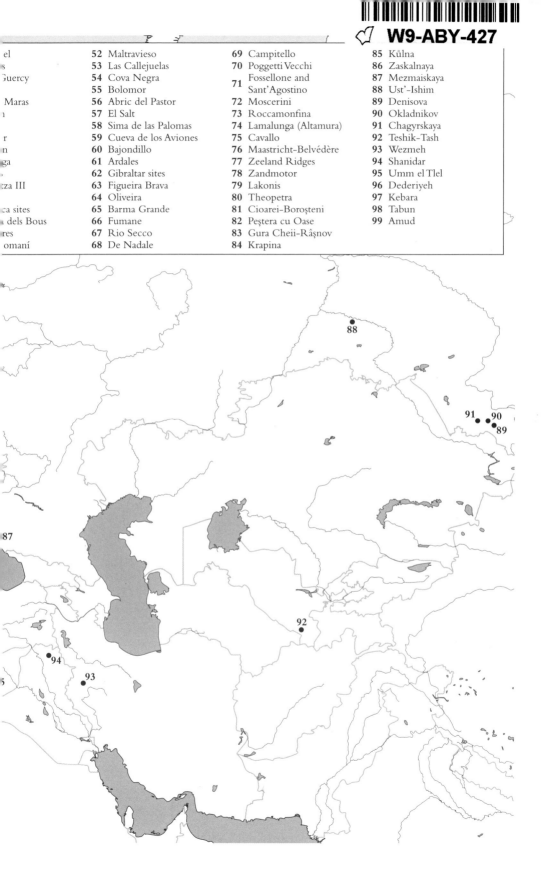

el
s
Guercy

Maras
l

r
n
ga

za III

ca sites
dels Bous
res
omaní

52	Maltravieso	69	Campitello	85	Kůlna
53	Las Callejuelas	70	Poggetti Vecchi	86	Zaskalnaya
54	Cova Negra	71	Fossellone and Sant'Agostino	87	Mezmaiskaya
55	Bolomor			88	Ust'-Ishim
56	Abric del Pastor	72	Moscerini	89	Denisova
57	El Salt	73	Roccamonfina	90	Okladnikov
58	Sima de las Palomas	74	Lamalunga (Altamura)	91	Chagyrskaya
59	Cueva de los Aviones	75	Cavallo	92	Teshik-Tash
60	Bajondillo	76	Maastricht-Belvédère	93	Wezmeh
61	Ardales	77	Zeeland Ridges	94	Shanidar
62	Gibraltar sites	78	Zandmotor	95	Umm el Tlel
63	Figueira Brava	79	Lakonis	96	Dederiyeh
64	Oliveira	80	Theopetra	97	Kebara
65	Barma Grande	81	Cioarei-Boroşteni	98	Tabun
66	Fumane	82	Peştera cu Oase	99	Amud
67	Rio Secco	83	Gura Cheii-Râşnov		
68	De Nadale	84	Krapina		

KINDRED

Also available in the Bloomsbury Sigma series:

KINDRED

NEANDERTHAL LIFE, LOVE, DEATH AND ART

Rebecca Wragg Sykes

BLOOMSBURY SIGMA
LONDON · OXFORD · NEW YORK · NEW DELHI · SYDNEY

BLOOMSBURY SIGMA
Bloomsbury Publishing Plc
50 Bedford Square, London, WC1B 3DP, UK
29 Earlsfort Terrace, Dublin 2, Ireland

BLOOMSBURY, BLOOMSBURY SIGMA and the Bloomsbury Sigma logo are
trademarks of Bloomsbury Publishing Plc

First published in the United Kingdom in 2020

Photo credits (t = top, b = bottom, l = left, r = right, c = centre)

Colour section: P. 1: © Inrap, Dist. RMN-Grand Oalais / image Inrap (t); © Peter Pfarr,
Niedersächsisches Landesamt für Denkmalpfl ege (b). P. 2: © Archive of the Royal College of Surgeons of
England. Photo: Paige Madison (t); © Philip Gunz (c); © Almudena Estalrrich (b). P. 3 © Fumane project:
Marco Peresani (t, c, b). P. 4: © Martisius et al. Science Reports 10: 7746 http://creativecommons.org/
licences/by/4.0/ (t); © Francesca Romagnoli; based on Figure 2 Quaternary International 407 (2016)
29–44 (c); © Jordi Mestre/ IPHES (b). P. 5: © Landesamt für Denkmalpfle ge und Archäologie
Sachsen-Annhalt / Juraj Lipták (t); © Rafael Martinez Valle (c); © D. Cliquet, Ministère de la Culture (b).
P. 6: © Thomas Higham (tl); © Erik Trinkaus (tr); © Francesco d'Errico (cl); © Graeme Barker (bl, br).
P. 7: © Jacques Jaubert (tl); © Public sector information licensed under the Open Government Licence
v3.0 (tr); © Christine Verna (cl); © Amud Cave archives, The Institute of Archaeology, Hebrew University
of Jerusalem. Photo: Erella Hovers (b). P. 8: © Tom Björklund (tl, b); © Field Museum Library (tr). In
text: Figure 1, p. 35: Based on elements kindly supplied by John Hawks, and on Figure 2,
Galway-Witham, J. et al.. 2019. Aspects of human physical and behavioural evolution during the last
1 million years. *Journal of Quaternary Science*, 34: 355-378. Figure 2, p. 51: Drawn after Figure 3, Caspari,
R. et al. 2017. Brother or Other: The Place of Neanderthals in Human Evolution. In Marom, A. &
Hovers, E. (Eds.) *Human Paleontology and Prehistory: Contributions in Honor of Yoel Rak*. Springer.

A catalogue record for this book is available from the British Library

Library of Congress Cataloguing-in-Publication data has been applied for

ISBN: HB: 978-1-4729-3749-0; eBook: 978-1-4729-3748-3

11

Chapter illustrations by Alison Atkin
All other illustrations by Marc Dando

Typeset by Deanta Global Publishing Services, Chennai, India
Printed and bound in Great Britain by CPI Group (UK) Ltd, Croydon CR0 4YY

Bloomsbury Sigma, Book Fifty-seven

To find out more about our authors and books visit www.bloomsbury.com
and sign up for our newsletters

Contents

A Note on Names

The scientific world of the nineteenth century is very different to that of the twenty-first century. It's not just about dramatic changes in analytical methods, but also profusion: there were many, *many* times fewer scientific articles published between 1800 and 1900 than in the past decade alone. When writing a definitive account of the Neanderthals, it's possible to cover key early prehistorians in some detail, in large part because there were so few of them. Moreover, in the nineteenth and earlier twentieth centuries, these individuals form part of the context for exploring how the first Neanderthal discoveries impacted science and society more broadly.

But from about 1930 onwards the numbers of people working on the subject really balloons, and therefore I took the decision to stop including named individuals and instead refer generically to 'archaeologists' or 'researchers'. This was about readability – I find that lists of names and laboratories tend to get mentally skipped over – but also brevity. Having been trained in science, where everything one says needs to be backed up by citations, this choice took a lot of thought. But for the different kind of writing required for *Kindred*, I wanted to make each word matter in telling the story of the Neanderthals themselves. There simply wasn't the space to mention the names and affiliations of researchers for every site or piece of information.

However, in no way do I want to imply that the contributions to what we know about Neanderthals over the past 80 to 90 years by those anonymised are any less important. Many of the people not mentioned individually in the text have been and remain my colleagues, and some are also good friends. Their names and publications can be found in the online bibliography accompanying the book (rebeccawraggsykes.com/biblio), but I want to specifically recognise them here, since without their dedication, grit, inspiration and literal perspiration this book would not exist.

Introduction

A sound out of time fills the cave: not the soughing and sighing of waves, for the sea fled as the cold bit and the mountains grimaced against icy armour. Now rough walls surround a soft ebbing breath, chasing a slowing pulse. At the end of the world, literally and figuratively, the last Neanderthal in Iberia witnesses a low, glinting sun across the distant Mediterranean. As a flint-dark sky lightens to grey dawn, soft coos of rock doves clash with the keening of lost gulls, crying like hungry children. But there are no more babies, no more of the people left, no one at all to join in watching the stars disappear; to hold vigil until the last breath leaves the air to cool.

Some forty thousand years later, oceans have risen again, salt tinges the air, and the walls of the same cave are ringing to voices and music – a requiem for a dream of the ancestors.

This is Gorham's Cave, Gibraltar, 2014. Archaeologists and anthropologists gather annually on this balmy southern tip of

Europe for one among many conferences on Neanderthals. But that year, something special happened. Among delegates visiting the great cathedral-like caves was musician Kid Coma, aka biologist Professor Doug Larson. He began to work the strings of a guitar, singing of the 'last man standing': some of the youngest-known Neanderthal archaeology comes from the Iberian Peninsula, and these caves. For a few minutes as his voice reverberated in the great stony chamber, professional concerns of presentations, hotly debated theories or intricacies of stone tool classifications were muted. Colleagues simply listened, and the human urge to connect with the ancient past took over. You can experience this strange, oddly moving moment, because someone thought to film it and it's now on YouTube.

That serenade to the graveyards of millennia throws a candid chink of light onto the people behind the science. Once the meticulous, objective scientific presentations are over, it's at the cafes and bars where less constrained – even passionate – speculation emerges between colleagues (who are also friends). Conversations range between 'dream' sites, to knowns versus unknowns; all dance around the question of whether we'll ever manage to glimpse the subtle reality of who Neanderthals were.

This book is a window onto those discussions. It's for those who've heard of the Neanderthals or not; for the vaguely interested to the amateur expert; even for the scientists lucky enough to research their ancient world. Because that's an increasingly immense task: convoluted paths through data and theory are criss-crossed by new discoveries, forcing diversions and even U-turns. The sheer amount of information is hard to process: few specialists have time to read *every* fresh article in their own sub-field, never mind the total scholarly output concerning Neanderthals. Even the most seasoned researchers can be left open-mouthed by new discoveries.

And this abundance of attention and analysis is because Neanderthals *matter*; have always mattered. They possess pop-cultural cachet like no other extinct human species. Among our ancient relatives (known as hominins), Neanderthals are truly A-list: big finds grab covers of major science journals and headlines in mainstream media. Our fascination shows no signs of lessening: Google Trends shows that searches for 'Neanderthal' have even overtaken those for 'human evolution'.

This degree of celebrity is however a double-edged sword. Editors know Neanderthals are potent click-bait and will tempt readers with sexed-up coverage, often angled towards some flavour of 'X killed the Neanderthals' or 'Neanderthals not so dumb as we think!'

Researchers' enthusiasm for sharing their work is tempered by frustration with consistent and contradictory spin, often framing them as boffins stumbling from one idea to another. Science manifestly operates by contention; however, new data and theories don't reflect the bafflement of researchers, but their enormous dynamism. Moreover, persistently clichéd 'Neander-news' means the average person never hears about some of the most fascinating modern findings.

The bigger picture too is hard to grasp, having transformed drastically since 1856, when odd fossils* from a German quarry were tentatively seen as a vanished species of human. Scholars began digging for more of these strange beings, and by the First World War, growing numbers of Neanderthal bones made it clear earth had birthed many siblings alongside us. Attention expanded to stone tools found in multitudes, and the first serious investigation of Neanderthal culture began. Time itself was key: by the mid-twentieth century, sites that had previously floated in time and were widely separated in space were connected through progress in dating techniques and geological chronologies. Fast-forward seven more decades, and it's upon these foundations that today we survey the grand prospect of the Neanderthal world, spanning thousands of kilometres and well over 350,000 years.

Yet twenty-first-century archaeology is worlds away from its beginnings, and might more closely resemble a Victorian futurist's fantasies. Early prehistorians had little more than stones and bones with which to reconstruct the ancient past, whereas today's researchers work in ways their forebears didn't know existed. Laser scans instead of ink sketches take the likeness of an entire site, as specialists study objects no one a century ago dreamed of finding. From fish scales and feather barbs to the micro-histories of individual hearths, our insights are as likely to emerge under the lens of a microscope as at the edge of a trowel.

---

* Fossilisation is the process by which bone becomes mineral.

We can almost spy over Neanderthals' shoulders, reconstructing the few minutes taken for a cobble to be efficiently reduced to sharp flakes 45,000 years ago. The static archaeological record itself becomes dynamic: we watch as tools move around sites and are taken away into the landscape. We can even trace them in reverse to the original outcrops. And incredibly intimate insights into Neanderthal bodies are now possible. Just considering teeth, we can scrutinise daily growth lines, assess diet from micro-polishes and even chemically 'smell' hearth smoke that infiltrated their dental calculus.

Out of this abundance of information has come a renaissance in Neanderthal research over the last three decades. A parade of astonishing findings have hit the headlines, and our basic understanding of where and when they lived, how they used tools, what they ate and the symbolic dimensions of their world has been revolutionised. Perhaps most astonishingly, once-rubbished stories of inter-species love are teased out from nondescript morsels of bone, and a teaspoon of cave dirt can produce entire genomes.

Whizzy machines allow us to extract terabytes of information from every conceivable substance, but all this is tempered by archaeologists' realisation that *how* sites form is crucial to understanding what they contain. Over millennia, vagaries of preservation, erosion and time mean everything comes to us as fragments. Recording the positions of artefacts is vital to understanding the integrity of each layer, before we get carried away with analysis. Broken and long-separated parts can be reunited, while soil structure, the lilted angle of flint flakes or weathering on bone splinters all contribute to deciphering site formation. It's from this tattered and sometimes jumbled archive that we must glean history.

So archaeologists still feel excitement while excavating, but the average dig produces tens or hundreds of thousands of carefully collected objects that must be washed, labelled and nestled in individual sealed bags. Coexisting digitally within massive provenance databases, they're a priceless resource allowing us to explore the intersections between geology, the environment and hominin action. Such caution has also altered how we deal with museum collections accumulated long ago. Increasingly, 'classic' sites – some visited by thousands of tourists each year – are revealing new and sometimes unexpected secrets through state-of-the-art reanalysis. It's the sum of all this that allows us to answer more accurately than ever before fundamental questions such as 'What did Neanderthals eat?'

Nonetheless, even a brief foray into the science of Neanderthal diet shows how deceptively simple such a query is. Not only due to the range of materials and methods available – examining proportions of animal bones, microscopic wear on teeth and stone tools, preserved food residues or chemical and genetic analysis of fossils – but because healthy suspicions about how sites form extends to forensically investigating diet. Even in places stuffed with animal remains covered in slice marks from stone tools, things aren't always clear-cut. For example, archaeologists have learned the hard way to take into account the role of other predators, and that body parts decay at different rates.

But each advance adds to the overall picture. It turns out that much more than just big beasts were on the menu, yet did *all* Neanderthals eat the same foods, in all times and places? Everything in Neanderthals' lives was interconnected, and entanglements with other Big Questions abound: How much food did their bodies need? Did they cook? How did they hunt? How big were their territories? What were their social networks like? Each question unfolds another layer of complexity.

Sorting patterns from the multitudes of artefacts and sites means looking up and out, bridging between places and times. Neanderthal life was four-dimensional, so as we reconstruct in phenomenal detail how they were hunting reindeer in one place, we must ask what were they doing elsewhere, and else-when? Many sorts of sites exist, from ephemeral scatters of stones haloing a carcass to masses of bones bedded in colossal ashy deposits: the ruined pyres of *hundreds* of beasts. Considering such different kinds of records brings us hard up against the capricious temporal cadence of the past: depending on how layers form, two equal depths of sediment might contain an afternoon, or 10 millennia. Dating individual objects is a powerful tool, but only if we're confident they've not moved between layers. And the information gleaned from individual artefacts, layers or sites expands outwards, connecting different scales of behaviour.

Such subtleties rarely feature in public discussion about, and understanding of, Neanderthals. Most people have rough ideas about them, but less so the scientific details. Moreover, they're largely set against a backdrop of ice and mammoths. Yet a whole other Neander-world existed beyond persistent stereotypes of shivering ragged figures in frozen wastes, barely hanging on until the arrival of *Homo sapiens* before dropping dead. Despite greater access than ever before to research as it happens – via social media-savvy researchers or livestreamed conferences – the tsunami of new data and complicated interpretation means balanced and truly up-to-date perspectives are hard to find. Genuine 'wow' finds do catch the attention of 24-hour news cycles and even researchers by surprise, but the 'bling' stories aren't always the most fascinating. Carefully argued theories and debates lasting decades make for poor headlines, but such stories contain some of the most surprising ideas about Neanderthal lives.

In fact, nuance underlies many of the most significant reorientations in understanding. Perspectives broaden in step with accruing data, and the gap between 'us' and 'them' continually diminishes. Many things we thought beyond Neanderthals' ken are today widely accepted by a slow aggradation of data: tools made of materials other than stone, use of mineral pigments, collecting objects like shells and eagle talons ... and by extension, engaging with aesthetics. Furthermore, diversity has emerged: Neanderthals today are less like identikit hominins than denizens of a world as wide and rich as the Roman Empire. Its huge scope in space and time means cultural variety, complexity and evolution. Varied and adaptable, Neanderthals survived in vanished worlds where kilometre-high glaciers met tundra, but also in warm forests, deserts, coasts and mountains.

Over 160 years since their (re)discovery, our obsession with Neanderthals persists. This is a love affair longer than a lifetime, but compared to the vast span of time they walked the earth – squinting against sunrises, sucking lungfuls of air, leaving footprints behind in mud, sand and snow – it's barely a shiver of the second hand on Time's great clock. How we think and feel about them is constantly evolving, from the average person googling 'Are Neanderthals human?' to those who work on their remains every day. Neanderthals are reimagined before our eyes, each discovery stoking anew our desires (and fears) about who these ancient people really were. Strangest of all is the afterlife they could never have conceived: entangled through nearly two centuries of science, history and popular culture, their story now extends into our far future.

The rest of these pages will paint a twenty-first-century portrait of the Neanderthals: not dullard losers on a withered branch of the family tree, but enormously adaptable and even *successful* ancient relatives. You're reading this book because you care about them, and the greatest, grandest questions they pose: who we are, where we come from and where we might be going.

Look through shadows, listen beyond echoes; they have much to tell. Not only of other ways to be human, but new eyes to see ourselves. The most glorious thing about the Neanderthals is that they belong to all of us, and they're no dead-end, past-tense phenomenon. They are right here, in my hands typing and your brain understanding my words.

Read on, and meet your kindred.

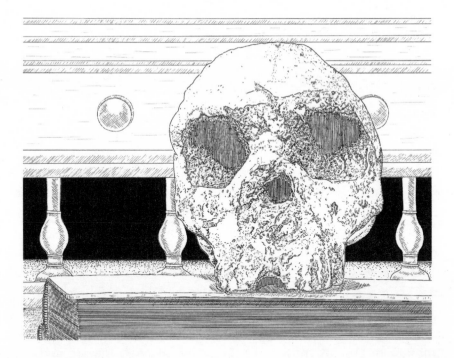

CHAPTER ONE

The First Face

Gritty roof-dirt scratches under your feet, for we stand atop a vertiginous space-scraper. Beyond any dream of Babel, this tower has grown up from the earth like a hyper-stalagmite, a metre for each year of humanity's history. Atop its three hundred kilometre-high roof, the International Space Station streaks overhead, almost faster than you can blink. Peer over the tower's edge, and along its full length you see a halo of light from thousands of openings. Towards the top are LED-lit apartment windows, but farther down — deeper in time — the quality of light shifts. Your eyes adjust as amber fluorescent bulbs give way to luminous gas lamps, then massed choirs of candles begin.

You're squinting now but perceive, even farther below, a softening. Old light from tens of thousands of clay lamps gleams out, their smoky trails wreathing the tower, yet we're still not all the way to the depths of human history. You take out a small telescope and as your pupils expand, greedy for ancient photons, you see the flickering of hearth fires begin some thirty kilometres down, and continue for ten times that depth, all the way back to three hundred thousand years ago.

Flames and shadows twist and arch, reflecting on stone walls, until there is only darkness and the years are uncounted.

Time is devious. It flees frighteningly fast, or oozes so slowly we feel it as a burden, measured in heartbeats. Each human life is marbled with memories and infused by imaginings, even as we exist in a continuously flowing stream of 'now'. We are beings swept along in time, but to emerge and view the whole coursing river defeats us. Not so much counting or measuring; today's science can calculate values to brain-imploding levels of accuracy, whether the age of the universe or a Planck second.[*] But truly comprehending the *scale* of time on an evolutionary, planetary, cosmic level remains almost impossible, as much as for the first geologists, staggered at glimpsing earth's true age. Connecting to the past beyond three or four generations ago – the boundary of 'living memory' most of us manage – is challenging. Relating to more ancient ancestors gets even harder. Old photographs embody how our perspective becomes fuzzier, and even this visual archive extends just a couple of generations farther back. Then we enter the realm of painted portraits, and another gauze layer of unreality settles onto the past. Comprehending the gobsmacking hugeness of *deep* archaeological time is much, much tougher.

Handy mental tricks exist to bridge this gap between our mayfly existences and the abyss of time. Shrinking the universe's 13.8 billion years to a single 12-month period puts the dinosaurs shockingly close to Christmas, while the earliest *Homo sapiens* arrive only a few minutes before New Year's fireworks. But plotting time on that relatable scale doesn't communicate the immense, yawning stretches of years. Surprising juxtapositions push it home a bit: for example, fewer years lie between Cleopatra's reign and the moon landings than between her and the building of the Giza pyramids. That's only the last few thousand years, whereas the Palaeolithic – the archaeological period before the last ice age – is even more mind-bending. Lascaux's leaping bulls are closer in time to the photos on your phone than to the panels

---

[*] The shortest measurable unit of time.

of horses and lions at Chauvet. Where do the Neanderthals fit in? They take us *way* back beyond fingers tracing beasts on stone walls.

While it's impossible to pinpoint the 'first' of their kind, they became a distinct population 450 to 400 thousand years ago (ka). The night sky then hanging over earth's many hominin populations would have been alien, our solar system light years away from its current position in a never-ending galactic waltz. Pause halfway through the Neanderthals' temporal dominion at around 120 ka, and while the land and rivers are mostly recognisable, the world *feels* different. It's warmer and ice melt-swollen oceans have flooded the land, shoving beaches many metres higher. Startlingly tropical beasts roam even the great valleys of Northern Europe. In total, the Neanderthals endured for an astonishing 350,000 years, until we lose sight of them – or, at least their fossils and artefacts – somewhere around 40 ka.

So far, so dizzying. But it's not just time: Neanderthals also spread across a remarkably vast swathe of space. More Eurasian than European, they lived from north Wales across to the borders of China, and southwards to the fringes of Arabia's deserts.

The more we find out about Neanderthals, the greater range and complexity we uncover. But following all this can get confusing: there are thousands of archaeological sites. So we'll hold on to anchors: key sites that offer touchstones in the Neanderthal story, while also looking outwards at the enormous scope of the field. Some – whether Abric Romaní in Spain or Denisova Cave in Siberia – tell us incredible stories of twenty-first-century discoveries. Others, like the Le Moustier rockshelter in the heart of the south-west French Périgord, offer chronicles of Neanderthal life woven through the history of archaeology itself. Two extremely important skeletons we'll meet later were found there, and it's also a stone artefact (lithic)* type site, where a particular Neanderthal culture was defined. Le Moustier has witnessed over a century of research, hosted a succession of scholars and even been a flashpoint for massing geopolitical anxieties just before the First World War. But neither Le Moustier nor France in

---

* 'Lithic' means stone, and researchers prefer the term 'artefact' to 'tool', which is more specific to something actually used in the hand.

1914 are where the Neanderthal story truly begins. We need to go back another five decades, to the 1850s.

Ground Zero

Everyone loves a 'how did you meet?' story. The knotty tale of our entanglement with Neanderthals is tousled by threads of intuition and perplexity: birthed by the Industrial Revolution, scorched by wars, glittering with treasures lost and found. From forgotten meetings tens of millennia ago when we saw each other as human, to the comparatively recent rediscovery of these ancient kin, our infatuation is perennial. Impatient for hoarfrost and mammoth breath, it's tempting to fire up a time machine and speed straight back into the Pleistocene.* But we need to start in the midpoint of this grand and convoluted history, before we can clearly see a beginning, or an end.

Let's journey just five or six generations back to witness the birth of human evolution as a science. Fundamentally narcissistic – a child of the Victorian worldview, after all – it's always been about asking who, and why, we are. Amid perhaps the greatest socio-economic upheavals the world had yet seen, nineteenth-century scholars struggled to wrap their minds around the strange bones coming out of European caves. But one thing was certain from the start: the Neanderthals detonated growing debates over what it meant to be human. There are few bigger questions, and beyond mere curiosity the answers matter deeply. Tracing how early prehistorians wrestled with categorising these confounding creatures helps us appreciate the many contradictory things believed about the Neanderthals, and explains preconceptions that still persist today.

This history begins in late summer, 1856. Quarrying to meet demand from burgeoning marble and limestone industries had progressively consumed the deep gorge south-west of Düsseldorf, a once-famous Prussian beauty spot. Towards the clifftops a cavity – known as the

---

* The Pleistocene is a geological division of time, and is the first epoch of the Quaternary, beginning around 2.8 million years ago, until around 11,700 years ago when the epoch we're living in – the Holocene – began.

Kleine Feldhofer Cave – was revealed, plugged by thick, sticky sediment requiring blasting. One of the quarry owners' eyes was snagged by large bones workers cast down from the cave mouth. Being a member of a local natural history association, he speculated that they could be old animal remains of scholarly interest, and so rescued a motley assortment – crucially including the top of a skull. The founder of the natural history club, Johann Carl Fuhlrott, visited and realised the bones were human. Moreover, they were fossils and thus must be very ancient.[*]

It seems that the Feldhofer discovery caught local imaginations as press reports appeared, and scholars further up the intellectual hierarchy began asking to view the mystery bones. At the start of 1857, a cast of the skull cap was sent to anatomist Hermann Schaaffhausen in Bonn, whose mind was thankfully open to the possibility of fossil humans. Eventually, a wooden box containing the real remains, chaperoned by Fuhlrott, travelled to Bonn on the barely 10-year-old railway. Schaaffhausen's expert eye immediately focused on the bones' abnormal bulk – especially the skull – while other features like the sloping forehead reminded him of apes. Given their patently ancient condition and origin in a cave, he was inclined to agree that they must be a primitive kind of human. That summer he and Fuhlrott presented their findings to the General Meeting of the Natural History Society of Prussian Rhineland and Westphalia. Just a few years after this unofficial debut into society, more serendipitously rescued bones would become the first scientifically named fossil human: *Homo neanderthalensis*.

The word 'Neanderthal' is so familiar today, yet its history is full of strange congruence. The Neander '*thal*' (valley) containing the bones' original resting place was named for a late seventeenth-century teacher, poet and composer, Joachim Neander. A Calvinist, his faith was partly inspired by nature, including the famous ravine of the Düssel River. Its geological wonders – cliffs, caves, arches – were so beloved by artists and romantics that it developed its own tourism industry. Joachim Neander died in 1680, but his celebrated hymns – performed three centuries later for Queen Elizabeth II's diamond

---

[*] Even in fossils 'just' a few tens of millennia old, textural differences are apparent.

jubilee – were an enduring legacy. By the early nineteenth century one of the gorge's formations was named Neanderhöhle after him, yet within a few decades the surroundings would have been unrecognisable to Joachim. Consumed by massive quarrying, the ravine disappeared and the new valley became known as the Neander Thal. Here's the weird bit: Joachim's family name was originally Neumann, later converted by his grandfather to Neander following the fashion for more classical names. Neumann – and Neander – literally mean 'new man'. Could there be any more fitting moniker for the place where we first discovered another kind of human?

Yet even if the anatomical case seemed obvious, proof that the bones really were incredibly old was needed. Fuhlrott and Schaffhausen returned to the quarry to interview the workers, who confirmed that the remains had lain about 0.5m (2ft) deep in undisturbed clays. Interpreted within a hybrid biblical-geological framework, for Fuhlrott this pointed to an age before the Flood, making the skeleton enormously ancient. It gave them confidence to publish the explosive claim that a vanished human species had existed before *H. sapiens*. More convergence: the same year, 1859, witnessed another convulsion of the scientific community with the natural selection theories of Darwin and Wallace. But it wasn't until around two years later that Feldhofer really hit the big time, when the fascinating biologist George Busk translated the original German article.

Little known today, Busk was at the heart of the nineteenth-century scientific elite, and like many contemporaries his interests were multi-disciplinary in a way virtually impossible now. A member of the Geological Society, President of the Ethnographic Society and by 1858, Zoological Secretary for the Linnean Society (the foremost learned society for biology), Busk added a commentary to his 1861 translation of the Feldhofer discovery. He pointed out that extreme human antiquity was well established by artefacts found elsewhere alongside extinct animals, and specifically compared the skull to chimpanzees. He also noted the urgent need to find another.

In fact, earlier, unrecognised discoveries already existed. Humanity had forgotten its long-lost cousins for millennia, then – something like buses – three appeared in the first half of the nineteenth century. The first came in 1829 at the hands of Philippe-Charles Schmerling.

One of a growing number of 'fossiling' hobbyists, he also had a medical background, and at Awirs Cave near Engis, Belgium, found parts of a skull. Together with ancient creatures and stone tools, it had lain sealed beneath 1.5m (5ft) of flowstone-cemented rubble.[*]

Despite its unusual elongated shape, the Engis skull didn't attract wider notice because it was from a child: like us, young Neanderthals had to 'grow into' their adult form. The adult Feldhofer skull was more obviously heavy-looking, and furthermore it came with other body parts.[†] Although the Engis child was to remain unclassified until the early twentieth century, happily for Busk someone else had already found another adult Neanderthal; and it came from British-controlled soil.

In 1848, while stationed in Gibraltar, the exquisitely named Lieutenant Edmund Flint came into possession of a skull. Once again, limestone quarrying – this time to reinforce British military fortifications – set the discovery in motion, but Flint's rank and personal interest in natural history ensured it was not disposed of.[‡]

The Rock spikes up from the peninsula like a vast hyaena's tooth, and its flora and fauna attracted the interest of enthusiastic natural historians among Flint's fellow regimentals; he was Secretary of their Scientific Society. Minutes for 3 March 1848 record his presentation of a 'human skull' from Forbes' Quarry, above the eighteenth-century artillery battery. No doubt the officers passed it around, gazing into the huge eye sockets, but despite being essentially complete (unlike Feldhofer) it apparently wasn't considered extraordinary. A coating of cemented sediment may have obscured details, but the inability to 'see' its exotic shape is noteworthy.

The Forbes skull sat unremarked in the Society collections until 1863. That December, Thomas Hodgkin,[§] a visiting physician with ethnographic interests, saw it amongst other objects. Perhaps primed

---

[*] This material is known as breccia.

[†] In total the original Feldhofer bones consisted of both thighs, the left hipbone, parts of the collarbone, a shoulder blade, most of the arms and five ribs.

[‡] It was very probably unnamed quarry workers rather than the Lieutenant who made the find.

[§] The describer of the condition Hodgkin's lymphoma.

by his friend Busk's translation of the Feldhofer report, he *did* see something remarkable in the skull, which at this point was probably in the care of Captain Joseph Frederick Brome, a respected Gibraltarian antiquarian and governor of the military prison. Passionate about geology and palaeontology, Brome had been sending finds from his own excavations to Busk for several years, and so the Forbes skull duly set sail for Britain, arriving in July 1864.

Busk must have immediately realised that the large nose and pushed-forward face were strikingly similar to features hinted at by the Feldhofer skull, which consisted only of the upper cranium plus a partial eye orbit. He also understood that these vanished people must have lived 'from the Rhine to the Pillars of Hercules'. Just two months later, the Forbes skull made its own scientific debut, although someone received a special preview. Thanks to the prodigious correspondence habits of Victorian gentlemen, we know that the Forbes skull had very likely been in the hands of Charles Darwin, conveyed by a palaeontologist colleague of Busk – Hugh Falconer – as Darwin's ill health prevented his travelling to the grand scientific unveiling. Darwin thought it 'wonderful', yet in keeping with his reticence on human origins there is no record of his scientific reaction to the Neanderthals.

Anxious to establish the skull's geological context, Busk and Falconer rushed back to Gibraltar before the end of the year. What they saw gave them confidence to publish that this was a second extremely ancient 'pre-human'. However, their intended species name of *Homo calpicus*[*] was not to be. William King, ex-curator of Newcastle's Hancock Museum and Chair in Geology and Mineralogy at Galway, had studied casts of the Feldhofer remains and, just as the skull from Gibraltar was docking in Britain, his suggested name *Homo neanderthalensis* was published. Following the 'first dibs' rules of science, this remains the one that we still use today.

But the appellation of these peculiar fossils was the least controversial thing. Assigning them as extinct members of our own genus, *Homo*, had profound implications that reverberated beyond the scientific

---

[*] Calpicus being a reference to the ancient Phoenician name for Gibraltar; had the earlier Belgian find been recognised, we might be referring to 'the Awirians'.

world. Dramatically at odds with nineteenth-century Western world views, the idea faced passionate opposition.* Scathing criticism rapidly appeared from August Franz Josef Karl Mayer, a retired anatomist colleague of Schaaffhausen, and a creationist.

Mayer claimed that the remains were simply from a diseased and injured – but otherwise normal – human. Somewhat later in 1872 the eminent biologist Rudolf Virchow examined the Feldhofer bones and agreed that their anatomical peculiarities could be explained if a lost Russian Cossack with arthritis, rickets, a broken leg and bowed limbs from his cavalry career had secreted himself in the cave and died. This sounds absurdly far-fetched today – and ironically underlines just how human-like the bones are – but Virchow was a widely respected medical pioneer in cellular pathology and designed the first systematic autopsies. Perhaps, then, it's not surprising that he was inclined to interpret the Feldhofer anatomy as illness and injury, even suggesting the formidable brows resulted from excessive frowning due to chronic pain.†

Yet Busk was also a medical man. Decades as a navy surgeon treating varied injuries, illness and parasites surely made him just as likely to see Neanderthals through a pathological filter, but this was tempered by a zoological background and experience in species classification.‡ Busk was certain no disease or physical trauma could account for the anatomy he saw, and noted with some satisfaction that those refusing to accept Feldhofer must admit there was little chance of a sickly Cossack expiring in Gibraltar. These debates smouldered on well into the twentieth century, yet in some ways Neanderthals weren't burning arrows shot out of the dark, entirely unexpected. Doubts had been coalescing in Western intellectual communities that the world might not precisely mirror biblical accounts.

---

* The editors of the original Feldhofer article anticipated this, adding a polite note pointing out that not everyone shared the authors' outlandish interpretations.
† Virchow once used his scientific research to defend himself after being challenged to a duel by Bismarck; Virchow was allowed to choose the weapon and selected two sausages, one of which contained parasitic larvae he'd shown could infect humans. Bismarck dropped his challenge.
‡ Busk performed specimen identifications from Darwin's 'Beagle' collection, and edited his and Wallace's papers on natural selection.

Diverse revelations since medieval times about nature – from unknown continents to the identification of previously invisible astronomical bodies – forced the restructuring of knowledge and philosophy. And while fossils had been noticed for millennia, by the eighteenth century biologists began treating them as once-living creatures that could be studied. Earth's deep places were increasingly explored, such as the great Gailenreuth Cave in Germany as early as 1771, adding to dawning comprehension of 'lost worlds' populated by extinct beasts. Theologically inspired cycles of disaster and renewal remained influential, but the unfamiliar nature of pre-Flood worlds was apparent by the early nineteenth century. Not only had Arctic creatures like reindeer once lived thousands of kilometres farther south, but the inverse was true, with hippopotamus bones found in decidedly non-tropical Yorkshire. Yet not everyone was convinced creatures truly evolved. Some – including scientists with religious leanings, like Virchow – even perceived moral risk in such theories, fearing it would lead to social Darwinism.

Nevertheless, as more fossils emerged, the case for another sort of human began to solidify. Just the year after King officially named the Neanderthals, a heavy, chinless lower jaw from Belgium found with mammoth, reindeer and rhinoceros was proposed to be from the same species. But it was another two decades until mostly complete skeletons were found. Again in Belgium, remains of two adults came in 1886 from the Betche-aux-Rotches Cave at Spy, showing that flat, long skulls, sloped-back jaws and robust limbs previously known from other sites all belonged to the same creatures. This cemented scholarly acceptance of Neanderthals as an anatomically defined extinct population. But the fossils are of course only half the story.

Time and Stone

Early prehistorians faced a fundamental problem: time. Lacking methods to tell exactly how old anything was, they relied on relative chronologies: fossils or artefacts found with extinct animals were obviously older than the current world. British geologist Charles Lyell knew that Earth's deep past must extend far beyond the biblical confines of a few millennia, and showed in his great work *Principles of*

Geology that — given enough time — simple, observable geological processes were entirely responsible for creating the world. A complete planetary history could therefore be deciphered through the principle of stratigraphy: since sediments accumulate on top of each other through time, greater depth must correlate to greater age. Lyell was intensely interested in Feldhofer, and in 1860 — even before Busk's translation — he visited to examine remaining deposits. Fuhlrott showed him the skull and gifted him a cast: Victorian-era data sharing. By then the cave itself was on the cusp of destruction, and Lyell's expert opinion was crucial to gaining scientific acceptance that it was truly ancient.

More than this, Lyell's concept of stratigraphy formed the bedrock of archaeology as a discipline. It could provide structure to deep time processes, establish relative ages across landscapes and illustrate how deposits *within* sites form. During excavation, variation in sediment colour or texture as well as the contents of each layer — artefacts and animal bones — are signposts for how conditions changed through time. For many decades, proof Neanderthals were as indecently old as many suspected rested purely on such reasoning. It took nearly a century for scientists to finally develop methods that could *directly* date things. Beginning in the 1950s with radiocarbon,* myriad other approaches have followed that are applicable to almost anything: bone, stalagmites, even single grains of sand.

Some categories of lithic artefacts can even be directly dated, though none of the early Neanderthal fossils seemed to be accompanied by cultural objects. In fact, we now know there *were* plenty of lithics at least at Feldhofer, but the discoverers weren't familiar enough with stone tools to tell the difference between naturally shattered versus intentionally knapped rock.

As with fossils, humans had long been interested in prehistoric artefacts before the first Neanderthals were found. In metal-focused societies, chance discoveries of hefty handaxes or delicate stone arrows required an explanation. People looked to both natural and supernatural

---

* Radiocarbon is probably the most familiar type of direct dating method to most non-specialists. Based on the predictable decay rates of the carbon-14 isotope, it can now be used to date organic materials up to about 55,000 years ago.

causes, calling them thunderstones and believing them capable of deterring lightning,* or weaving tales where they were elfshot: weapons of the 'Little Folk'. Historians on the other hand understood such objects within available chronologies. One of the first recorded descriptions of a prehistoric stone tool comes from 1673, when a triangular-shaped artefact was discovered near 'elephant' bones at Gray's Inn Lane, London. Despite understandings of geological time beginning to crystallise around then, the find was nevertheless interpreted as a Roman elephant attacked by a Celtic warrior. The notion that such an object was made by hands *thousands* of generations before Rome was founded simply wasn't in anyone's range of possibilities. Yet a century or so later, understanding had developed enough for deeply buried handaxes to be described as likely from 'a very remote period indeed, even beyond that of the present world'.† The true importance of lithics for understanding ancient humans was, however, still to come.

The first known person to intentionally excavate Neanderthal artefacts, albeit unknowingly, was the Frenchman François René Bénit Vatar de Jouannet. Between 1812 and 1816 he excavated the rockshelters of Pech de l'Azé I and Combe Grenal, south-west France, finding burned animal bones and remnants from lithic production. Crucially, he noted they were embedded in obviously ancient flowstone, but because even the Engis skull wouldn't be found for over a decade, he had no concept of Neanderthals or indeed any extinct hominins. His best guess for the chronology of the artefacts – 'very old Gaul-ish' – was surprisingly similar to the Gray's Inn interpretation nearly 150 years earlier.‡

After de Jouannet, evidence mounted that such finds could be stuffed neither into historical nor biblical chronologies. In south-east France, antiquarian Paul Tournal had been digging up cave bear and

---

* This isn't as outlandish as it seems, since in the right silica-rich sediment lightning can produce a mineral called fulgurite.
† The words of John Frere, who in 1797 discovered lithic artefacts in association with extinct animals in Norfolk, Britain.
‡ He was working just before the 'three ages' of stone, bronze and iron were proposed in 1817 by Christian Jürgensen Thomsen.

reindeer bones alongside clearly human-made artefacts in the Bize caves, leading him to propose in 1833 an 'anté-historique' age. Around the same time, knapped flints entombed deep in river gravels of the Somme valley, northern France, were being found by the French archaeologist Jacques Boucher de Crèvecoeur de Perthes. It was hard to imagine they could have arrived there recently, nonetheless even evidence of elephant and rhinoceros fossils garnered little scientific acceptance. It wasn't until around the same time that news of the Feldhofer find began spreading that things changed.

We meet here once more Hugh Falconer, who would bring the Forbes skull to Darwin. Like Busk, he remains little known today, but was central to the origins of human evolution as a science. After years in colonial India where he'd pursued palaeontological interests, by 1858 Falconer was excavating Brixham Cave, Devon, finding lithics and extinct fauna sealed under a stalagmite floor. The same year he visited de Perthes' gravel pits, and convinced of their great age then advised the geologist Joseph Prestwich to take the trip. By chance, Prestwich met stone tool expert John Evans there – together with Charles Lyell, who'd made his own de Perthes pilgrimage – and in 1859 they published expert opinions verifying that the time of the lithics and extinct beasts truly lay together in the deepest past. As far as the 'scientificos*' were concerned, the matter was settled, but sceptics persisted: was it possible the toolmakers, albeit ancient, had lived *after* creatures like mammoths were already dry bones?

Absolutely incontrovertible – and utterly spine-tingling – testimony soon proved humans had indeed witnessed extinct beasts in all their vital, hairy glory. More than 560km (350mi.) south of the Somme gravel pits is the village of Les Eyzies-de-Tayac, at the confluence of the Beaune and Vezère rivers. Today in January it's quiet enough to hear peregrines cry over the massive cliffs rearing above, but summer sees narrow sun-baked pavements heaving with tourists, for the village is the capital of a prehistoric wonderland, surrounded by hundreds of caves and rockshelters in spectacular limestone gorges and plateaux. After sampling truffle omelettes at the Café de La Mairie, visitors

---

* Darwin's correspondent, the botanist and explorer Joseph Hooker, uses the term 'scientificos', versus 'plebs'.

amble up to the National Museum of Prehistory, built round a ruined chateau hunkering below the limestone overhang. Elaborate fireplaces remain, a strange echo of the prehistoric ashy layers stacked metres deep beneath. From the old ramparts, a huge art deco sculpture of a Neanderthal stares inscrutably out: like the statue's secret thoughts, this landscape has hidden many things.

Les Eyzies' relative isolation ended in 1863 when an ambitious railway linking Paris to Madrid opened a branch into the Périgord, initiating its transformation from sleepy hamlet to epicentre of debates over Western civilisation's origins, and eventually a World Heritage listing. To follow the trail today, near where the railway line arcs gracefully southwards from the station, you can rent a canoe and paddle up the snaking route of the Vezère. After a few kilometres and opposite a hilltop chateau is the La Madeleine rockshelter. Famous medieval remains receive the tourists today, but adjacent is a prehistoric site, still concealed by vegetation, much as it was in 1864.

That summer Falconer was at the scene, visiting an archaeological collaboration between two passengers who arrived on shiny new trains the year before. A British financier, Henry Christy's wealth allowed him to assemble 'one of the choicest private archaeological collections in Europe',[*] giving him unusually good knowledge of stone tools. His French partner, Édouard Lartet, was already something of a celebrity prehistorian, digging ancient sites since the 1830s.[†] Based on rumours of a local *vicomte*'s collections and finds in a Parisian antiquarian shop, they began collaborating in the Vezère valley. Initially investigating the upper shelter at Le Moustier, on their return one day they noticed across the river another large shelter, visible only because it was winter and the branches in front were bare.

Known as La Madeleine, this site turned out to contain hugely rich archaeology made by early *H. sapiens* tens of millennia after the Neanderthals. Nevertheless, it included an object critical to acceptance of their place in our evolutionary past. Until then, those sceptical of

---

[*] From Falconer's memoirs (p. 631).
[†] Lartet was originally trained in law, but allegedly developed a passion for palaeontology after receiving a mammoth tooth from a farmer in payment for his services.

deep human antiquity had explained away carved reindeer antler objects found elsewhere in France as resulting from already fossilised material collected and engraved much later. That argument collapsed at La Madeleine when Lartet and Christy's workers turned up a shattered piece of mammoth tusk with markings on it. That very day, Falconer – the world's most eminent fossil elephant expert – happened to be visiting. As the soil was brushed off the ivory, he immediately saw that sweeping engraved lines formed the distinctive domed head of a mammoth, complete with carefully rendered shaggy fur.* This single artefact proved that humans *had* lived alongside extinct species, and that all the 'rejectamenta' of their lives being hauled out of caves across Europe really were from a prodigiously ancient world.

The La Madeleine discovery laid the final foundation stone of today's discipline of human origins. It would take another 50 years or so for prehistorians collecting lithics to really begin grasping who made what, and when. But they had already crossed a Rubicon between two cosmologies: the old view of a universe made for us, and a new world where we were the children – with many sisters and brothers – of the earth itself. The path into the latter is where the rest of this book will take us, to learn how Neanderthals morphed from scientific oddities to the strangely immortal, oddly beloved creatures we've both discovered and also somehow created. But first we need a family portrait, to help place the Neanderthals into their frankly immense evolutionary context.

---

* Eighteenth-century Russian permafrost finds had already shown that mammoths were hairy.

CHAPTER TWO

The River Fells the Tree

Close your eyes, shake off your shoes. Dulled solar redness glows beyond eyelids, grass prickles toes and dust lies beneath your soles. Warmth brushes your arm as a hand slips around yours; somehow you know who it is. Open your eyes, and below a sky both bright with sun and somehow star-speckled black, your mother stands before you. This is the place out of time, where all humans find each other. Rustling footfalls approach, and another woman steps forward: your maternal grandmother. Maybe you spoke to her last week, or twenty years ago, or maybe you only know her from blurred photos. She joins her hand to your mother's, and then turns her head: behind, threading over an endless plain is a line of more women, bonded by hand and gaze.

Your eyes lose count but you sense them by the hundred, the thousand. Faces become unfamiliar with distance, despite you somehow knowing cheek curves, curls of hair, or a shift of the hip. Beyond, the thread continues to meet the horizon and your gaze rises up to the milky spindrift above: there, many tens of millennia away, even the stars have shifted. Then you feel it like a lightning strike passed through forty thousand hands: endless cycles of love and loss thumping through breasts and bones for five hundred thousand years into your blood, your heart. Dizziness swirls, but your mother's hand squeezes, and just then, through blinking eyes, you see it. Spreading out from this single thread of maternal ancestry is an immense human tracery, fretworks of (im)mortality blending towards the distanced-blued plateau at the edge of time. They're all here, the others. They always have been.

We are the embodied heritage of all our mothers. The predecessors of your eyes focusing on these words first saw light over 500 million years ago (Ma). The five dextrous fingers moving these pages have clutched, grasped, scrabbled for 300 million years. Perhaps you can hear music, or a recording of this book right now; that ingenious triple-bone ear structure began listening for sounds of love and terror while we scuttled beneath saurian feet. The brain processing this sentence had ballooned almost to its current size by 500 ka, and was shared by the Neanderthals.

Situating us and them within a deeper biological and evolutionary context drives home what we share. It also reveals how off-target nineteenth-century visions of Neanderthals as the missing link between us and other apes were. Fossil primates were already known: in 1836 none other than Édouard Lartet found an ancient monkey. Later – the same year Neanderthal bones were blasted out of Feldhofer Cave – he uncovered the first European ape, named *Dryopithecus*. Despite this, fossil humans were still a shock.

Today the situation is profoundly altered. While details remain debated, our family tree is more crowded than scholars like Busk or Darwin ever imagined: there are over 20 identified hominin species from just the last 3.5 million years. Its roots go deeper too. Transforming small scuttling mammals into hominins and eventually Neanderthals took an extraordinarily long time. Immense forests 25 Ma bristled with monkeys as the split leading to the apes was already underway. Early ambassadors of these tail-less primates, the apes *Proconsul*, were already playing away from the trees in East Africa. Then, as the Great Rift Valley cracked open, a massive global cooling began, and apes embarked on huge diversification and dispersals. Evolving into at least

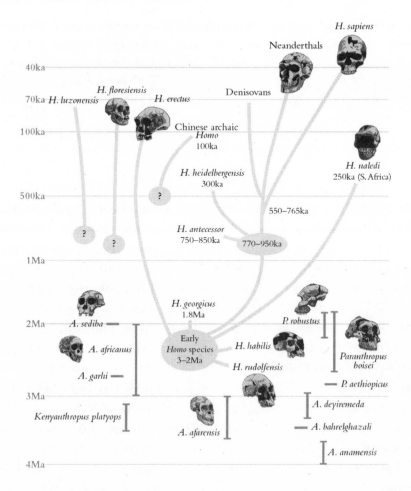

Figure 1 *Neanderthals' evolutionary context as a member of the hominin family.*

100 species between 15 and 10 Ma, the clever fingers of *Dryopithecus* and others searched out foods in wet forests and open lands alike.

From this point there's increasingly detailed fossil and genetic evidence showing when and where our fellow great apes began going their own ways. Orangutans in Asia shared their jungles with the enormous *Gigantopithecus*, whose roaring and chest-slapping must have reverberated through misty dawns.* Shifting back to Africa, around 10 Ma first gorillas separated off, then chimpanzees, and

---

* Despite weighing something like 400kg (880lb), they probably lived relatively peaceably like gorillas and may have survived until comparatively recently.

around this time we begin to see creatures walking on two legs. Not all were hominins – directly ancestral to Neanderthals and us – but it marked a watershed.

A sort of 'mosaic' evolution is found in the rare hominin bones from between 7 and 3 Ma, often with confusing combinations of primitive and advanced anatomy.* There's flat-faced *Kenyanthropus*, and the many australopithecines: true 'proto-humans' who strode fully upright and were growing larger brains. And someone by 3.3 Ma was responsible for the Lomekwian: the simplest stone artefacts. This is probably the beginning of an intensifying feedback cycle between meat and lithics: casual carnivore tendencies likely went back much earlier, but sharp cutting edges are essential for accessing most of the flesh and fat on large carcasses.

It's not yet clear from which prior hominin group the *Homo* genus emerged, but the first certain common ancestor the Neanderthals shared with us enters the stage from around 2 Ma. This was *H. ergaster*,[†] and already by 1 Ma these archaic humans were certainly living as 'true' hunter-gatherers, with much greater technological sophistication than earlier species. They made the first finely shaped lithics known as bifaces – tools formed on both sides[‡] – and carried them ever farther in the landscape as part of lives marked by greater planning and growing social networks.

H. ergaster had essentially human bodies. They were tall, capable runners with no hints of tree-clinging feet, and their tucked-under faces, shrinking teeth and correctly proportioned limbs all flag them as direct ancestors to Neanderthals and ourselves. Most striking are the ballooning brains: these were the canniest, most versatile primates yet to have walked the earth. They certainly moved outside the great African continent, although fossils and simple tools from a previous 'super-archaic' Eurasian population are known from 2 Ma.[§]

---

* This is typified in *Ardipithecus ramidus*, a bipedal Ethiopian hominin that still retained grasping foot bones pointing to tree-clambering.
† In Africa this was known for decades as *H. erectus*, but that name is now reserved for its representatives in Asia.
‡ These tools are also often called handaxes.
§ Fossils and lithics dotted across Eurasia date between 1.8 and 1 Ma, and tools have now been found in China dating before 2 Ma.

But where exactly did the Neanderthals come from? The oldest hominin remains in Western Europe are from the Sima del Elefante fossil site at Atapuerca, Spain, around 1.2 Ma, but they're *far* older than the most ancient bones that look Neanderthal. A younger Atapuerca locale known as the Gran Dolina contains bones dating around 850 to 800 ka that may well represent a population ancestral to both Neanderthals and *H. sapiens*, or at least a close sister group. Named *H. antecessor*, these hominins weren't just in Iberia but were also surviving less-than-balmy climates in Europe's far north-west. This was demonstrated in 2013 by remarkable finds at Happisburgh, on Britain's eastern North Sea coast, where squalls and surging tides uncovered 900,000-year-old ancient clays. Their strangely pockmarked surface turned out to preserve dozens of footprints: the traces of a small group of hominins moving upriver from a massive estuary, where the Thames emptied from a now-erased northern course.* Within just a fortnight this incredible site was erased by the sea, but 3D recording revealed there had been at least one adult with a group of youngsters, from teenagers to small children. The latter must have struggled to stay upright as they negotiated sucking, slippy muds squeezing up between their toes, and preserved pollen grains confirm this wetland was surrounded by cool pine and spruce forest.

Traces of soft bodies so deep in the past are incredibly rare, and their immediacy is in contrast to the dry fossils with which researchers must try to define ancestry for Neanderthals. Genetics tells us they emerged as a lineage around 700 ka, and though the Gran Dolina people lived only around 100,000 years prior to this, they don't look very alike. It's possible that more than one sort of hominin lived in Europe at this point, but many bones over the next few hundred millennia somewhat resemble contemporary fossils from Africa, including a massive lower jaw found in Germany in 1907 that was named *H. heidelbergensis*. While these hominins were long claimed as likely predecessors to Neanderthals, more recent work at a third Atapuerca locale, the Sima de los Huesos, has refined the picture.

---

* The entire course of the Thames was shifted southward by later, much more extensive glaciers around 450 ka.

How the 28-odd hominins – many in exceptional condition – ended up deep inside the 'Pit of Bones' here is something of a mystery. But it's their age of around 450 to 430 ka and anatomy that make them prime suspects for being true proto-Neanderthals, confirmed in 2016 by DNA analysis.*

Why does knowing Neanderthals' deep evolutionary history matter? The misconception that they represent a literal bridge to the apes is still common, despite the millions of years that lie between us both and our closest primate cousins. In purely anatomical terms, we can 'see' Neanderthals emerging slightly earlier at the Sima de los Huesos than the oldest *H. sapiens*-like fossils from Africa around 300 to 200 ka, a temporal gap filled by thousands of generations. But in broader evolutionary terms, they're one of the very youngest hominin species, and *extremely* similar creatures to us. Just as important, getting a handle on where they came from shows how evolution didn't follow an arrow-straight Hominin Highway leading to ourselves. Instead, many simultaneous pathways existed, some finishing in dead ends, others like Neanderthals developing their own unique bodies and minds that were a match for our own. And they didn't stand alone: as we've been discovering in the past decade or so, the *Homo* lineage itself has other stories to tell. The 'hobbits' of Flores, Indonesia, are one case, perhaps going back as far as 700 ka and surviving until around 50 ka. In 2013 more unexpected skeletons emerged halfway round the world in South Africa. Named *H. naledi*, these hominins had some very primitive features and were expected to date somewhere in the millions of years. Instead, they turned out to have been living just 250 ka, making them contemporaries of the Neanderthals and our own early representatives.

But among all the recent human evolution discoveries, perhaps the most startling for understanding the Neanderthals is that they could and did interbreed with us. It now looks like for most if not all earth's current human population, our deep maternal ancestry – an archive of bodies and blood pulsing through to the present – included Neanderthals. This revelation has led to a profound refocusing that

---

* The Sima de los Huesos fossils remain the oldest hominins to have provided genetic material from anywhere in the world.

in one fell swoop shifted them from a dead-end, primitive family branch to bona fide forebears who contributed to what and who we are right now.

This is the new stage upon which we must turn to look again at the archaeology of the Neanderthals. Like revolutionaries, they've uprooted the old dynastic tree in whose crown we proudly perched. Instead, our deep history is more like a shower of leaves fluttering down onto a great river. Some follow fast rivulets, others slow trickles. They split, join and pool in hollows until the waters overspill and reunite with deep channels cutting into the earth.

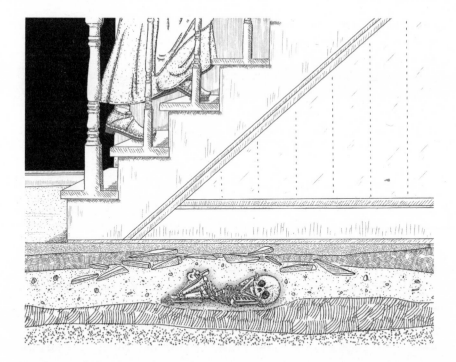

CHAPTER THREE

Bodies Growing

Dawns rise high over the cliffs, and green tinges the branches when her people feel the urge to move again. The furred who live nearby have grown wary, leave no prints to follow. The tiny body that she pushed herself inside out to birth stayed thin, sucked only weakly. Eventually it stopped, went rigid like dried tendon. Yet still she carried it, as skin shrank around stick-limbs and stretched over shoulder blades. Now the group gathers things to leave, and she feels the pull to join them. But putting down her burden is frightening. Inhaling her infant's heady birth-smell, still caught in its dark hair, she crouches at the shelter edge. The clutched bundle is lowered to the ground, and for the first time lies apart from her body. The others approach curiously, reach out to prod, pull, stroke; to know. Yet beasts will come later and this precious thing must be protected. She scoops a hollow, slides the already dusty form down, covers it. Embraced by the soft dirt full of their stony leavings, she draws back and joins the people to walk on.

Skies flicker as the days, years, centuries pass. Soils condense and hold the delicate bones tight. Others come and go above but eventually the vibrations of footfalls stop. Even icy tendrils from heart-aching cold cannot penetrate to the

tiny skeleton. Tens of thousands more winters, then deep thumpings murmur
downwards. The weight lessens. Voices come: new people are building a house,
where none have lived for so long. Patten-shod feet clatter up and down wooden
stairs over the small remains, a lullaby of life above death. In a blink the house too
is gone, then sediments shiver and shift as hands pull at the dirt. Clods crumble,
and a chink of early summer sunlight grazes the white shatter of eggshell-thin
skull. A voice shouts out, 'Arrête! Os!' Rough yet gentle hands – like those that
last touched this lonely little one – reach down, after so long, to pick it up.

It's 19 May 1914. In the space of a month, a car will take a wrong turn in Sarajevo, a gun will fire and two aristocratic deaths will snowball into 200 million. On the other side of Europe in the French Périgord, a life lost 40 millennia earlier comes to light. Denis Peyrony's diary for the day's work at the rockshelter of Le Moustier records the discovery of pathetically tiny bones. One of the twentieth century's most respected prehistorians, he's relatively unsung today, unlike his younger and flashier colleague François Bordes.[*] Born to a farming family the same decade that the Feldhofer and Gibraltar Neanderthals hit the news, Peyrony grew up close to the land. Alongside becoming the Les Eyzies village school teacher, he was fascinated by the deep past.

By 1894 he'd begun collaborating with Louis Capitan, a pathologist turned prehistoric anthropologist, and seven years later they discovered the astonishing ice age paintings of Font de Gaume. As the spring of 1914 ended, Peyrony was a seasoned excavator of Neanderthal sites, including skeletons at the great La Ferrassie rockshelter. When remains were uncovered that May at Le Moustier, he immediately recognised them as a very young baby. Miraculously untouched by construction and demolition right above it, the infant now known as Le Moustier 2 would be lost and found once more over the next 80 years. It's one of many individual Neanderthals with fascinating histories of discovery.

All hominin bones, whether fossilised or not, are special. Bodily representatives of lives lived tens or hundreds of millennia ago, their

---

[*] Also a science fiction novelist, Bordes even has a tram stop named in his honour, directly outside the Université de Bordeaux where he worked for decades.

immediacy captivates. But also their rarity: we have *millions* more artefacts made by Neanderthals, than bones from the hands that once touched them. Nonetheless, collectively, we know them more intimately than any of our other close relations. The handful of remains a hundred years ago whispering stories of another kind of human have today swelled to thousands of fossils from many, many sites. They represent a couple of hundred individuals, from newborn babies to old adults, who – while not decrepit by today's standards – were probably elders in their own society. This rich sample allows us to reconstruct the biological character and diversity of the Neanderthals.

Even with such impressive numbers, every skeletal part is still worthy of velvet-cushion reverence. They're conserved and transported in locked cases like diamonds, or sacred relics. Their priceless value lies in being treasure troves of data on individual lives, and simultaneously acting as windows onto entire populations. Specialists apply a vast range of techniques from biochemistry to high-tech visualisations, examining whole bodies or zooming down to the near daily layers inside teeth. Through the DNA they contain, Neanderthal remains are also our direct connection to these vanished people.

We may be twice-removed from their dry bones – by time, and the glass of museum cases – but encountering them, it's hard not to feel a shiver pass across our own still-vital skin. Perhaps especially so when in miniature: a child's life too abruptly ended, no matter how long ago.

Growing Up Neanderthal

That any bones at all survive over such immense timescales is astonishing, and even more so the fragile bodies of babies. This is particularly the case at Le Moustier, a rockshelter sited where a pier-like limestone ridge juts out between two valleys. Over more than a century these cliffs have seen as many theories about Neanderthals swirl around them as floods of the river. Le Moustier has suffered the growing pains of prehistory as a discipline, being found before it was truly understood how excavation both reveals and destroys the archaeological record. All the clever techniques in the world for studying individual artefacts matter little if a site has been emptied without recording a crucial dimension: the patterns telling us where things came from, and in which order things happened.

Archaeology distinguishes between different parts or features of a site. One level up from individual artefacts is the assemblage, essentially the smallest identifiable grouping of finds that seem to belong together. Typically, assemblages belong to layers, sediment deposits picked out by excavators according to colour, texture or their archaeological content. The sequence of layers is called the stratigraphy. It's the archive of things that have happened in that place, whether human detritus or naturally accumulated rockfalls, muddy sludges or windblown dusts. Digging means removing layers, and those lower down will be progressively older.

There are often complications: erosion, localised inversions or even disturbance by later prehistoric activity. Identifying mixing or movement between layers is crucial, and achieved by carefully examining not just artefacts but also the soils and spatial relationships between things. To paraphrase Carl Sagan, to understand what Neanderthals were doing at any particular site, you must first reconstruct its entire history of formation.[*] This is taphonomy, today recognised as the most important part of archaeology.

Le Moustier had not lain undisturbed between Lartet and Christy's digging and Peyrony's work. A Swiss archaeologist named Otto Hauser was also active in the Périgord, including at Le Moustier from 1907 onwards.[†] By a miracle, neither his attentions nor eighteenth-century building work had disturbed the tiny bones just 25cm (10in.) below the surface; a hair's breadth in geological terms. We'll return to Hauser later, but after his departure Peyrony took over, finding intact deposits beneath a demolished house. It was here the infant had lain hidden; a ghost under the stairs. Despite having excavated several Neanderthal skeletons already, Peyrony recorded virtually no details about the find, beyond asserting the existence of a pit.

But he immediately sent the remains to Parisian anatomist Marcel Boule, already an authority on Neanderthals. Just a week later his

---

[*] His famous quote involves the history of the universe and baking apple pies.
[†] It's the lower shelter at Le Moustier that Peyrony and Hauser were digging; Lartet and Christy worked on the upper shelter, but little information survives.

opinion was received, confirming that the remains were of a newborn. At this point, incredibly, the skeleton disappears from records. Peyrony's journal never mentions it again, and within two months, fieldwork was abandoned as the First World War enveloped Europe. It was assumed for many decades that the baby was another casualty, whether lost or destroyed, of the long years of conflict.

In fact, some of the remains had been waiting safely, albeit unrecognised, just a few kilometres from the site. In 1913, the year before he'd found the skeleton, Peyrony founded a magnificent museum in Les Eyzies. During an inventory of its vast collections some 80 years later, bones labelled 'skeleton' turned up among the Le Moustier boxes. Clearly those of a single newborn, researchers dared to hope it might be the lost Neanderthal baby, last known to have been in Paris. Six months of exhaustive analysis revealed that sediments still encasing some of the bones and containing tiny lithic fragments were identical to those from Le Moustier.

Some of the skeleton must therefore have stayed behind in the Périgord, and over time been somehow forgotten.* But what happened to the parts sent to Paris? Their tale was one of mistaken identity. Peyrony's attention in 1914 was divided between *three* Neanderthal sites, all with skeletons: Le Moustier, and two other rockshelters, Pech de l'Azé I and La Ferrassie. He was sending much of this to Boule, some still inside sediment blocks to be excavated in the lab. Decades later it was noticed that two bones from a supposed double infant burial at La Ferrassie were suspiciously different in colour and condition. Twentieth-century analysis confirmed that adhering sediments and the tiny shards of flint they contained matched not those at La Ferrassie, but Le Moustier. What's more, the thigh and upper arm were the exact parts missing from the Le Moustier baby. Boule's laboratory, fairly heaving with unlabelled Neanderthal remains and likely hectic at the war's outbreak, was the perfect setting for them to become confused and wrongly curated with the La Ferrassie infant.

---

* Even Peyrony omits them from some of his Le Moustier publications, and they were last mentioned in 1937, the year after he retired.

This strange meeting of two little lost souls separated in life by many millennia continues today, since the Moustier baby's limbs remain in Paris more than 160km (100mi.) from the rest of its body.

With Neanderthal bones today treated as priceless objects, such a history seems astonishing. But the tiny skeleton's rediscovery wasn't just a happy coda. The tragic inverted existence of fossil infants, with vastly longer afterlives than their time under the sun, nonetheless offers opportunities. To understand whether Neanderthals physically and cognitively developed as fast as our own children, we need to know what the starting point was. Their fragile remains also remind us that each Neanderthal lifespan was unique, and sometimes interrupted on the great path from birth to creaky-jointed old age.

Let's meet some Neanderthal youngsters who now rest in museums around the world. The Le Moustier baby would have fitted into any newborn's first outfit, but there are children of many ages. Imagine a group photo: barely sitting up at the front are 7-month-olds alongside slightly older crawlers, plus fidgety toddlers and an unruly gang of 3-year-olds. At the back stand children ranging from 4 years up, losing their last babyish features. They come from Spain, France, Israel, Syria; even an 8-year-old from Uzbekistan.

Only DNA can identify if children were female or male, but their age can be worked out from teeth and bones. It's this that suggests Neanderthals grew up at a slightly different tempo than *H. sapiens*, but in varying ways.

Teeth are mostly mineral, making them like proto-fossils that survive decay when bones do not. When researchers count the inner growth lines, known as 'perikymata', they find that formation rates were on average a day faster in Neanderthal children. Similarly, some Neanderthal children lost their milk teeth anywhere up to one to three years faster. But perikymata and tooth development in others match typical rates today. This is shown by one of the most complete young Neanderthals, found in 1961 at Roc de Marsal, a few hours' walk downriver from Le Moustier. Skeletal age estimates came out between 2.5 and 4 years old, but synchrotron radiation microtomography – a type of super-intense X-ray scan – found more advanced molars alongside front teeth lagging behind comparably aged children today.

A similarly contradictory impression emerges from the body of a young boy from the cave of El Sidrón, north-west Spain. His back teeth were less developed than the perikymata implied they should be, and some of his bones looked more like those of a child two or three years younger. Maybe he was just a tiddly kid, but this all shows how Neanderthals had their own range of variation and complexity in development.

Intriguingly, the El Sidrón boy's brain was also a bit behind for his apparent age, and understanding this aspect of growing up is especially important. One fact that sticks in people's memory – probably because it's unexpected – is Neanderthals' supposedly larger brains. Lacking any mummified or frozen corpses, we can't examine them directly; however, brains do leave an impression on the inner skull. Once studied by plaster casts, modern scanning technologies recreate them using inverted 3D models: vanished grey matter reappears in ghostly digital form, even down to a snaking artery that once pulsed with blood. It turns out that their apparently bigger brains are actually due to sex-biased samples: when only males are compared, the difference is much less, highlighting the probability that most complete Neanderthal skeletons belong to men.*

From birth their skulls started out pretty similar in size to ours, but had you cradled the Le Moustier baby's downy head, its shape would have felt a little unexpected. Combining scans of this and another newborn's skull shows how the middle parts of their faces were already slightly pulled out, and they lacked the cute chins of our own babies. There's much debate over the way their brains developed during the crucial early years, and some size projections look remarkably like our own, albeit growing slightly faster. Yet the structure itself didn't develop more rapidly. That tells us Neanderthal babies were hitting the magical milestones of smiling, grasping and babbling at roughly the same time ours do. Small differences eventually add up, though, so physiological childhoods may have finished earlier, leaving less time to learn complex social and technological skills. But what was going on with brains is balanced in other body parts.

---

* Men on average have bigger heads than women.

Bones to Bodies

It's quite amazing that, in total, while the remains from less than 0.01 per cent of all Neanderthals who ever lived have made it through the mill of time and taphonomy, they represent somewhere between 200 and 300 individuals. The majority comprise the odd bone or jaw fragment with teeth valiantly holding on, but between 30 to 40 are much more complete skeletons, and must have originally entered the dirt whole. We'll consider the debates over burial in Chapter 13, but whatever their backstory, each skeleton means a chance to 'know' an individual intimately. And even the bits and pieces are important, helping us explore populations: injury patterns, ages at death and whether men and women used their bodies differently.

One site with an extremely rich fossil record is the Krapina rockshelter, Croatia. It produced over 900 bones from somewhere between 20 and 80 individual Neanderthals.* However, even based on the lower figure, something like three-quarters of skeletal parts are missing. Undoubtedly, its rapid excavation at the end of the nineteenth century is part of the reason, yet the site of Spy was found not much before and has far more complete bodies. In fact, many of the Krapina bones were fractured by Neanderthals themselves, and probably never deposited as entire skeletons. In contrast, El Sidrón was discovered in 1994 – almost a century after Krapina – and is the most abundant Neanderthal fossil locale yet known.† Careful excavation recovered over 2,500 remains, but from just 13 Neanderthals: 4 women, 3 men, 3 teenagers, 2 children and 1 baby. Their bodies had also been broken up, but obviously had originally been more complete.

These cases show that no two fossil sites are alike. Interpreting them requires caution, in particular when trying to look at mortality patterns. The age distribution within human populations tends to

---

* The result varies depending on which calculation method is used because of the highly fragmented state of the bones.
† Echoing Feldhofer, the cavers who first found bones at El Sidrón thought they may have been from soldiers who hid there during the Spanish Civil War.

reflect changing lifetime health risks: there are lots of children, fewer adults and some old people. But fossils aren't necessarily a mirror image of populations. Just as certain sections of society were excluded from churchyard burials, the archaeological record shows that not all Neanderthals were equally likely to be preserved, and this varies for every site.

Bearing that in mind, what we have is still incredibly diverse and enough to mean that our understanding of what they were made of – literally and figuratively – is quite exhaustive. More than ever, we can reconstruct what made them different to us, and even how they experienced the world.

Stand face to face with a Neanderthal, and they'd be recognisable as a kind of human, but decidedly unconventional. Somewhat shorter than average, with broader chests and wider waists, their limb proportions were also slightly different. Beneath massively muscled thighs were thicker, rounder and slightly curved leg bones; nonetheless, unlike countless inaccurate reconstructions, they absolutely walked as upright as us.

Zoom in, and almost everywhere there are anatomical idiosyncrasies; some more obvious, others subtle. As a member of *H. sapiens*, you're your own handy anatomical model: pinch your chin, and under wobbly flesh and muscle you'll feel a bony core. Nearly all Neanderthals lacked this, even as babies. Feel your head: it's tall but globular; your face is short and tucked under the forehead. Though they shared our massively inflated brains relative to other hominins, Neanderthal skulls are shaped very differently. Lower crowns gave them a more aerodynamic, sculpted look, finished by an obvious bump just above the neck.* Larger and deeper-set eyes gazed out from a face whose nose and mouth seemed pulled forwards, but with swept-back cheekbones. Framing all this were magnificent arched brow ridges; not centrally separated like your brows, and much more imposing. But the brain inside – controlling those eyes staring intently back at you – was just as big and deliberating as your own.

---

* This is known as the occipital bun.

The differences go beyond the surface. Feel where your jaw meets your head, and pretend to chew; this moving joint in Neanderthals was shaped quite differently, with a shallow, asymmetric gap and an extra bulge of bone. Slide your tongue to the back of your mouth where your wisdom teeth are (or were); most *H. sapiens* teeth go up against the jaw arch, but in Neanderthals they're pulled further forwards, creating a gap. Perhaps they could wriggle their tongue into that space, and they would also have felt the slightly curled-back edges of their large front teeth with a 'shovel'-like shape. Inside their jaw, their back teeth were different too, often with massive fused roots. Even newborns' tooth 'buds' are distinctive enough to identify them in the absence of other bones.

Reach out in greeting, and you'd see that while your thumb tip is shorter than the second bone, in Neanderthals – even infants – both are nearly the same length. And the hand that clasps yours in an emphatically firm grip is wider, with more flared fingertips.

The panoply of variance across their bodies isn't, however, an indication of being more primitive, in the widely understood sense.[*] Both we and they inherited some ancient features in common, but their lineage kept others we lost, while the reverse is also true. It's more that Neanderthals and *H. sapiens* reflect two diverging pathways of being human, each with their own oddities. The narrow chests, features of the inner ear, or teeth that are unique to us are just as 'weird' as Neanderthals' foibles in the broader context of hominin evolution. Nonetheless, explaining *why* these differences exist, and what they meant for how Neanderthals lived, has remained a key focus of research.

Our inquisitive minds love to discover reasons for everything, but in fact evolution via natural selection is simply about reproductive success, not forging an uber-adaptation. Explanations for Neanderthal biology often focus on possible advantages, but the reality is more complex with multiple influences at work. Building bodies is an interconnected process, and altering one part can cause

---

[*] 'Primitive' in evolutionary terms simply means a feature with very ancient roots shared between species from the same ancestral group.

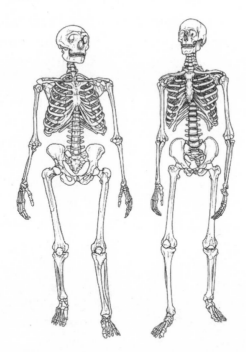

Figure 2 *Skeleton of an average Neanderthal (left) vs. an average recent* H. sapien *(right).*

transformation elsewhere. Genetic mutations are just random copying errors, and can sometimes result in anatomical features that, if they don't negatively affect survival, persist in small, isolated populations.

While a genetic blueprint is extremely important, how hominins live also profoundly impacts bodies, from bones right down to the cellular level. The surrounding environment and regular activities can both leave permanent traces; think how extreme athletes' muscles over time can change their skeleton.

Untangling the influence of genetics versus behaviour is crucial for our understanding of Neanderthal anatomy and lifeways. For example, were the differences in limb lengths inherent, due to use, or both? This is why Neanderthal babies and children are so important, as well as those in the awkward teenage years. One particular individual who helps us understand this second developmental period of life has an especially compelling story: the *first* skeleton found at Le Moustier.

Ice to Fire

After Lartet and Christy's 1860s excavations at the upper rockshelter, the cliffs lay quiet until the early twentieth century, when two of the more bizarre 'Neanderthal afterlife' tales began. One was the baby from Peyrony's work, which is officially known as Le Moustier 2 because another individual had already been found six years before. This Le Moustier 1 skeleton was also swept up in the tides of war and for decades also believed destroyed, but originally found by Otto Hauser, not Peyrony. He began digging the large lower shelter in 1907, working between the buildings. It wasn't until the following spring that a spade wielded by Jean Leysalles[*] bit down through thick lower leg bones, and unlike Le Moustier 2 (which remained hidden) there is a far more detailed record of what they found.

Over several days more bones emerged until on a rainy night the skull was uncovered. It took several months, however, before everything was removed. The delay led to claims Hauser was staging things for wealthy visitors, but in fact the motivation, together with covering and re-excavating the bones, may have been to ensure the remains were protected and witnessed by experts. Hauser had picked Hermann Klaatsch – an anthropology professor and a world expert on Neanderthals[†] – for the team, and invited international scholars to attend the final 'raising' on 12 August. However, only German colleagues turned up. It largely fell to Klaatsch to remove the bones as Hauser took photographs; a unique archive for the time. After attempted reconstructions of the skull in front of the village cafe while children watched,[‡] all the remains were securely nailed in a wardrobe before being boxed up and sent to Germany. Thus began nearly a century of extraordinary travels.

---

[*] Leysalles owned a local cafe where Hauser's team stayed beneath rockshelters at Laugerie, across the river from Les Eyzies.

[†] Klaatsch had studied first-hand the Feldhofer, Spy and Krapina collections, plus other hominin finds.

[‡] A then 7-year-old girl, Mme Guimbaud, recalled watching through the fence and witnessing the skull knocked off a table at one point; it broke and the reconstruction had to restart.

Hauser had arranged a lucrative sale to the Berlin Museum of Ethnography, where for decades the skeleton was displayed as a prize object. That repose ended early in the Second World War, when as an irreplaceable treasure it was secreted within a massive bunker in the 'Zoo flak tower'. One of several fortified structures housing anti-aircraft systems and a vast air raid shelter, it also functioned as secure storage for priceless cultural materials.

Towards the end of the war the Nazis tried to shift the hoard, and some things made it out, but much did not. As Berlin fell in May 1945, the Zoo tower hosted a last stand and suffered heavy bombardment, together with the remaining animals of the zoological gardens.[*]

Only a few hundred beasts out of thousands survived, and alongside the Neanderthal in the tower's darkness, a strange Pleistocene rearguard – lions, hyaenas, an elephant and a hippopotamus – were left waiting for the Red Army. Ravaging the city, the Soviet Trophy Commission also looted nearly 2 million objects from the flak towers and elsewhere across Germany. At some point the Le Moustier 1 skull was carried off on another train, stacked alongside Old Masters and golden treasures from Troy, all headed for Moscow.

Over a decade afterwards, from behind the Iron Curtain the skull returned to Berlin. Its Russian sojourn had protected it, but the rest of Le Moustier 1 left behind hadn't been so lucky. Shortly before the war's end, over 2,000 Allied aircraft had begun a massive bombardment, devastating the museum where, incredibly, the skeleton had still been on display. The headless body must have lain there as the walls shook and were engulfed by the inferno. It became buried for a second time, lost among a vast mass of rubble and melted artefacts, until the whole mess was painstakingly excavated 10 years later.

But it took another three decades for reunification. Returned war booty was in disarray and the skull was only recognised with meticulous cross-referencing of old photos and catalogues. After the Berlin Wall was torn down, restoring sundered families and friends, in 1991 Le Moustier 1's remains finally came together once more.

---

[*] The Zoo tower withstood everything the Soviets could throw at it except huge amounts of dynamite.

Scientific pilgrims began arriving to study this famous relic, and 99 years after its original discovery, the first definitive study of Le Moustier 1 was finally published. Probably a boy aged between 11 and 15 years of age, he's the most complete adolescent Neanderthal known. His skull had the classic long, narrow form, its widest point towards the back, but he looks to have been in the middle of growth spurts. His face was expanding faster upwards than forwards, so he lacked the distinctive space behind his wisdom teeth, and neither his brow ridges nor nose were as imposing as an adult's. Thanks to Le Moustier 1 we know that Neanderthal teenagers had their own awkward teenage phase; perhaps excesses of hormones meant he was also prone to spots and a quick temper.

As if being shuttled across Europe, bombed and burned was not enough, his skull had endured five physical restorations, some more gentle than others. But twenty-first-century tech allowed a more accurate virtual effort, using a mirror image to 'de-skew' parts warped from sediment pressure. The result revealed a face that, while still immature, was already dominated by huge eye sockets, and looks quite different to any living teenager. Interestingly, though, his brain was already comparatively large, so he may have ended up a particularly large adult.

A last mystery remains: at some point between the war's end and the reuniting of the skeleton in the 1990s, a front tooth and some facial bones went missing. Did this happen in Berlin as the muddled war booty was unpacked again? Or perhaps earlier, when crates from the Zoo tower were opened somewhere in the Soviet Union? One might imagine the skull being damaged as it was assessed by soldiers surrounded by bullion and oil paintings. We'll never know, though the idea of a lost Neanderthal tooth still lying in the darkness of a Russian salt mine is rather enchanting.

Faces and Senses

Neanderthal skulls are captivating, but even without warping, the complex functions of their structures are hard to reconstruct. In general, unpicking the reasons for anatomical differences between them and us is an enormous can of worms. Skull geometry intersects in complex ways, and researchers are only just beginning to understand the genetics and biochemistry behind bone growth. Probably some of the overall skull shape is down to random drift in genes over thousands of generations, but it's always features that might have had evolutionary advantages that grab attention, particularly in an ice age world. However, these days the dominance of glacial conditions as drivers of their physical evolution is far less obvious. Instead, their bodies may have been moulded in large part by how they lived.

Starting at the top of the body, we can explore how ideas have altered. Explanations for those huge brows have ranged from structural support for the big face, to acting as natural sun visors. One recent, slightly left-field theory claims that Neanderthals relied on them to communicate, in a similar way to baboons signalling status by waggling brightly coloured brows. But modelling revealed that massive bony ridges actually make that harder, and chimpanzees show that there's plenty of other ways to convey meaning with the face and body.

Next come the eyes: just how did Neanderthals see the world? Their sockets were larger than any *H. sapiens* past or present, and bigger eyes potentially mean more photo-absorbing retina and greater light sensitivity. Why would they need this? Assuming Neanderthal heartlands lay in western Eurasia, this region is at a much higher latitude than most of the African continent, dealing with less light and especially dim winters. Northerly animals *do* tend to have bigger eyes, and on average, even people from higher latitudes have eyeballs up to 20 per cent bigger than those from near the equator. Expanded eyes would require a larger visual system, and this area, housed in the distinctive occipital bun, is clearly bigger in Neanderthals.

Better vision in low-light conditions might have usefully extended the day, but even allowing for slightly larger Neanderthal brains, it could have left less computational capacity for other things. The frontal cortex in particular deals with social interactions, and its size seems connected to larger social networks. Our own brains are particularly inflated in this area compared to Neanderthals. But on the other hand, minds are famously flexible, physically adapting after serious injuries by shifting tasks between zones, and even growing new tissue in frequently used areas.* Without directly observing Neanderthals in an MRI scanner, it's hard to be certain if their big eyes and greater visual neuron volumes held back other cognitive and social capacity.

Whether or not they had owl-like vision, Neanderthals probably shared our weird (among apes) white eyeballs and palette of iris colours. Reconstructing individuals' pigmentation, however – whether in eyes, hair or skin – is surprisingly tricky. Many genes are involved, interacting in different ways to produce a fiendish number of combinations. Similar to our own evolutionary history, very dark-skinned Neanderthals are unlikely, because even with continuous sun exposure, getting enough vitamin D at the higher latitudes they lived in would be impossible.

Neanderthals therefore very probably developed lighter colouring, but DNA shows it wasn't identical to the various biological mechanisms operating in people today with Eurasian ancestry. Basic genetic comparisons indicate that the combination of red hair and freckles is *possible* in some Neanderthal individuals, but we can't be totally sure those genes were expressed exactly the same as in us. What's clear, though, is that their population was also diverse: the ginger-freckle marker is found in some Spanish and Italian Neanderthals, whereas other analysis indicates that individuals from Croatia had darker skin, eyes and hair.

Whatever the colour of their eyes that gazed at herds on the horizon, paying attention to the auditory world was just as crucial to survival. High-resolution bone scanning technology shows that Neanderthals' tiny inner ear bones and soft tissues beyond neither mirrored ours nor the common ancestral form. Could Neanderthals have heard differently? Surprisingly, functional modelling suggests that these parts still transferred and amplified soundwaves exactly as in your own ears.† Evolution seems to have adjusted their shape in parallel

---

* For their professional qualification, London black cab drivers must memorise over 20,000 of the city's streets, causing the rear of their hippocampus – which deals with spatial memory – to significantly enlarge.
† At least, in the lower frequencies that are possible to reconstruct.

with changes to the skull, while keeping them tuned into the same sorts of sounds as us. And there's a fair amount of evidence that for humans, to a large extent this means sounds we make using vocal communication.

If their vision was perhaps more acute and their hearing just as sensitive to voices carried on the breeze, what was their experience of smell? In 2015 a perfume called 'Neandertal' was released,[*] claiming to be inspired by the 'hot flint aroma' produced from making stone tools. Remarkably, this isn't just sales-speak: knapping flints does produce a distinctive scent. It's often compared to a fired gun, and is exactly how astronauts described the smell of moon dust. Around half of the talcum-fine lunar surface is asteroid-pulverised silica: the main ingredient in flint, quartz and other commonly knapped rocks. Strange to think that moon-tang would have been more familiar to a Neanderthal than to Neil Armstrong.

However, while Neanderthal visual systems were enlarged compared to ours, their olfactory bulb – the brain region that deals with scent – was reduced. Interpreting that as lesser sensitivity, however, requires caution, and it's here that once again genetics comes in.

Though not identical, there's clearly some overlap in scent-detection genes between us and Neanderthals. One molecule, androstenone, is intriguing. It contributes to the 'perfume' of human sweat and urine, and among the roughly 50 per cent of living people who can detect it,[†] strong dislike is one reaction. If some Neanderthals could also detect this odour, maybe it had a useful function. Androstenone affects hormones and emotions in people, but it's also given off by wild boar, and smelling the porcine version has a dramatic effect on dogs. Perhaps it's involved with hunting: being able to smell a herd over the hill or detect that an animal had passed by would have had strong advantages. Whatever the specifics of particular odours, though, Neanderthals very likely experienced smells – pine resin, horse sweat, old smoke – as a strong trigger for memory.

Sucking in scent through their prodigiously sized noses raises the question of why they were so huge. Massive nasal apertures dominate

---

[*] Available in a handmade biface-shaped bottle; 90ml (3fl. oz) cost just shy of £200.
[†] Specifically, the very similar pheromone androstenol.

the middle of their faces, and in the flesh they'd have had a hyped-up King Charles II–esque profile. Microscopic study of their skulls found comparatively huge numbers of bone-growth cells in the mid-face, showing how much the whole area was pushed forwards. Biomechanical models, however, don't support theories that this snouty face gave increased strength for intense chewing (although the next chapter will discuss how they definitely used their teeth for more than eating). In contrast, the smaller, tucked-under appearance of our faces are caused by bone-absorbing cells, and unexpectedly, we actually have a stronger bite.

The nose itself functions just as much for respiration as for smell. Modelling of nostril airflow in Neanderthals reconstructed the soft tissue for the skeleton from La Chapelle-aux-Saints, France, confirming that the entire set-up was nearly a third larger than living humans. In general, one function of the nose is to 'condition' air by warming and moistening it before it reaches our sensitive lungs. That can be especially important in arid and cold conditions, and in some ways Neanderthals' large internal nasal structures resemble reindeer and saiga antelope, which have extensive mucous membranes to reduce dehydration and heat loss. Fascinatingly though, the internal structures in Neanderthals appear to be *worse* at air conditioning than our own (though better than *H. heidelbergensis*). What those cavernous nostrils could do, however, was control air flow, allowing Neanderthals to snort in air at almost twice the rate we can.

Well over 150 years of staring at Neanderthal remains at increasingly minute scales means we know an astonishing amount about them, sometimes in mind-boggling detail. Tracing how they grew, developed and sensed the world finds remarkable congruence with ourselves. Blinking against the low glare of winter sun, keeping an ear open for the sound of children playing, or wrinkling noses against woodsmoke were shared experiences of humanity across the millennia.

Nevertheless, Neanderthals *were* different anatomically, in a multiplicity of ways. Interpreting features large and small across their bodies means rethinking evidence for evolutionary adaptations to their very particular world. We're still unravelling the functions of things like bigger eyes, but other aspects – such as noses – may be much less to do with arctic-ready adaptations than once believed. Instead, running a fuel-hungry body with a high-impact lifestyle may have been the biggest survival challenge they faced.

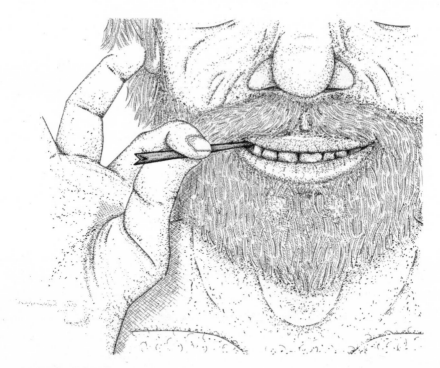

CHAPTER FOUR

Bodies Living

Soles slapping along
Feet flying beyond the striders and shufflers
It's good to run!
Puffing lungs, cheeks hot in the wind
Berries! Swift fingers pluck.
The hill now
Short legs fall behind.
Until there, the dark skull-eye of the rock,
Watching for the people.
Tired, empty stomach,
Tall ones bend down, give a fat-chew.
Teeth work, copying the jaws all around the fire circling to soften hides.
Clear twilight, time to eat:
Small hand still learning the cut–chew, cut–chew,

and, like everything, marks are left behind on skin, enamel, bones.
then
Over-bright eyes full of hearthglow,
Lids droop, head sinks into a lap of dreams.

Neanderthals have long held the title of Most Hench Hominins. Though shorter than us, they weighed about 15 per cent more, were bulkier and had thicker, heavier bones. While far from hardcore bodybuilders, their bodies were strongly muscled. Traditionally, the explanation for this was basically 'ice ages'. Since the nineteenth century, biologists have known that cold-adapted species – often in higher latitudes – tend to have larger bodies, but shorter limbs. Such stumpy proportions mean less surface area and better heat retention. But seasonality is also involved: body mass is linked to the length of the growing season, because this dictates food abundance and availability, and being bigger means creatures can have more fat stores for times when food is scarce.

Living humans appear to roughly follow these geographic and seasonal bodily patterns, with Europeans tending to be stockier and have thicker bone shafts than those of African backgrounds.* On the face of it, Neanderthals also fit the trend, and since initially their bones were mostly found in clearly glacial contexts, this idea was hugely influential. Long-lasting and severe bodily stress from living in cold conditions causes biochemical reactions that involve the production of growth hormones, and as we saw with noses, the majority of Neanderthal anatomical idiosyncrasies were long viewed through this lens.

But robust bodies aren't unique to them, nor to cold climates. Older hominins, and even very early *H. sapiens*, were sturdier and had thicker bones than living people. Moreover, recent research shows that Neanderthals' more compact size and shape would have bought

---

* However, the ancestors of most people with European backgrounds have not been in northern latitudes for more than a few millennia: older hunter-gatherer populations in Europe were largely replaced by Neolithic farmers from the Near East.

barely 1°C of extra cold resistance, and their large brain size doesn't fit a thermal trend either. To be fair, some nineteenth-century scholars like biologist Thomas Huxley saw Neanderthal brawn not as a mirror of brutality, but resulting from highly mobile lifestyles. Backing up his prescient views, the past few decades have seen a significant shift towards more nuanced explanations, often thanks to anatomical research and biomechanical models.

It's the impact of extremely demanding lives that's looking increasingly important. Neanderthals had to balance conflicting needs: more massive bodies coped better with their intensive lifestyles, but needed a lot of fuel. And extra calories require more oxygen to convert into energy. Therefore, respiratory efficiency became crucial, exemplified by huge noses siphoning in air and larger chests to accommodate more capacious lungs: every breath they inhaled was deeper. Furthermore, experiments show that increased exercise makes not just the limbs of young animals more robust, but their whole bodies. Skulls get heavier, brow ridges larger, muscle attachments bigger. And body proportions change: children growing up at high altitudes, whose metabolisms must work harder with reduced oxygen, can develop slightly shorter legs. That all sounds very familiar to what we see in Neanderthals.

From finger to toe, their skeletons show clear evidence for thicker bones and larger muscles, making them at least 10 per cent beefier than even similarly stocky *H. sapiens* populations. This was definitely genetic, since it's visible in babies, but even youngsters also had physically tough lives. Le Moustier 1's legs were already well developed before his teens because of enormous activity.

Average Neanderthal leg-to-arm strength ratios were even greater than cross-country competitors running 160km (100mi.) per week. But it's not necessarily only about distance: Neanderthal relative limb bone thickness more closely resembles prehistoric and recent *H. sapiens* populations who habitually travelled over extremely rugged landscapes. And their strength wasn't only in legs, with arms as powerful as many of today's athletes.

So they had the bodies to cope with rough country, but climate is still likely part of the equation. It's becoming apparent that some complex feedback processes were at work in shaping Neanderthal bodies: intensely active lifestyles in challenging landscapes, with cold phases fine-tuning particular features. Adaptations developed during glacials might have persisted even during warmer times, sometimes helpfully, in other cases perhaps creating their own challenges.*

That Neanderthals moved around an awful lot has never really been controversial, but the *manner* in which they did so was often debated. The knuckle-dragger cliché was at the margins of discourse from the start, despite anatomical evidence since Spy in the 1880s that they walked as upright as we do. As well as the leg bones from Le Moustier in 1907, a couple of years later and just a few kilometres west, Peyrony and Capitan uncovered the La Ferrassie 1 (LF1) male skeleton.

Still one of the most complete Neanderthal skeletons known, it was missing only a kneecap and small hand and foot bones. Though short at 1.6m (5ft 2in.) he was beefy, weighing probably about 85kg (190lb), and patently a fully upright walker. But it was the La Chapelle-aux-Saints skeleton – found in 1908 – that had huge influence. Boule inaccurately reconstructed the legs and spine as stooped, and this image was then projected to millions via an illustrated reconstruction in 1909 that – down to the prehensile toes – was decidedly ape-like.

Today there's no question that Neanderthals were fully upright, but it's possible that walking alongside them might have left you slightly out of step. Some anatomical differences indicate a gait that wasn't identical to ours, and being shorter means they likely covered ground around 4 to 7 per cent more slowly. However, recent biomechanical analyses don't point to drastically less efficient locomotion, especially when compared to roughly contemporary early hominins. On that basis, Neanderthal women used just 1 kilocalorie (kcal) more energy when walking, and if their overall heavier bodies are taken into account, their legs come out *more* efficient. While an image of them as tireless striders fits the skeletal evidence, running doesn't seem to have been a Neanderthal forte. With reinforced foot arches to cope with their greater bulk, sprinting and especially endurance running would have been less efficient.

---

* Unexpectedly, overheating during the very balmiest climatic periods might have been a real issue.

Perhaps Neanderthals might have lost to any *H. sapiens* in a 5,000m track race, but on the other hand their Achilles tendons made them much more sure-footed on uneven ground.

Biosocial Beings

So far, Neanderthals seem to fall somewhere between hardcore hill-walkers or trail-sprinters: huge lungs puffing, chunky thigh and calf muscles flexing as feet pound terrain. But what were those strapping arms all about? They had a seriously powerful wrist twist, and would've made champion arm-wrestlers. But most of the strength was in the upper arms, a pattern unlike any recent *H. sapiens* populations. And there are interesting asymmetries: we know from lithics and wear patterns on their teeth that Neanderthals were right-handers like us, and the dominant side was between 25 to 60 per cent more developed. That's close to what we see in cricketers or tennis players and implies strenuous, habitual activity, often assumed to be spear hunting. Fossils including a 200,000-year-old isolated arm from Tourville-la-Rivière, France, confirm that some Neanderthals were making upwards and rotating movements similar to throwing, and as we'll see later, actual javelin-like spears are preserved. But overall, their shoulder mechanics aren't as well suited to overarm movement as ours, and the asymmetric patterns in arm muscle development also don't match this.

Another possibility exists: electrode monitoring experiments reveal a better match for Neanderthal-style muscle flexing that's not spearing, but one-handed scraping. We know they were scraping a number of materials including wood, but the working of animal skins may have been the main task causing their right arm asymmetry. Chapter 10 explores in detail what we know about Neanderthal hide working, but fundamentally it's very intensive work. Each animal skin can require more than 10 hours of scraping in multiple phases, and so even if they only processed half the hides they obtained, a Neanderthal might end up doing 100 hours of scraping annually.[*]

---

[*] This is based on estimates from different North American Indigenous cultures including the Huron using about 30 skins per year for a family.

However, the electrode experiments threw up something else. It turns out that when spearing, it's not the dominant elbow joint that takes the strain but the opposite one, as it helps guide the shaft. Exactly this pattern of left elbow asymmetry is seen in Neanderthals, caused by huge tension while the arm was straight and extended. So their bodies may well record hunting, but in an unexpected way.

Should you pluck up the courage to offer your hand to a Neanderthal, would it be crushed? Only if they chose. Differences in bone anatomy and significant hand muscles gave them formidable power, but without sacrificing dexterity. Recent analysis doesn't support claims that they were less nimble-fingered, but their hands do seem built for dealing with massive forces and transmitting them into the arm. They possessed huge strength when grasping in the palm, and big muscles combined with large finger tendons ensured a steely grip. Their odd-looking wide fingertips – probably visible in the flesh – were likely adapted to hold things extremely tightly, with only a minimal loss of precision for very fiddly tasks.

Biomechanics suggests that it might have been knapping driving at least some of these anatomical features. The base of the thumb comes under greatest strain, matching exactly where their anatomy was suited to coping with strong forces. And when using lithic tools, it's the outer edges of the thumb and fingers that need to be strong, again mirroring their hand anatomy.

Even if their deft fingertip manipulation *may* have been marginally less than our own, the archaeology shows that they were perfectly able to make and use miniaturised artefacts. It may be that the greater finger grip strength and thumb flex compensated, allowing them to tightly hold teeny things.

But individual Neanderthals show variation, potentially related to lifetimes of doing different tasks. Biosocial archaeology explores skeletons according to age and sex, as a means to work out patterns in who did what. La Ferrassie 2 is especially well studied as one of few relatively complete skeletons identified as female. Well, *probably* female: sex identification relies on shape and relative sizes of particular bones like the pelvis. One of the most complete and famous females was found in 1932 at et-Tabun Cave, Mount Carmel, in then-Palestine. In remarkable serendipity, three women archaeologists were involved in unearthing her bones. The first fingers to touch Tabun 1, holding up a tooth to the sun, belonged to Yusra, a local expert fieldworker. Working alongside her was Jacquetta Hawkes, newly graduated archaeologist invited to the Tabun excavation by its director, eminent prehistorian Dorothy Garrod.* Sex identification isn't 100 per cent certain without DNA testing (which has yet to be done on any Neanderthals from the Near East), but assessments of how less heavily built a skeleton is can also help. Found just 50cm (20in) away from LF1, LF2 is also obviously adult, but clearly much less robust. Even so, the average size difference between living men and women is pretty similar to what we can perceive in Neanderthals: European males bulked out between 77 and 85kg (170 and 190lb), females 63 and 69kg (140 and 150lb).

Also like us, sometimes Neanderthal men and women tended to use their bodies differently. Legs overall were equally strong but women showed some asymmetry, with more buff thighs than lower legs. Differences in the amount of walking versus running might explain this, connected to the sort of terrain covered, but it's hard to model specifics.

There's also sex dissimilarity between upper and lower arm bones. LF2's biceps probably weren't as strong as the average Neanderthal man, or perhaps surprisingly even women from a range of H. sapiens populations. But her lower arm had a strength measure more extreme than any comparative group. This must reflect repeated, particular actions, but intriguingly in general Neanderthal women don't display asymmetry between right and left arms like the men. Whatever they were doing with their lower arms, it mostly involved both hands. Double-handed hide working is a distinct possibility, potentially during a particular preparation phase, based on studies of some hunter-gatherer women.

Another part of the body with interesting stories to tell is teeth. Crucial for working out age, they also record how their owners used their mouths for things other than gnawing or chewing. In many cultures where only a knife is used to eat, people slice off choice

---

* Garrod had already excavated another Neanderthal in Gibraltar, and seven years after the Tabun find became the first woman professor at Oxbridge.

chunks while holding the food in their mouth. As the edge drags against enamel, particularly with a stone tool, minute scratches get left on the teeth. Such marks are visible in Neanderthals and provide key evidence for rates of handedness,* but also social differences. Recent work comparing adults by sex – including some from El Sidrón – found that scratches tend to be more numerous and longer in Neanderthal women.

Other dental stigmata exist too. Think how useful your teeth can be: tugging at an especially tight knot or carrying things when hands are full. Ethnographic data tells us that mouths can be vital tools either to hold or process things by masticating. This might be animal products like sinew or plant materials. It's long been obvious that Neanderthals' front teeth are extremely worn from use like this, even exposing dentine. In particular they seem to resemble hunter-gatherer societies that used their mouths for hide working: clenching the teeth like a vice through which to drag the hide and soften it up, or process sinews. There are also sex differences here: some women seem to have much more intensively worn front teeth. The closest match is historic Arctic hunter–gatherer societies such as the Inuit, Yupik, Chukchi or Iñupiat, where women spent much of their time working hides. However, the pattern isn't identical: Neanderthal women were using their *upper* front teeth far more, and lacked heavily worn back teeth from chewing skins. Either they had a very particular method, or there was another task happening that is as yet unidentifiable.

To add to the impression that some tasks tended to be done by one sex more than another, Neanderthal women have a higher frequency of chipping damage on their lower front teeth, while in men it's on the upper set. There aren't any ethnographic clues for how these particular asymmetries were produced, but since these patterns are roughly similar across Western European sites, we might be looking at widespread commonalities in activity organisation.

A note of caution is needed, however, since the sample size for female skeletons is limited, plus the way we interpret things can be

---

* The scratches follow the direction the knife sliced, so their angles reflect handedness.

tinged with bias. For some it's easier to imagine Neanderthal women working hides than the possibility that asymmetry in men's arms might be better explained by one-handed scraping, rather than spearing. Moreover, we have little idea how they defined their own categories of gender, which goes beyond the spectrum of biological sex variation. Their social distinctions need not have been binary nor mapped directly onto anatomy.

Certainly, child bearers carried – literally – an extra lifetime's biological load, some of which might be possible to tease out from the bones. Maybe some of the locomotion difference in women came from routinely carrying youngsters, and a good proportion of the hide working that left its mark on arms and teeth was probably for children's clothing and wraps to carry infants.

Just like living people, however, individuality shines through. Some Neanderthals enjoyed or were better at particular tasks than others. They would've been likely to do those more often, creating the conditions for proto-craft specialists to emerge. Highly unusual tooth damage may reflect this. One man from L'Hortus Cave, France, had abnormally intense chipping on just one of his front teeth from use over a long time period. Meanwhile, the El Sidrón 1 man had major damage, but on *both* front teeth. It's not clear what caused this; retouching* lithics by biting is one possibility, since while sounding unlikely, it's known from a few hunter-gatherer cultures.

Probably the most numerous social category in any Neanderthal group was children. Born stronger than us, intense activity further toughened their little bodies. Even before the age of 10, in Uzbekistan the Teshik-Tash child's legs must have walked huge amounts, while Le Moustier 1's teenage arms were almost as muscly as an adult's. Youngsters' teeth also show them practising or joining in with adult tasks: at Sima de los Huesos, older children and teenagers had already begun to wear off their enamel. But even the littlest ones here and elsewhere have some distinctive clamping wear, suggesting that hide working was one thing they started to help with early on.

Overall, children's tooth micro-wear increases with age, but it's more complex than just greater amounts of mouth use. Micro-scratches

_____

* Secondary knapping.

in the young boy from El Sidrón were not only fewer but also diagonal, rather than vertical. This means he'd learned to eat like a grown-up using a lithic, but wasn't really doing a lot of other tasks with his mouth. There's a hint of the social settings where he and other children may have been learning and copying, since their overall tooth damage pattern on average resembles women's more than men's.

Neanderthal children certainly learned by doing, and from birth had front row seats for most of the tasks they needed to master as adults, whether slicing fat off muscle, eating around a hearth or walking the land. There was probably some teaching for particularly complicated things, but Western standards of appropriate child safety and supervision aren't shared by all societies. In many hunter-gatherer cultures youngsters will play with sharp tools, sometimes wielding them even before they can walk, and independently forage together. But busy childhoods brought with them a high cost, which some of the youngest paid.

Burdens in the Bones

Had the seventeenth-century philosopher Thomas Hobbes known about Neanderthals, he would have undoubtedly included them in his famous description of hunter-gatherers as having 'poor, nasty, brutish and short' lives, in 'continual fear and danger of violent death'. This prejudice has often been applied to Neanderthals and hominins, but drew extra support from their own bodies. Most complete skeletons bear marks of at least one affliction, whether illness or injury, and sometimes a veritable 'series of unfortunate events' befell them. Yet at the same time, modern research is tending to show that while Neanderthals certainly had it tough, it wasn't necessarily worse than for other humans living in such challenging environments.

A case in point concerns teeth growth interruption lines, quite common in Neanderthals and long claimed to show they suffered periods of childhood starvation. All the El Sidrón individuals had them, formed from toddlerhood to about 4 and even as late as 12 years old. Some there and elsewhere like Le Moustier 1 have multiple such phases. But it's not universal, and others had none. Better biomedical understanding now suggests that while they *can* be due to malnutrition, mostly they record systemic bodily stress such as a serious viral illness or infection.

Moreover, Neanderthals weren't more afflicted than other human groups. Samples from prehistoric Inuit sites show longer-lasting series of lines from early childhood, whereas Neanderthal children seem to be affected from toddlerhood onwards. This is perhaps related to the introduction of solid foods, a classic time for more exposure to germs. Interestingly, early *H. sapiens* samples show more lines in young babies, which could point to even greater health stress than Neanderthal infants suffered.

Even without growth lines, it's safe to say that most Neanderthals could have done with a visit to the dentist. Like many prehistoric societies, poor oral health was common and many must have been in varying amounts of pain from extreme tooth wear. Heavy calculus caused receding gums, trapped food and eventually led to abscesses in some individuals. Micro-wear scratches on one young adult from El Sidrón show over time that he switched his preferred hand for eating, potentially connected to a nasty root abscess.

Others suffered development conditions: Le Moustier 1 had an un-erupted canine, and an unfortunate individual from Krapina had two mal-positioned back teeth, one of which was impacted and probably very painful. Yet like us, Neanderthals tried to be their own hygienists. Distinctive grooves show that some practised habitual tooth-picking, especially on sore areas.* Rates differ between sites however, potentially indicating varying states of health or perhaps even social traditions. The chances of archaeologists finding actual toothpicks seems infinitesimal, yet at El Sidrón a conifer wood fragment was embedded in calculus right next to a tooth with pick-grooves.

Twenty-first-century analysis can even reveal things that aren't visible on bones: again at El Sidrón, DNA in calculus came from a gut parasite that causes nasty diarrhoea.† Overall, while they don't appear especially sickly compared to other hunter-gatherers, Neanderthals

---

* Experiments to accurately identify this type of wear involved spending hours picking teeth in 18th-century jaws from a Madrid ossuary.
† The particular species *Enterocytozoon bieneusi* is common in pigs and can be caught by eating meat contaminated by faeces.

endured a smorgasbord of other health problems. Some conditions are actually rare today: the Zeeland Ridges skull fragment is the only truly 'submarine' Neanderthal yet found, and bears the mark of a genetic condition. Dredged up in 2001 from some 30m (100ft) deep, 15km (10mi.) offshore from the Netherlands, a large ulcer caused by a deep cyst is clear. Certainly visible during life, it may have caused its bearer little bother, but such ulcers can lead to problems with balance and headaches, or more seriously, bleeding on the brain, convulsions and seizures.

Other conditions, which today have varying symptoms, are seen in multiple Neanderthals. Three individuals – two from the same site – bear distinctive bony growths on the spine and elsewhere.[*] They may have been pain-free, or caused backache, mobility problems or even the seizing up of joints entirely. Interestingly, it's most common today in males, and connected to meat- and fat-rich diets seen in historical cases, including the Renaissance Medici dynasty of Florence, Egyptian pharaoh Ramses II and medieval monks and merchants who enjoyed high-calorie food.

Several Neanderthals elsewhere had different bone growths on the skull.[†] Probably caused by hormones including higher oestrogen levels, it can lead to headaches, thyroid issues and substantial weight gain. Older women have a higher lifetime exposure to oestrogen, perhaps explaining why we see it on the Forbes' Quarry skull from Gibraltar, which belonged to a woman aged at least 40. But men with low testosterone are also at risk, putting the two known male Neanderthal sufferers in company with the eighteenth-century castrato singer Farinelli and modern prostate cancer patients.

Despite common belief that Neanderthals barely made it past their twenties, the Forbes woman is among a number of middle-aged or older Neanderthals. Even if they grew slightly faster as children, this would have had a minimal impact on total lifespan, so biologically there's no reason septuagenarian Neanderthals couldn't have huddled by hearths. The apparent rarity of individuals over 50 is common across the archaeological record of all periods, since accurately

---

[*] Diffuse idiopathic skeletal hyperostosis.
[†] Hyperostosis frontalis interna.

identifying age beyond that is tricky* and elderly bones tend to be more fragile.

So what did decades of hard living Neanderthal-style do to you? An especially famous old geezer lived at Shkaft Mazin Shanidar (Great Shanidar Rockshelter), Iraqi Kurdistan. Initially dug between 1951 and 1960 and currently being reinvestigated, this truly spectacular site produced more than 10 mostly complete skeletons. The very first found, known as Shanidar 1, was probably well into middle age and had overcome an astonishing number of physical difficulties.† Some time before adulthood his right upper arm suffered a terrible multiple fracture, and shrivelled after healing badly. Incredibly, it looks as if the lower part was amputated, presumably by someone else. While he survived and coped with these injuries, his right shoulder blade was malformed, and it may also have led to his collarbone being both unusually small and developing a serious bacterial infection, perhaps also damaged during the original injury.

But his troubles didn't end there. He had one of those probably painful bony growth conditions and poor hearing that meant he most likely found it hard to follow his fellows' communication.‡ To top everything off, he also survived multiple head injuries. A horrendous wound crushed the upper left side of his face, distorting the bone all around his eye and cheek. This may have been the same dreadful event that fractured his arm, but several other blows heavy enough to affect the bone definitely happened at later times. Whenever it took place, the large injury left him not only with massive soft tissue damage but probably also partially sighted – if not blind – in one eye.

In spite of living with chronic pain and his many challenges, Shanidar 1 adapted to daily life in the group. He continued to use his dominant arm despite missing a hand, perhaps even knapping using

---

* Tooth wear is one measure, but beyond a certain point it isn't reliable.
† This individual was immortalised as the warlock-like Creb in Jean Auel's novel *Clan of the Cave Bear*, which inspired a generation of prehistorians including myself.
‡ His partial deafness was caused by benign growths inside the ear; in medical literature they're known as 'surfer's ear' from being associated with swimming in cool water, but in fact they can be caused by infection, injury or even prolonged cold wind exposure.

an altered technique. Though by death he displayed a marked limp from advanced arthritis, well-developed leg bones point to an adulthood just as mobile as other Neanderthals. After coming through so much, however, slower reactions may have been his ultimate downfall, since some evidence points to him being caught in a rockfall.

Shanidar 1 is the most battered Neanderthal known, but he's far from alone in having endured more than one physical complaint. Decades before his discovery, another skeleton had earned the moniker 'Old Man'. Just four days before Le Moustier 1's skull was finally removed in 1908, three priests who shared an enthusiasm for prehistory were surveying caves near La Chapelle-aux-Saints, Corrèze. Within one cavity, set like an eye socket in a low hill, they exposed a body lying on its side, knees drawn up. Buzzing with adrenaline at their astonishing luck, they rapidly exhumed and packed up the remains, before writing for advice the same night to eminent scholars. Boule stepped up, and the remains were sent to his laboratory. Upon their arrival, the true significance of the La Chapelle-aux-Saints skeleton as the first almost complete Neanderthal was clear.

Though slightly younger than Shanidar 1, nonetheless a hard life was etched into the Old Man's bones. In addition to similar hearing loss, massive dental wear made him prone to excruciating abscesses and about half his teeth were gone; extreme even for older hunter-gatherers. Bone degeneration, which eventually would have made walking painful, is visible across his body. While some was perhaps due to injuries, much resulted from routine massive exertion, probably hauling heavy loads of stone or animal carcasses. On the other hand, unlike Shanidar 1 and other Neanderthals, his only clear injury was a long-healed broken rib.

In fact, a contrast could easily be made with La Ferrassie 1, who was also lying in Boule's laboratory at the same time. Probably younger than the Old Man at between 45 to 50 years old, LF1 had suffered more injuries (though fewer than Shanidar 1). Relatively common today and not especially severe was a fractured collarbone, perhaps leaving his shoulder somewhat lopsided and affecting arm use. But more seriously, LF1 had broken the top of his thighbone in the hip joint. Such an injury is quite rare and usually caused by a heavy fall while the leg is twisting.

Maybe muddy ground while LF1 was hunting turned a slip into something more dangerous, or perhaps he was even knocked over by the quarry itself. In any case, we know it happened decades before he died, and he never walked the same, eventually curving his spine. Extra discomfort came later from arthritis, followed by a serious condition causing agonising swelling of joints, fingertips and toes.* By his end, LF1 may have been in permanent pain.

Perhaps it's not surprising that some older Neanderthals weathered wretched health. But more unexpected are quite severe injuries in some juveniles. Le Moustier 1 is a case in point: he sustained a nasty broken jaw that healed badly, and probably caused asymmetric wear on his teeth from prolonged difficulty eating. As well as potentially affecting verbal communication, this tells us it happened some time well before his death, at between 11 and 15 years old.

And even younger children were battered about. Less than a kilometre from Forbes' Quarry in Gibraltar is the Devil's Tower fissure site, explored in 1925 by a young Dorothy Garrod. What she found, nearly a decade before she excavated Tabun 1, was the remains of a child no more than 5 years old with a broken jaw. Even more shocking, it had happened at least a couple of years before, and he had also sustained later, potentially fatal skull fractures. Would such a young child have been involved in risky activities like hunting, or are we looking at accidents while unsupervised? Yet *two* serious injuries seems particularly unlucky; the other possibility of course is that somebody hit him.

The risk of injury and death would be amplified if the danger came from within. Pernicious beliefs that Neanderthals were prone to violence have persisted; however, unambiguous evidence of assaults is rare. There is quite a high rate of head injuries, but in almost all cases it's uncertain how they were inflicted. Medical research shows that blows from fighting tend to either be facial or land above the level of the ears. And since 90 per cent of perpetrators will be right-handed, they're almost always on the left side. Among various head injuries at the Sima de los Huesos, one individual stands out, having been struck

---

* Hypertrophic pulmonary osteoarthropathy; mostly seen today in cases of lung cancer.

twice with the same object from different angles. That's hard to explain accidentally, but the weapon could've been a hoof rather than a handaxe. The sheer size of Shanidar 1's massive injury means he was either battered by something enormous or struck many times.

Similarly, a skull fragment from Krapina bears a gigantic sunken fracture just behind the right ear. This is the most serious cranial trauma known in any fossil hominin, but it's actually *too* big for most assault injuries, far larger than typically caused by hand-wielded weapons. Remarkably, this grave injury eventually healed, although the skull's owner may have been left with brain injuries and their long-term impacts. Overall, Neanderthals at Krapina do have quite a high rate of crushed skull injuries — some showing massive inflammation — but few are in the 'assault zone'. Accidents do seem likely explanations for most, which actually matches what we know from some hunter-gatherer populations, where falls commonly cause serious injuries.

Out of thousands of fossils, there are only two strong cases for Neanderthal-on-Neanderthal assault. One is another Shanidar adult whose chest was stabbed so deeply the wound slashed across two ribs; yet they survived. The ribs healed and, remarkably, grew around part of the weapon that remained inside. Based on the shape of the gap, this would match a lithic flake or point; nonetheless, it's just possible a terrible accident rather than intent was the cause. Perhaps during the last frantic seconds of a hunt, a spear thrust brought down not the beast but a fellow hunter.

The final example, however, is really 'beyond reasonable doubt'. In the late 1960s partial remains of a Neanderthal that we'll hear much more about in Chapter 15 were excavated at La Roche-à-Pierrot near Saint-Césaire, south-west France. Probably a woman, 3D reconstruction of her skull revealed that what seemed at first to be a warped fragment edge was actually part of an appalling wound over 7cm (2.7in.) long. It was located right on the top of the skull, and in forensic terms closely resembles injuries from sharp, straight-edged objects. This mystery object struck the Saint-Césaire woman's head — either from in front or behind — so violently her scalp ruptured and bone beneath shattered. Yet again, however, traces of healing show that even such violent trauma was survived.

So at least some violence was happening. Does that mean Neanderthals were routinely murderous? Probably not. Sick or injured Neanderthals may have been more likely to die in rockshelters and caves and therefore end up preserved, while living in such places might itself have been a hazard. Miners for example had high rates of head injuries until safety reforms brought in protective headgear.[*] Neanderthals weren't tunnelling with explosives, but they were lighting fires beneath stone ceilings, causing rapid thermal changes that make rockfalls a real risk.

Looking at large samples also shows that severe injuries weren't universal: from 279 upper limb parts at Krapina, just 3 arm bones and 1 collarbone are damaged. None of the 170-plus leg bones are affected. Certainly, older Neanderthals everywhere indicate that health got worse with age, but that's common to all humans living tough lives.

Moreover, comparing early *H. sapiens* sites is informative. The site of Mladeč, Czech Republic, produced remains of at least nine people dating around 36 ka, just a few millennia after the last known Neanderthals. Nearly all had poor health indicators, whether severe growth interruption lines on teeth, hearing impairment/deafness, infections, a benign tumour, bone degeneration, gum disease and potentially scurvy or meningitis. In addition to a broken arm, a single male skull known as Mladeč 1 bore *three* injuries, very likely from assault. Farther east and a few thousand years later, at Sunghir, Russia, there's even a cut-and-dried *H. sapiens* murder case. The throat of a richly buried adult skeleton had been violently gashed, very probably fatally.

Youth for early *H. sapiens* people wasn't any less tough than for Neanderthals either. Another spectacular burial at Sunghir is of two children buried head-to-head. Both had more than one phase of tooth growth interruptions, and one's thigh bones were extremely short and bowed, probably from a genetic condition. The other's facial bones were also abnormal, and probably made eating difficult: they had no tooth wear, suggesting that special soft foods were provided. We can even find a match for the battered little

---

[*] Some historic data comes from Cornish miners, one of the last regions in Britain where trepanation was routinely used for cranial injury.

Devil's Tower boy in the early *H. sapiens* skeleton of a 4- to 5-year-old at Lagar Velho, Portugal. As a toddler he'd suffered a severe facial blow and a healed serious arm injury.[*]

On balance, we may actually come out as *more* violent than Neanderthals, because nowhere is there evidence they killed youngsters. That's not the case at the early *H. sapiens* site of Balzi Rossi, north-west Italy, where a child very probably perished after being stabbed or shot in the back. A stone tool fragment was still lodged in one vertebra, and while it's possibly some kind of horrific accident, the weight points towards social conflict. Such aggression in our own species, even between hunter-gatherers, is certainly well documented, and clearly accelerated over the past 40,000 years. In contrast, we see no such phenomenon through the hundreds of millennia Neanderthals existed.

Many Lands, Many Lives

The bones of each and every Neanderthal hold unique stories. At grand scales across regions or through geological time, there are subtle anatomical differences. Two Neanderthal infants born probably 30,000 years and thousands of kilometres apart – Le Moustier 2 in France and Mezmaiskaya 1 in Russia – shared distinctively thick bones. But in other ways, such as arm proportions, they were slightly dissimilar. And even Neanderthals who lived in the same areas at roughly similar times were far from clones.

Variations in bodily form that appear more common in some times and places are visible in adults too. For example, Northern European Neanderthals' faces stuck out a bit more, giving them larger gaps behind the back teeth. And between Le Moustier 1 and the Krapina Neanderthals, living hundreds of kilometres and around 80,000 to 90,000 years apart, there are small but noticeable dental differences.

In some places we can see highly localised anatomical foibles. At La Quina, south-west France, three adults and a teenager all share a particular skull feature that's very rare elsewhere. This implies

---

[*] Not long before death his teeth record several growth interruptions within a few months of each other, suggesting serious illness.

long-term processes where sub-populations became genetically isolated enough that random mutations were distilled, just as happens in people today for various reasons. The La Quina Neanderthals lived at the end of an intense glaciation, which may have shrunk populations, causing isolation and high levels of inbreeding.

Sometimes regional climate might have directly influenced anatomy. Southern European Neanderthals were to some degree insulated from the cold (although not always from aridity), and strikingly, those from the Near East have less heavily built, thinner bodies. But if Neanderthal physique was affected by physical activity, this difference might also be reflecting how local ecology influenced levels of mobility.

That's backed up by recent studies on limbs. While males from Europe had more developed lower legs, those in the Near East had stronger thighs, pointing to variation in either how much they moved around or the kind of terrain. In women, although the sample size is very small, the difference is even greater. But both men and women from the Near East had beefier arms.

Tooth anatomy itself doesn't show clear regional trends, but wear patterns definitely do. More than 40 Neanderthals at over 20 sites ranging from Pontnewydd, Wales, to Shanidar, Iraq, clearly show that the environment affected not only what they ate, but how they used their mouths as tools. Those from regions or periods with more open vegetation like steppe had higher levels of tooth clamping. The most obvious reason may be that Neanderthals in colder conditions needed more clothing, and spent much more time working animal hide.

There are smaller-scale patterns, however, that are harder to tease out, but which could mark regional traditions in technology or tasks: in particular, Italian Neanderthals have more wear than Western Europeans. And strikingly, after 60 ka no Near Eastern Neanderthals display this type of wear at all. Combined with the evidence from limbs, it suggests that those living in this warm, arid but plant-rich area hunted, foraged and processed materials in unique ways.

But there are always exceptions. Some individuals from warmer and more vegetated environments were using their mouths in ways that look exactly like others from steppe-tundra contexts. Perhaps in clement conditions, when they were under less intense survival pressure, Neanderthals were more likely to diversify their skills, with their bodies taking on the imprint of their craft expertise.

The earthly remains of Neanderthals have furnished us with the most remarkable, intimate details about their lives. Each dawn brought a new day of arduous work, but their existence wasn't for the most part more challenging than typical hunting and gathering peoples. They knew both pains and pleasures great and small. Long-distance trekkers, they ate up tough terrain despite a short stride. Their arms and hands were immensely strong, yet capable of fine dexterity. And just like us, there were many ways to be a Neanderthal. A walking tour of their world would have meant meeting groups who looked and probably sounded quite different, whose idea of 'normal life' might have been just as unfamiliar to each other as to us. Moreover, within their biological diversity, every individual took their own path.

As ancient genetics develops further, mechanisms and adaptations behind Neanderthals' unique biology will become clearer. But perhaps the biggest revolution has been the toppling of 'ice age' explanations for why they looked and lived how they did. Their bodies were honed for and by hugely demanding lifestyles, whether or not they hunched against glacial winds. Living through extreme cold climates may have simply buffed up the already sleek engine they ran on. And when we fully survey the vast swathe of climates and environments that Neanderthals experienced, their story becomes even more unexpected.

CHAPTER FIVE

Ice and Fire

Grey wolf-light filters through trunks as the autumn dawn curls round acorns. Warm fur fluffs against the chill, and the macaques' tawny bodies unfurl to reveal snug youngsters sheltered from the dew. Their morning cacophony interrupts the magpie chatter and jay screeches that thread through this forest ablaze with ember colours. The limestone bluff it blankets is home, yet birds and monkeys alike are wary of the rockshelter. Rich pickings can be found, but danger also hides. Braver individuals drop towards the overhang's shadows; they know the panther is not here, because the people are. Or were yesterday; the smoke has gone stale.

Opportunists, the macaques greedily suck gristly leavings and marrow scraps lacing bone splinters. Sounds float from the mist-bowl valley below: crashing and deep rumbles mark elephants moving to the river. Faint cronking drifts down from a circling crane cyclone high overhead. Above their spiral – higher than mountains – a flock of winter thrushes travels west, seeking berry-laden woods across the sea. Many hours' flying later, they reach an island that will one day be famed for its bone-white southern cliffs. But this land belongs, for now, to the beasts.

The thrushes alight among trees galleried along the land's greatest river. Tan waters run thick and slow around belligerently peering periscope eyes. Water cascades off a grey bulk rising like a barrel-bodied submarine, and as it chews, surveying the water, the hippopotamus's ears flick. Beyond charging distance, water buffalo loiter in the muds, weed hanging from crescent horns adorned by egrets. They ignore the lion hunkered down on the high bank, waiting for the rising sun to sear away the fog just as the fallow deer take the chance for a morning draught. But nowhere on this river – nor anywhere on the island its tributaries drain – does woodsmoke wreath the dawn air.

An unimaginably far time in the future, winter thrushes gather once more, and golden morning light again gilds the great river. But its spreading reach is now squeezed by a great city, bridges spanning the water like corset stays. The lions have become immobile, mute on massive plinths. Before them are arrayed not herds, but a massed multi-coloured crowd, flowing against nudging traffic. Below the statues and the tumult, lost streams thread through the urban rhizome of wires, drains and tunnels. Behold, the stony sepulchre of Before-London: a city reclining on vast gravel beds, packed down and padded by rotted black-brown dirt. Here lie graveyards of entire vanished worlds: speckled by massive bones, shot through with muddy tatters of long-dead flowers, studded with still-iridescent beetle wings.

In twenty-first-century Europe, the raw stares of giant beasts are mostly filtered through glass and bars. Thames-side 'watering holes' are crowded by Westminster politicians, the banks of the once-sprawling river last brindled by megafauna tracks more than 2 millennia ago. London's lions today are moulded in bronze and sport bouffant manes that their predecessors lacked. Its hippos swim in concrete tanks, and even the deer are herded for royal pleasure. But look across the English Channel and the beasts are staging a comeback: brown bear roam Pyrenean ridges, boar saunter through Berlin suburbs and wolf prints will soon be on North Sea beaches.

Mostly we marvel at our remaining wild fauna from a safe distance, so imagining this crowded continent filled with even larger creatures is difficult; far harder is conjuring up entire vanished environments. Most writings about Neanderthals come through an ice-blue tint, their furry compatriots largely pictured as arctic-adapted creatures. The early discoveries from caves or gravel pits were mainly species such as reindeer or woolly versions of other beasts like mammoth and rhinoceros. The concept of a frozen Neander-world took hold, but understanding their authentic experience means deconstructing a simplistic, singular 'Ice Age' and exploring the many different worlds they lived in, each with its own menagerie.

Other, rarer nineteenth-century sites had strange mixes of creatures: at Victoria Cave in Yorkshire,* along with hyaena was hippopotamus and a weird straight-tusked elephant. Not only had artic species once lived further south, but tropical creatures had also roamed far up into Europe. Although geologists understood that the deep past encompassed vanished environments, true comprehension of earth's immense age stole up like a cloud-cast winter dawn. By the 1880s there was good evidence that the vast expansions of polar ice that scoured much of Northern Europe had been dovetailed by 'interglacial' phases, as warm as today.

It took another century to uncover the true complexity of palaeo-climate. Beginning around 3 Ma, global cycles of cooling and warming ramped up, directed by earth's eternal waltz around the sun. The details of how planetary elliptical orbit, tilt and wobble affect our climate are fiendish but predictable; essentially, the amount of sunlight largely dictates air and sea temperatures. These are the engines driving polar and mountain glacier ice expansion and melt, the effects of which spin out to create climate change.

The proof lies in sediments under the ocean and in Greenlandic and Antarctic ice. Cores drilled from deep down contain extraordinarily ancient records of palaeo-climate, revealing change in global temperatures over more than 100,000 years, at a millennial-scale resolution. By dating and comparing them with other, shorter records – pollen sequences in lake beds, dust accumulations blown from ancient tundra, flowstone in caves, or tropical coral reefs – it's possible to finely calibrate palaeo-climate over massive spans of time.

The pattern is remarkably consistent: the undulating cycles of cooling and warming varied in intensity and some lasted longer than others, but earth's heart beats out the same tempo whichever

---

* Similar to Lascaux, Victoria Cave was found when a dog got lost down a hole.

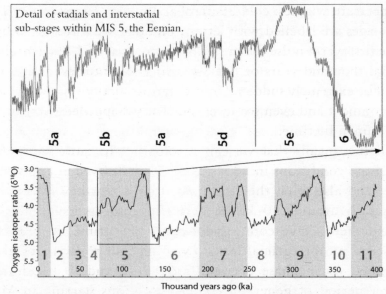

Marine Isotope Stages 11–1: changing concentrations of Oxygen isotopes (^{18}O and ^{16}O) which tracks global climate; data from both deep sea and ice cores.

Figure 3 *Palaeoclimate during the time of the Neanderthals, including glacial periods and the Eemian; a world 2–4°C warmer than today.*

record we look at. The long-term climate cycles have been labelled by researchers using a system known as Marine Isotope Stages (MIS; after the ocean cores). We're living in Stage 1, a warm – or interglacial – period following Stage 2, the last cold phase, which ended around 11.7 ka.

Stage 1 marks the geo-chronological Holocene boundary; everything before that (almost a hundred cycles) until nearly 2 Ma is the Pleistocene. Swoop back in time, and you can see that warm climate peaks are odd numbers (MIS 3, 5, 7 …), while the cold troughs are even numbers (MIS 2, 4, 6 …). Even if elements were appearing earlier, distinctively Neanderthal bodies and culture became clearly expressed not during a glacial period, but the clement MIS 9 interglacial after 350 ka. Moreover, looked at over the full span of time between around 400 to 45 ka, contrary to the clichés, Neanderthals actually lived through more interglacials than glacials.

Another important advance in our understanding is that every climate stage was unique. Each contained smaller, shorter temperature wiggles: cold 'stadials' and warm 'interstadials'. These sub-stages are labelled with letters; one that we'll return to below is the first warm pulse during MIS 5, known as 5e.* They might last several thousand years or just a handful of centuries, and changes could be extremely sudden. Drastic transformations in temperature, environment and even sea level sometimes happened over the span of a human lifetime.

All this means that we can reconstruct in quite outstanding detail the conditions in which Neanderthals lived at any point in time, and also what the world was like when they disappeared. This falls around 40 ka in MIS 3, meaning its climate and environments have come under special scrutiny. Despite being classed as an interglacial in its own right, it's really more like an extended warm phase between around 65 and 30 ka, during a longer period of generally colder conditions starting in MIS 4 until the end of MIS 2.

In comparison to more ancient, true interglacials, MIS 3 for most Neanderthals was never really toasty. Summers north of the Alps were about equal to those in the Scottish Highlands today, followed by downright soggy autumns. We often envision Neanderthals hunched against bitter snow flurries, but standing in sheeting rain is just as accurate. Winters, though, were definitely colder, with snow probably on the ground for many months. But still, Eurasian MIS 3 environments were far from glacial wastelands. Instead, what makes this climate cycle distinctive is its instability, temperatures see-sawing rapidly up and down.

Beyond the Ice

If Neanderthals squelching around in mud as well as ploughing through snow is unexpected, even bigger surprises exist. The most recent proper interglacial, MIS 5, was *even warmer* than today. As the preceding MIS 6 glacial ended, temperatures rose rapidly, peaking

---

* The letters go backward, so 'e' is the earliest of five interstadials and stadials in MIS 5.

around 123 ka as the sub-stage MIS 5e, known as the Eemian.* This remains – so far – the warmest period hominins experienced across Eurasia. It lasted roughly 10,000 years,† short on geological scales but equivalent to some 500 generations.

So what was this balmy world like? At least early on, it was sunnier than today. Earth's position relative to the sun was slightly different, bathing the planet in more summer solar radiation. This pushed average temperatures 2 to 4°C higher, with palpable effects. Alpine caves at today's snowline were warm and moist enough to see stalagmites grow, and vast forests spread across the continent. Most dramatically, melting polar caps and glaciers sent sea levels up some *8m (25ft)*.

As it got hotter and hotter, beaches crept higher and a kaleidoscope of tree species replaced each other. Pollen records show that birch and pine gave way to oak-rich woodland, speckled by elm, hazel, yew and linden, eventually maturing into dense hornbeam forest. These lasted until the arrival of spruce, fir and pine, heralding the start of the succeeding MIS 5d cooler stadial. Neanderthals living at different points in time across this 10,000-year mutating woodland heard very different dawn choruses. Squabbling crossbills and crested tits yielded to crass jays and mellifluous nightingales, until cold mornings saw clattering capercaillies sending breathy vapour into biting air.

The Eemian's other fauna also subverts traditional ideas about Neanderthals. Alongside aurochs and horse – who enlarged their diet from grass to browse – were wild boar, roe deer and their spotted fallow relatives.‡ Beavers feasted on saplings, flooding valleys and creating rich new habitat where turtles swam. In a bizarre ecological shift, these reptiles were hunted by badgers.

Other big beasts arrived with the warmth: water buffalo, straight-tusked elephant and hippopotamus. But one southern climate migrant is especially interesting: monkeys, specifically Barbary macaques.

---

* Eemian is a Western European palaeo-climate term based on an ancient pollen type site; many other regional names exist – for example in Britain, the Ipswichian.
† Vegetation in Southern Europe reacted faster, so it lasts several millennia longer than in Northern Europe.
‡ All 'wild' fallow deer in Europe today are the result of historical introductions.

This species today is confined to isolated parts of North Africa, especially mountain forests. Yet during the Pleistocene they were far more widespread, and occasionally turn up in earlier and later Neanderthal sites. The shallow cave of Hunas in Germany is probably Eemian, and contained macaque remains from the same layer as a Neanderthal tooth and lithic artefacts. These fellow primates must surely sometimes have encountered each other. Barbary macaques are mostly plant eaters, but will take insects and even meat like young birds and rabbits when times are tough. They love to scavenge human trash today, and it's just possible they picked over Neanderthal leavings too.

The Eemian sounds like a lush paradise, and though Neanderthals didn't have to worry about frostbite, it wasn't exactly Club Tropicana. Deciduous forests are surprisingly difficult places in which to be a hunter-gatherer, since much of the plant stuff that's edible takes time and energy to make it so. Easily foraged things like nuts and berries tend to be seasonal. Large game was around, but far harder to find in forests.

For a long time the lack of sites led to doubts that Neanderthals had really adapted to the Eemian, but in fact later erosion probably scrubbed away most sediments of this period. Today about 30 locales are known, though very few are caves or rockshelters. Mostly they are preserved in buried lakebeds, or at carbonate-rich surface springs. In that world, sticking close to water makes sense, as the one resource all prey need.

Newer research also shows that the Eemian wildwood wasn't a gigantic, unbroken canopy. Two deep lakebeds at Neumark-Nord, eastern Germany, have stunning preservation, and by sampling every 5cm (2in.) for tiny plant, insect and mollusc fragments, researchers found that the shores were surrounded by mixed vegetation. Alongside light woodland were scrubby hazel thickets and dry, grassy areas with tormentil, mugwort and daisies underfoot. This harlequin environment attracted a spectrum of animals: creatures of the forest like boar and straight-tusked elephant, alongside grazers like bison or aurochs. Some herbivores like horse were there too, shuffling between ecological niches, but skeletons and even tracks show all species were drawn to the lakes.

How did the Neanderthals fit into this leafy world? The Neumark sequence largely covers the initial hot, drier phase of the Eemian when dense forest had yet to develop. This period had more diverse animal and plant species, and it's then we find most of the archaeology. Later as the trees close in, the lake shrunk, prey probably became sparser and Neanderthals began melting away. They didn't entirely vanish, however: one upper layer formed during a time of dense woodland includes over 120,000 animal bone fragments. It's a uniquely rich record of intense butchery for the Eemian, and proves that Neanderthals clung on even as the sunlight was filtered through an ever-thicker leaf filigree, and furry flanks became harder to spot behind the great trunks.

While Neanderthals were stalking deer at Neumark-Nord, westwards across the Channel there seems to have been no one to bother the beasts. Britain actually appears virtually devoid of hominins for 150,000 years, between the end of MIS 7 and the start of MIS 3. A potential window for recolonising would have come just as the deep cold of MIS 6 receded, but a gigantic natural disaster may have literally got in the way. Fed by glacial meltwater from the entire ice cap and swollen by river outflows draining much of Europe, a vast lake formed behind a chalk ridge running across from eastern Britain to France. The soft rock couldn't take the pressure and collapsed, releasing a monstrous flood that ravaged the Channel floor. Seismic surveys have uncovered the scars from gargantuan cataracts that bulldozed out massive valleys now buried under sea sediments. So stupendous was the inundation that its closest match is the gorges that slice halfway round Mars. Neanderthals would have heard the unearthly roaring kilometres away, while mammoth herds at far greater distances would have felt infrasound waves rumbling through the ground.

Doggerland – the now-submerged region between Britain and the Continent – was left barren and totally denuded: a wasteland. Deep gorges, treacherous landslides and vast areas of rock and gravel lay between Neanderthals and the uplands of Britain. This may have been enough to put them off, but it seems that the beginning of MIS 5 also saw a rapid rise in sea levels. Britain became cut off before hominin populations had time to reach it. The warm-loving species like elephants and hippo that did make it over may have been those that could safely cross wetland, swollen rivers and even short stretches of sea. It took another 60,000 years for the climate to cool and sea levels to drop enough so that Doggerland re-emerged at the end of MIS 4. As the mammoth-steppe environment stretched out once more from the Atlantic to the Pacific, Neanderthals and horses returned to their most north-westerly domains. The only hint that hominin pioneers may have crossed earlier comes from a couple of probable lithics in south-east Britain during a cold phase late in MIS 5. However, if they're genuine, sea levels were probably not yet low enough to walk over, so by what means Neanderthals arrived is unclear.

Climate Crisis

As resolution in palaeo-climate and environmental research has increased, it's been discovered that even the Eemian contained brief but drastic climate change. The warmest temperatures and highest sea levels existed for just 4 millennia between 126 and 122 ka until gradual cooling began, but this was the calm before the storm. What came next is known as the Late Eemian Aridity Pulse (LEAP) and was a time of crisis. The evidence for LEAP comes from lake sediments inside anciently flooded volcanic craters. Extremely fine annual sediment layers called 'varves', just 1mm (0.04in.) thick, accumulated over time, and it's in this muddy archive at about 118.6 ka that something odd appears: for precisely 468 years, dust rained down. Researchers counted over 50 thick dust pulses, each testament to a landscape plunged into sudden cold and drought. Drastic loss of vegetation led to huge soil erosion and massive dust storms tore through the air. Other sources record this immense climate shock, from flowstones suddenly halted in their growth, to pollen cores where warm forests vanish and tundra appears in less than a century. Repeated layers of charcoal are visible: it was so dry, bushfires were igniting.

We can only imagine what Neanderthals made of the ravaging of their familiar forests over just a couple of generations, or the frightening and unpredictable weather. The LEAP ended as fast as it began. Yet although temperature and humidity rallied briefly and gave time for some warm-loving trees to recolonise, other regions never recovered.

Instead, coniferous forests developed, marking the beginning of a temperature decline that would last until the end of the Pleistocene. True tundra took over in Northern Europe around 115 ka, and swelling polar ice caps began sending out massive flotillas of icebergs reaching as far south as Iberia. The MIS 5 interglacial was on its last legs, gnawed by increasingly rapid snaggle-toothed successions of warm and cold, tracking its failing pulse. Even so, the Neanderthals kept going. Throughout the last gasps of the interglacial, the number of sites increases, and there's growing technological inventiveness.

Ice Age

Neither forest, heat nor dust were enough to do Neanderthals in. But what about the true ice ages during glacials? Their most intense phases had average temperatures of around 5°C lower than today. That was enough for a vast ice front hundreds of metres thick to creep down from the poles. The extent of the ice caps varied during each glacial, but at the peak of MIS 6 they came as far as the British Midlands and across to Düsseldorf, Germany.* During the last deep freeze of MIS 2, even the French south-west – where summer temperatures now reach 40°C – was a land of permafrost and polar desert. As well as the cold, oceans locked up as ice caused tremendous drops in global sea level, sometimes below 100m (300ft). This created one of the few benefits to glacial living: vast new lands fringed by rich estuaries.

Yet even when ice caps were restrained, times were still extraordinarily harsh. Weather patterns were probably peculiar, with snow and ice storms of a scale that we've never seen. And as well as frigidity, glacials brought aridity. Biting dry air combined with frozen-solid groundwater in permafrost zones made dehydration a real risk.

All this caused colossal environmental impacts. Across much of northern Eurasia pine forests evaporated. Periglacial tundra extended

---

* The mother of all glacials was before the time of the Neanderthals, during MIS 12: the ice front in Britain extended just north of London, shoved the Thames down to its current course and obliterated an ancient river that drained much of the Midlands called the Bytham.

south of the ice cap, forming a variegated carpet of hardy mosses, lichens and dwarf trees that might reach past ankle height if they tried hard. Southwards, this softened into steppe-tundra, resembling parts of Siberia today, but containing a mix of species with no modern analogues. Winds swept a mosaic of aromatic herbs, grasses and scrub; verdant green in spring, the land glowed like flame and blood in autumn.

Microhabitats existed with lusher vegetation, and pollen and charcoal tell us that *some* trees clung on. Whippy birches crowded by rivers, and even interglacial stragglers – oak and lime – stowed away in sheltered gorges. Eastwards towards Asia the steppe was mottled with taiga: boggy coniferous forest favoured by elk,* but tough to travel through. Even in the more insulated south around the Mediterranean, plant communities changed in response to drier conditions.

Neanderthals mostly avoided truly arctic conditions. For example, in northern France the rich archaeological record of late MIS 5 only really diminishes once harsh, persistent cold in MIS 4 kicks in. Some may have moved south, others died out, but the rare MIS 4 sites are probably linked to brief warm spikes in temperature. Woolly rhinoceros and mammoth also left the harshest tundra to arctic specialists like reindeer or Arctic fox. The most hardcore are musk oxen, adapted for bitter cold and deep snow, and which only expanded southwards during extreme glacials. Fascinatingly, occasional sites where lithics are found along with musk ox bones do exist. They testify that Neanderthals were resilient enough to cope, at least temporarily, with the ultimate in challenging ice age environments. But Neanderthals were much happier in steppe-tundra exemplified by MIS 3, populated by herds nearly as rich as those in the great African grasslands today. Current more nuanced understanding of Pleistocene climate and environment certainly casts the 'hyper-arctic' explanations for Neanderthals' anatomy in an even less certain light.

---

* Elk in Europe is what's known as moose in the US; however, we have hardly any evidence Neanderthals hunted them.

Coast to Peak

Neanderthals rode earth's climate rollercoaster for hundreds of
thousands of years, coping with the extremes in weather that went
with it. What's more, myth-busting about their world goes beyond
temperature to their vast environmental range. They lived far beyond
the European regions where they were first discovered, and long
believed to be their heartlands. Exploring the breadth of their
geography shows that though they happily adapted to steppe-tundra,
in ecological terms Neanderthals should equally be thought of as
creatures of Mediterranean woodland. Peninsulas like southern Italy
stayed warm enough for hippo to survive even after MIS 5, and as far
as we can tell, such landscapes were always a Neanderthal stronghold,
right until the end.

Let's begin at Europe's south-west tip: Gibraltar. Here the great
Rock juts into the Mediterranean, the only place in Europe where
wild macaques still clamber on rocks. DNA shows they're not ancient
relics but historic introductions from Algeria and Morocco; North
Africa is visible southwards through sea-haze.* But Pleistocene primates
in the form of Neanderthals did live here. Though tiny – just 6km
(4mi.) long – Gibraltar is a rich microcosm of habitats, nurtured by
uniquely stable conditions even during the worst climatic upheavals.
Olive groves, geckos and even tree frogs have survived for many tens
of millennia, untroubled by the harsh, arid conditions endured farther
north in Iberia. For Neanderthals this place was prime real estate:
abundant, diverse resources and caves bathed in the rising sun.

Humans have always been drawn to the Rock. Neolithic and
Roman artefacts come from some of the 200-plus caves, though from
the eighteenth-century, use of this hulking mass of stone became
more intense. Today, military fortifications jostle for space with a
cable car and the nature reserve where macaques hassle tourists for
snacks. But beneath all this, tunnels 10 times longer than the entire
promontory honeycomb the limestone. Military digging began in the

---

* Supposedly there was a superstition that should the macaques leave, the British
would lose Gibraltar, leading Churchill to order for their population to be
replenished during the Second World War.

1700s, eventually leading to the discovery of the Forbes skull. And much later, the First World War stationing of Dorothy Garrod's mentor, prehistorian Henri Breuil, set in motion her finding of the Devil's Tower child.* But it wasn't until the 1980s that the large caves further along the cliffs were investigated, revealing what life on 'Costa Del Neanderthal' was like.

Only during high sea levels like today does the dawn light scatter off the Mediterranean into the lofty Vanguard and Gorham's caves, but the ocean has always been important to Neanderthals living here. When the shore was closer, they collected seafood and made the most of rarer, bigger finds like fish or marine mammals. During glacials the coast retreated up to 5km (3mi.) away, exposing a dry dune plain in front of the caves, but even then Neanderthals brought back shellfish, including large mussels gleaned from rocky estuaries a considerable walk to the east.

As we'll hear in Chapter 8, the few sites where we see Neanderthals as coastal foragers on both the Atlantic and Mediterranean seaboards are probably the tip of the iceberg. Many hundreds more would have been flooded by rising oceans: there must be drowned caves whose current tenants in the form of crabs and suspicious moray eels are matched by buried seafood remains. Submarine archaeological surveys are now beginning to search Europe's rocky margins, but already we must rethink Neanderthal feet as not only striding across steppe but leaving prints on sandy strands.

The beach now in front of the Gibraltar caves is mixed with fine blasting debris from military tunnelling that included creating vast water tanks for the Second World War garrison. Supplies for the town remain a concern, and infrastructure works in the 1980s more than 250m (650ft) up the cliffs uncovered a small cavity. A stop-off for Neanderthals hunting high up on this mini-montane environment, their main quarry were ibex, relatives of goats but a lot bigger. Their remains gave the site its name – Ibex Cave – yet it seems Neanderthals were here not to stay or even butcher carcasses, since there are very few lithics. Instead, the main attraction may have been the expansive

---

* Breuil noted lithics on the slope while out walking, and suggested that Garrod excavate.

views of the plain below. They took what ibex they caught back down
to Gorham's and Vanguard caves, perhaps sliding down an enormous
dune of windblown sands that banked up against the cliff: a good
shortcut when hefting weighty haunches of meat.*

Ibex are found at dozens of other Neanderthal sites in craggy
environments. Nimble, cocky and extremely strong, they must have
required particular hunting techniques, and special care to avoid the
enormous curving horns. Perhaps even harder to catch were chamois,
but we find them too, including some 1,400m (4,600ft) up at Las
Callejuelas, west-central Spain. It's cold and very dry even today, but
what about even higher altitudes? No problem: Neanderthals in the
Alps, Carpathians and other mountain ranges were climbing at least
up to 2,000m (6,500ft). Outside full interglacials, these lofty landscapes
must have been marked by swollen glaciers and snowy slopes for most
of the year.

Why did they choose such extreme places? Following species like
red deer that seasonally move to high grazing is one possibility. But
mountains themselves may have been attractive, offering abundant,
high-quality stone: a resource Neanderthals must have been constantly
alert for. They may even have traced the sources of river cobbles
upstream, as later prehistoric peoples did.

Other mountain specialities might have been on the menu, such as
hibernating bears, but elsewhere it seems that Neanderthals were
active up at impressive elevations because they were simply comfortable
there. Noisetier Cave in the French Pyrenees is at over 800m (2,600ft)
elevation and has no obvious montane resources that might have
attracted Neanderthals there between 100 and 60 ka. The red deer
and mountain sheep they were hunting were also common at lower
altitudes, and neither is there any particularly good stone nearby.
Instead, this looks just like many other caves across their world:
regularly, if briefly, used as a place to stay. Either Neanderthals here
were permanent mountain dwellers, or were using it as a stop en route
to other regions. If that's the case, we're looking at crossing the
Pyrenees. That's backed up by lithic sourcing studies proving

---

* For nineteenth-century officers on Gibraltar, the dune was a favoured haunt of
botany enthusiasts as well as a route for military deserters.

Neanderthals undertook such journeys over high passes in the Pyrenees, the Massif Central and other mountain ranges.

Whether they were up among spindrift-shrouded peaks or searching among seaweed, there appear to be few landscapes that really put Neanderthals off. Even deserts feature on the great map of their realm; the complete opposite of what we'd expect for arctic specialists. The warm, rocky Mediterranean ecology that stretched between Gibraltar and Turkey eventually transforms into even drier environments towards Central Asia, all of which are rich in Neanderthal fossils and archaeology. Ecologically they could shift gear, adapting to whatever was on offer, from palm dates to olives, tortoise to gazelle; even on the fringes of Arabia, giant camels.

The only sort of country in western Eurasia lacking evidence so far for Neanderthals is true wetlands. To do more than briefly visit such places, serious investment in either watercraft or raised structures like trackways and platforms is needed. But, never say never. Perhaps somewhere a surprise waits, in northern peats deep below the sleeping moss-tanned bog bodies of the Iron Age.

Visions of Neanderthals struggling through deep snow, breath puffing in frozen air, have been remarkably persistent since their discovery. Yet our ice age blinkers have concealed their innate adaptability. Polar desert was never their true home, though *in extremis* they may have coped for a while. Mostly they avoided extreme cold, and appear to have been most successful in more temperate climes, whether surrounded by grassy plains or wooded glades.

Even cold-adapted animals such as woolly mammoths were ecologically malleable, originating during the MIS 6 glacial but found later alongside warm-loving straight-tusked forest elephants.[*] It's our own pigeon-holing tendencies that limit Neanderthals to glacial worlds, when the reality is far more diverse. But they had something other creatures didn't, allowing them to cope with all but the worst the Pleistocene could throw at them: a complex technological culture.

---

[*] These in contrast became extinct at the end of the warm MIS 5 interglacial.

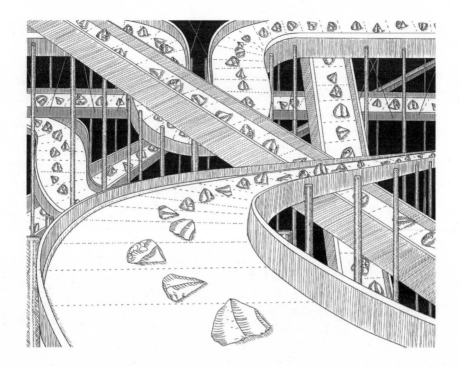

CHAPTER SIX

The Rocks Remain

Sun glints splinter from a sea swollen by vanished polar caps. It is eighty million years ago: today's continental jigsaw is half-finished, the Pyrenees and Alps yet to heave their heights up. Instead, where Europe will one day be, a sub-tropical kaleidoscope of archipelagos emerge and disappear as the oceans rise and fall. Titanosaur footsteps shake terra firma, while on briny coasts the immense wings of pterosaurs lift them up above the sea. Their shadows rush over turquoise waves as a dark mass heaves beneath. A swarm of ammonites explodes, torpedoed by a mosasaur, and shell fragments glitter as they spiral slowly down. Soft muds blossom as the splinters land on the sea floor, an oozing wasteland. It's replenished by a never-ending drizzle of broken sponges, molluscs and decaying forms of uncountable plankton.

Spin the earth like a marble: continents creep, muds thicken and squeeze, cementing to limestone. The newborn rock sweats out silica gel into voids, some inside ancient burrows, others filling miraculously un-shattered shells. Place a finger on the marble to slow the spinning planet. Seas have drained, sierras risen and now vast ice caps pulse at the poles. As age upon age passed, immense pressures bore down, congealing the silica and germinating microscopic crystal lattices that evolved, shifted state. Became flint. Far above, hoofed legs run across the land, furred pelts are mussed by breezes. Waxing and waning climate cycles erode the limestone, while tectonics and rivers saw it down into winding, terraced canyons.

A stormy afternoon a hundred thousand years ago: the grey-green thundery squall hurls itself up one of the gorges. Rain-soaked rock gives way, slithers down and disgorges a stony pearl among the rubble, black as the clouds. The flint splashes into the river below, joins the slow-rolling bed of cobbles. For fifty thousand years it surges with floods, freezes into iced eddies, pauses for centuries on gravelly bars. One spring the now-rounded flint lies on a small bank, glistening after a rain shower. Above it, smoke blues the sky, trailing from a wide limestone shelter. People clatter down to the river, glancing as always at the stones, and the cobble's shine catches an eye. It's hefted up and down; tapped with another stone to call forth its clear voice. A few blows reveal good insides smooth like old fat, and soon it will be slippery with bright blood.

Stone tools were the atoms of Neanderthal life. They connected every aspect of their world, and are the fundamental units with which archaeologists try to reconstruct their cultures. Known to researchers as 'lithics', each possesses a unique story arcing from rocky formation to the day a Neanderthal picked it up, and onwards to rediscovery with the scrape of a trowel. Its geological heritage, whether under the sea, beneath mountain roots or as flowing lava, dictates its character. It's also what caught the hominin gaze tens of millennia ago, but today the eyes of museum visitors tend to slide over these objects.

Kept in glass cases, they're hard to relate to when so few people have ever held one, never mind making and relying on them for daily survival. Their stark beauty may be appreciated, even exhibited in galleries as art, but to most they remain mute. In truth, from individual objects to whole assemblages, lithics are extraordinarily rich sources of insight about Neanderthal lives.

For early prehistorians, the most pressing concern was classification. Having little direct experience of the manufacture or use of lithics, they concentrated on appearance. Arranging objects according to visual similarity and very basic technological features allowed them to create 'typologies'. One of the first to do this was de Jouannet, who was not only digging up lithics very early on but also trying to understand them. He assumed that they became more refined over time, and in 1834 produced a chronological typology placing knapped* (chipped) stone objects deeper in time than ground or polished tools.

Over several decades others sketched out cultural classifications that are essentially the same as those used today: in 1865 antiquarian John Lubbock coined the term 'Palaeolithic' for the most ancient prehistory, and later Lartet and Christy proposed a tripartite subdivision.† They perceived that lithics from some sites including Le Moustier were chronologically 'in-between' those from the Somme gravels – which they termed Lower Palaeolithic – and blade-based tools from places like La Madeleine, labelled Upper Palaeolithic. Concurrently, the French prehistorian Gabriel de Mortillet used the lithics from Le Moustier as a 'type site', and gave the first name to a Neanderthal Middle Palaeolithic culture: the Mousterian.‡

Yet the first Neanderthal fossils to be found seemed to have no accompanying artefacts (lithics at Feldhofer went unnoticed, only rediscovered in quarrying waste in the 1990s). For some 30 years the makers of the Mousterian were a mystery, and equally nobody knew if Neanderthals had material culture. The first observed association between skeletons and Middle Palaeolithic artefacts was at Spy, and it took far longer for prehistorians to comprehend that Neanderthals manipulated stone in remarkably sophisticated ways.

Unsurprisingly, twenty-first-century lithic experts are a world away from typologists. Nothing is ignored, even tiny chips and splinters, and examining an assemblage can mean hundreds of hours of work. Yet if recording the ten-thousandth artefact becomes tedious through repetition, one only need recall the preposterous privilege it is to simply hold such objects. Excavation and recording are increasingly digital and automated, but focused attention is still crucial in analysis.

---

* 'Knap' can mean a small hill in English, while '*knopp*' in some northern European languages means to strike, cut or even eat; similar words exist in Irish and Scottish Gaelic.

† Christy's fortune funded a luxuriously illustrated 17-volume series of their work.

‡ Mousterian is anglicized; the French is *Moustérien*, a simpler version of de Mortillet's original, somewhat clunkier *Moustierien*.

Each object becomes imprinted in memory as its creation is mentally reconstructed by reading technological stigmata on its surface.

Here's where learning knapping mechanics is necessary. The object to be worked, such as a cobble, is known as the 'core' and is struck with something harder. The parts that come off are called flakes; how this happens in practice is down to a mix of skill, geology and physics. How hard and where one hits determines what the flake will be like. Kinetic energy spreads out in a cone shape from the point of percussion, and its edge is what forms one side of the flake. This process often leaves visible ripples on both the core's negative 'scar' and the mirror image of the flake's inner surface. By looking for these and other distinctive knapping features on cores and flakes, researchers can reconstruct the method of knapping and to some extent its sequence, sometimes with a single artefact.

Neanderthals as cavemen thugs whacking stones together is very far off the mark. How rocks react to knapping reflects their structure: the more homogenous and fine the particles are, the more predictable fracturing will be, and the sharper the flakes' edges.* Neanderthals gauged these characteristics by sight, touch and even sound: high-quality stone like flint will often ring when struck. By using a variety of different knapping methods and techniques, they could control the size and shape of a product from any given core, even with poorer-quality rock like quartzite.

More artisans than klutzes, they appreciated the right tools for the job. Selecting hammers – the things that struck the cores – was crucial. Small cobbles have the necessary mass to hit hard for big flakes, but for more delicate work pebbles are better. And using 'soft' rather than hard hammers produces different effects. Elastic organic materials like antler and bone, or even less dense rock like limestone, spread out the kinetic energy and produce thinner, longer flakes. This was crucial when shaping was the goal, and for secondary knapping (retouching). Tools – lithic artefacts that were used to do other things – were often retouched, sometimes to give a particular edge, but often to resharpen them: flakes dull very fast even when cutting meat.

---

* The finest and sharpest natural rock is obsidian, a volcanic glass that cools so rapidly no crystal structure can form; though brittle, it can slice between molecules.

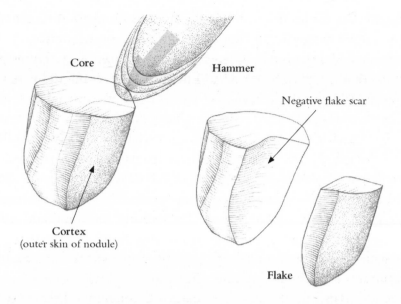

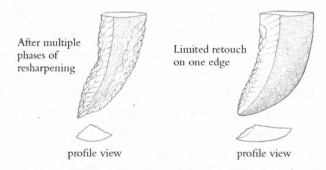

Figure 4 *The mechanics of stone knapping, and terms for lithic artefacts.*

Neanderthals had patently mastered the basics of fragmenting stone, but where do they fit into the broader evolution of lithic technology? Going back 3.5 Ma, the australopithecine makers of the oldest known artefacts – crude flakes smacked off blocks – would have viewed Neanderthal knapping with just as much awe as that practised by early *H. sapiens.* It took until roughly 2.5 Ma for hominins to develop geometrical concepts, allowing them to begin truly controlling stone fracturing. The first 'centripetal' cores appeared, with flakes carefully and sequentially removed from around the perimeter, leaving wheel-spoke patterns.

By 1.8 Ma the ability to mentally divide volumes is shown by the most visually iconic of all Palaeolithic artefacts: bifaces. The manufacture of these two-sided tools (also known as handaxes) was possible due to growing use of soft hammers, allowing their surfaces to be shaped by shallow flaking.

Taming Your Rock

Neanderthals inherited these already ancient ways of working stone and also made bifaces, but went a step beyond in managing material masses. Variegated systems for much more systematic, precise ways of getting large flakes off cores began to emerge, and it's these that truly define the Middle Palaeolithic. It first appears in Africa by around 500 ka, probably made by populations ancestral to *H. sapiens*, but in Europe it really explodes just as we see Neanderthal anatomy fully materialise between 400 and 350 ka. What makes Middle Palaeolithic flaking different is that it further developed the conceptual division of stone blocks, with cores treated as two halves. By shaping the base and preparing special side zones to be struck, it was possible to guide how flakes came off the upper surface, controlling their shape and size.

The first identification of this flagship Neanderthal technology was at the hand of Victor Commont, a prolific early twentieth-century amateur prehistorian who noted distinctive large flakes and cores turning up in quarries. The method is known as Levallois, after his hunting ground: a rapidly expanding Parisian suburb pockmarked by industrial activity,* and today Europe's most densely populated area. By the time of Peyrony and Bordes, the presence of Levallois technology was noted in Mousterian as well as other Neanderthal cultures too. Sometimes the stone blocks being worked were huge, and initially most prehistorians paid attention to a type of Levallois where after each 'main' flake was removed, the upper surface and sides of the core needed re-shaping.

That fact led to Levallois being regarded for some time as somewhat profligate, but decades of careful work has revealed it to be much

---

* Levallois-Perret hosted the Eiffel company headquarters, builders of both the famous tower in Paris and the American Statue of Liberty.

more sophisticated and flexible. By taking off small, preparatory flakes in different patterns across the upper surface, Neanderthals created outlines that directed the kinetic energy of subsequent removals. Through varying the preparation phase, they could produce massive flakes, long blades or even triangular points, sometimes making several in sequence before needing to reshape the surfaces. They also sometimes swapped patterns on the same core.

What transformed our understanding of Levallois, and other Neanderthal technologies, was when archaeologists began refitting knapped artefacts back together. Truly 'slow science', it is meticulous, time-consuming work – an immense 4D jigsaw – and demands well-preserved sites. But it's worth it: the next best thing to actually looking over Neanderthals' shoulders. For the first time it was possible to reconstruct the thought processes and choices made by individual Neanderthals, revealing dynamic responses to each block of stone.

The benefit in technological terms of Levallois and other 'prepared core' methods is that Neanderthals now had reliable ways to get particular products, especially large, thin flakes. Unlike bifaces they weren't much good for truly heavy-duty stuff, but for the same weight of stone, highly portable Levallois flakes gave far greater amounts of cutting edge. Neanderthals were skilled enough to use prepared core technology on all kinds of rock, from very hard volcanic stone to tiny pebbles. Where there was good flint, such as in Britain or northern France, they sometimes made extraordinary giant flakes and points 10 to 15cm (4 to 6in.) long.

The other advantage that flakes of various kinds had over bifaces is that they were far more easily retouched.* Although as a technique it was far more ancient, retouching really defines the Middle Palaeolithic. Sometimes Neanderthals retouched a flake in order to adjust its edge to match a particular task: blunting for scraping, creating notches or serrations for shaving and sawing. However, it's now understood that much – or even most – retouching was about resharpening edges. Fresh flakes quickly lose their razor-sharpness, but it can be maintained by shallow, thin removals using soft hammers along the edge. However

---

* To repeatedly alter the edge of a biface, you must also keep thinning the volume or it becomes too steep-edged to knap, whereas a flake is already thin.

the impact of systematic resharpening went further than individual artefacts. It expanded the scale at which Neanderthals were active. Longer-lasting flakes that were easily carried allowed them to move around over greater distances. Proof that this was going on comes from refitted Levallois knapping sequences where the flakes are missing having been taken elsewhere, and from geological sourcing. Universally, Levallois and retouched tools were the most far-travelled objects. These new ways of engaging with stone meant Neanderthals were ranging across the landscape farther than any hominins before.

While Levallois is often presented as the gold standard, it's far from the only thing Neanderthals did. It would require an entire volume to describe the whole gamut of knapping methods – known as 'techno-complexes' – they invented, but homing in on two just from Western Europe demonstrates quite how diverse their lithic world really was. Known as Discoid (since some cores are disc-shaped) and Quina (named for the type site), they were just as systematic as Levallois, but aimed at producing flakes with a sharp edge directly opposite a blunt one (a 'margin'), providing built-in natural ergonomics. Yet in all other ways they were quite different technological beasts.

For a long time both were viewed primarily as adaptations to bad stone, but since the 1990s this view has transformed, and they're now understood as techno-complexes in their own right. Looking first at Discoid, it turns out to be a masterclass in economy. The cores need only a few initial preparatory flakes to create a decent knapping angle, and every flake thereafter is the 'good stuff': sharp but easy to handle and ready-to-use. What's more, each removal creates the surface for the next strike, with no need to stop and refashion the surface. Far from simple or unsophisticated, Discoid technology provided Neanderthals with a zero-waste system not far off production-line efficiency.* Moreover, it's just as flexible as Levallois, since differently shaped products could be created, from oblong pieces to pointed ones, each with its handy blunted back like a penknife.

Thanks to 3D refitting methods it's possible to follow an individual Neanderthal using Discoid knapping on one day somewhere around

---

* While the preparatory flakes taken off during Levallois core shaping can be used, they clearly weren't the key objective.

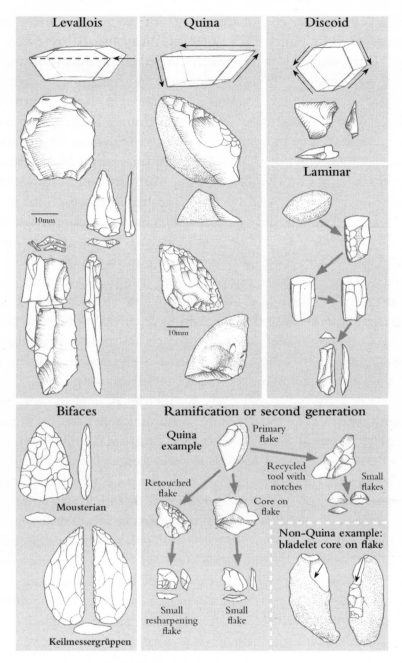

Figure 5 *Some of the lithic technologies made by Neanderthals, showing different knapping concepts and 'second generation' production.*

46,000 years ago in the southern Italian Alps. Up a narrow road winding past cliffs, Fumane Cave sits quite unassuming behind security fences. We'll return to other aspects of its exceptionally important archaeology later, but the Discoid lithic assemblage of Level A9 contained a special find. Excavators uncovered a cluster of artefacts nestled together across just a few centimetres, all made of the same distinctive grey stone. By combining hand and digital refitting methods, an exceptionally complete Discoid core sequence was reconstructed.

After sourcing a flint cobble from the nearby stream, a Neanderthal had sat down and knapped over 60 flakes in 10 phases, until just a small lump remained. The 14 pieces missing from the reassembled jigsaw were almost all the best, with long, fine edges opposite blunt backs. None of the other 8,000-plus lithics in the layer matched, so they must have been taken out of the site.

The grey flint core and its flakes stood out because of their colour, but technologically everything else in Level A9 told the same story. Neanderthals here were totally focused on Discoid technology, but not uncreatively. As knapping progressed and each core got smaller, they altered their technique to produce differently shaped flakes.

Working out what Discoid was especially good for isn't straightforward. Based on use-wear,* its strong, short and thick flakes were often applied to hard materials like bone and wood, but Neanderthals were also happy to use them for butchery.

What makes Discoid truly distinctive from Levallois or Quina is that the flakes were very rarely retouched. It's no coincidence that broadly, they also tend to be made using nearby stone, within 15km (10mi.). In contrast, in the average Levallois and Quina assemblage, a number of artefacts will be made from rock from distant sources. That tells us two things. First, Discoid technology was specialised, yet somewhat disposable: the flakes weren't intended to last long and be transported elsewhere. Second, this kind of techno-complex would only have suited Neanderthals who knew the rock resources extremely well, and weren't regularly moving long distances.

---

* Just like with teeth, using lithics leaves wear on the surface. By comparing artefacts with experimental collections, it's possible to identify the substances processed.

The third key techno-complex made by Western European Neanderthals was the Quina.* Prehistorians originally focused on its distinctive, steeply retouched scraping tools, but in recent decades attention shifted to how it worked as a system for making big flakes perfectly suited to resharpening. Unlike Levallois, Quina flakes bit away at the cores, rather than shaved across them. In this sense it's conceptually more like Discoid, but rather than fat chunks, Quina flakes are like badly sliced bread, one edge thicker than the other. When the stone nodules used were cylindrical, the impression is more of sausage slices.

This made the Quina similarly efficient to Discoid, with little initial shaping or ongoing core maintenance needed. What mattered to Quina-making Neanderthals wasn't the overall shape of their flakes, but their features. The Quina was all about getting the longest, thinnest edge possible opposite a thick, blunt margin, requiring a particular, very hard way of striking.[†]

The point was to produce perfect flakes that could withstand intense and repeated resharpening. Even the retouch – almost always forming a scraper edge, rather than a serrated one – was peculiarly specific, using a unique motion that essentially tore the stone off. In some sites limestone hammers as well as bone retouchers were favoured, but everywhere the intensity of resharpening was clear: sometimes four or more retouch phases are identifiable. Each time the edge steepened as it moved back towards the thicker margin.

For Neanderthals, Quina combined low waste with abundant, ready-to-use flakes that could withstand heavy use and resharpening. It's fundamentally about anticipating future tool maintenance, both within sites and during longer movements across the landscape.

However cleverly Neanderthals made flakes, for many decades it appeared that they were unable to produce lithic blades, which define the subsequent Upper Palaeolithic culture made by *H. sapiens*. But the

---

[*] Pronounced 'keen-ah', the Quina type site is actually a number of locales spread along several hundred metres of river bank, and located two valleys north-west of Le Moustier.

[†] Hitting very hard means knapping errors were quite common, and sometimes Quina cores broke.

reality is more nuanced. Neanderthals did develop blades from around 300 ka, producing large, wide examples as part of the Levallois system. Later, they also began to experiment with true blade – or laminar – technology: defined by products twice as long as they are wide. They used pre-formed linear ridges running from one (or both) ends of a core, which could direct the kinetic energy and ensured that the shape of each removal would be highly elongated. This is a systematic, nearly continuous process, with each blade setting up the spot where the next strike would be.

There was still a distinctly Neanderthal 'flavour' to this technology, however: unlike Upper Palaeolithic knappers, they used stone hammers instead of bone, and generally prepared cores much less. But these weren't sub-standard, and could be impressively large: refitting at the early site of Tourville-la-Rivière has found blades over 10cm (4in.) long.

The most striking Neanderthal 'blade culture' took place in post-Eemian MIS 5 in north-west Europe, where blades were quite common in some sites over about 20,000 years, often alongside Levallois flaking. Yet this phenomenon doesn't persist. Blades do appear again elsewhere, but they're never dominant – nearly absent in some areas like Iberia – and quite variable. Neanderthals at Fumane Cave were making blades within Levallois-based knapping, but shifted the precise technique over time. And sometimes they went micro. Around 80 to 70 ka in the deep sequence of Combe Grenal, south-west France, up to a fifth of all artefacts were related to laminar knapping. Some were really tiny, under 3cm (1.2in.) long, known rather sweetly by archaeologists as bladelets.

It's long been assumed by post hoc reasoning that laminar technology must be better since later *H. sapiens* made more of it, but what did it actually offer? Experiments suggest blades aren't vastly more economical than flakes, or better for slicing. Moreover, they can barely be resharpened and by themselves are no good for long-term use.

What they lack in robustness is made up for in their standardised, rectangular shape. It's highly likely that at least some – especially bladelets – were used in composite tools, which as the next chapter will show, was definitely within Neanderthals' capability. Being easy to slot in and out, just like blades on a craft knife, they provided sharp edges of a different sort. Since thousands of years separate the different laminar phases at Combe Grenal and many other sites, this implies that Neanderthals may have invented the use of blades multiple times.

Varied methods of making flakes – and to a lesser extent blades – dominated Neanderthal technology in many ways, but the older legacy of bifaces was not forgotten. Though typically rare in the earlier Middle Palaeolithic, from about 150 ka there's a resurgence in bifaces as part of growing technological diversity. Yet this doesn't happen everywhere, and neither were Neanderthal bifaces technologically identical to those of the Lower Palaeolithic. Just like their more ancient counterparts they used bifaces as multi-purpose tools with edges that could effectively pierce, slice or scrape materials. Use-wear shows that they were used on materials from meat to wood, but it also shows that this happened through many phases of use, sometimes with more than one material on the same artefact.

That's because, just like some kinds of flakes, Neanderthals were resharpening them an awful lot. But they used a different technique, often striking very shallow flakes across the tip or down the main working edge. It was possible in this way to rejuvenate the margin many times while keeping the angle relatively acute. This meant that bifaces could be nearly as long-lived and far-travelled as things like Levallois flakes or Quina scrapers.

The discovery of some biface 'workshop' sites, often near very good flint sources, shows this in practice. A single layer at the Pech de l'Azé I rockshelter contains nearly *25,000* distinctive biface shaping flakes. The average knapper produces fewer than 50 during initial production, so over 500 bifaces must have been made here. Yet very few were actually found: clearly, after making them, Neanderthals were taking them elsewhere.

In other places it's possible to see what happened on these biface journeys. In 2002 at Lynford Quarry, eastern Britain, beneath metres of gravel, Middle Palaeolithic artefacts were noticed coming from black organic sludges. This sediment came from a small river course that, 60,000 years ago, was on the margins of a great plain now submerged under the English Channel. It produced thousands of lithics, including around 50 bifaces. Some had been quickly made on cobbles from the nearby river, but most had been knapped elsewhere using beautiful black flint before being used at Lynford and then discarded.

Why would Neanderthals leave so many bifaces, most still useable? In fact, to experienced knappers they're quick and easy to make; much less bother than a Levallois core. For Neanderthals who mastered knapping as youngsters and knew the location of good stone like the back of their hand, lugging bifaces wasn't always the wise choice; far better to carry more meat, fat or other materials.

Nonetheless, where decent stone was scarce, bifaces were hoarded and only discarded as heavily resharpened hulks at the end of their life. We can even find the in-between places where solely resharpening flakes remain as witnesses to Neanderthals staying perhaps just for a night, sharpening their tools and carrying them on further.

Rocky Generations

Whether making bifaces, blades or flakes, something many Neanderthals had in common was their fondness for a good second-hand deal. Recycling could be simple, but a pioneering refitting study from three decades ago shows how in other cases it was quite elaborate. Similar to the distinctive Discoid core at Fumane Cave, an excavator at Coustal Cave, south-central France, spotted a cluster of jasper artefacts, an unusual stone that wasn't available locally. Almost all were within 1m² (1.2yd²), and when reassembled demonstrated an extraordinary transformation in reverse. A Neanderthal had originally brought a long serrated flake tool into the site and likely used it, then converted it to a core and back again to a tool through *eight* stages. This single object highlights how Neanderthals could effortlessly shift in treating artefacts as one category or another, altering the knapping methods used. But it's not an anomaly. Refitting at Combe Grenal shows how a different Neanderthal rescued a botched blade core by serrating the edge, turning a mistake into a useful tool.

What happened at Coustal was probably the work of the tool's original maker, but in other contexts it's possible to see that considerable time passed before recycling happened. Le Moustier contains striking evidence of this habit of repurposing already ancient artefacts. Recent reassessment of bifaces from the base of one layer found distinct colour differences showing that, rather than being badly made, they'd been scavenged from the underlying level and recycled into cores. Despite being totally focused on Discoid technology, it's inconceivable these later Neanderthals didn't recognise the bifaces as having been tools, even if they were only interested in them as an easy source of good flint.

Recycling artefacts is actually very common across many sites, and it seems that just as the eyes of archaeologists are drawn magpie-like to glinting lithics, Neanderthals would have homed in on exposed artefacts inside caves or in open-air locales.* Such encounters may well have been the genesis for an appreciation of old objects not just as sources of stone, but as tokens of time, history and even the presence of 'those before'.

While Neanderthals' habit of recycling lithics has long been known, something much more extraordinary has only recently been appreciated. Known as 'ramification', it's thanks to refitting that researchers uncovered the existence of 'hidden' sub-systems of knapping. Somewhat like tributaries from a river, Neanderthals were taking the flakes they made with primary methods and then further reduced them into a 'second generation' of tiny flakes.

This is easiest to do on thick artefacts, and sizeable things like large Levallois or Quina flakes with fat bases or edges were ideal. In some contexts ramification is so systematic that Neanderthals were obviously not just viewing their lithic technology as about fragmenting stone, but as a means of producing portable reserves of rock that could then be treated as mini-cores.

There were a multitude of second-generation approaches, some of which were aimed at making miniature versions of the original flakes. By taking out notches from Quina flakes or tools, Neanderthals obtained little second-generation flakes that had exactly the same properties as the first-generation products knapped direct from the core: a sharp edge opposite a naturally blunt one.

Remarkable research combining refitting and use-wear at the site of Jonzac not far from Saint-Césaire highlights the astonishing consistency in some second-generation systems. Originally exposed when nineteenth-century quarrying cut through deep archaeological layers against a limestone cliff, one layer contains masses of animal

---

* Burrowing by creatures like hyaenas probably also uncovered older material.

bones, mostly reindeer, which had been intensively butchered. Quina lithics were used to cut up carcasses and scrape flesh off skins, typically being resharpened at least once before then being put to use for very heavy work, probably chopping bone. This led to splintered edges, at which point some then had notches removed so they could be used again, while others became hammers or anvils.

But that's just the main sequence; there was also a second-generation cycle using all the smaller flakes that came off during shaping and resharpening of the tools. Neanderthals were using roughly half, some of them retouched. But most fascinatingly, there was further selection in what the second-generation flakes were used for. Those flakes that came from tool shaping were employed for butchery and to clean fresh skins, just like the original Quina scrapers. In contrast, flakes that came from resharpening were *only* used to cut meat, and sometimes themselves resharpened. Even the bits that came off when battered Quina tools were notched were requisitioned, but again only for cutting meat.

These specific patterns may be unique to the groups that visited Jonzac, but they exemplify how Neanderthals in all places and times were hyper-aware of the material potential of everything they made. This 'waste not, want not' attitude has been identified across virtually all techno-complexes: back at Fumane Cave, in one layer thrifty Neanderthals were taking bladelets off the edges of flakes. Occasionally there's even a third generation. At Combe Grenal, after turning Discoid flakes into micro-cores for making diminutive points, they then knapped off tiny blades and bladelets.

In some settings it looks like Neanderthals used ramification to manage scarce high-quality stone, but it could also function as a time-saving device, making the most out of decent pieces from among plentiful but bad rock. Yet it's intriguing to note that stone availability isn't always the motivation. Instead, as at Jonzac, specialisation in activities underlies why Neanderthals made very small artefacts. Though some probably could be used in the hand, others like petite Levallois flakes just 2cm (0.8in.) long from Pech de l'Azé IV would surely have been hafted, and it's worth noting that *all* bladelets made by Neanderthals were produced within second-generation systems. This strongly implies that bladelets were not accidental knapping by-products, but were integrated within technological systems: part of the diverse range of artefacts they were aiming to make from the outset.

An Embarrassment of Riches

Over the past several decades, the overarching story has been that Neanderthals used stone in more systematic, complicated and nuanced ways than was ever suspected. But explaining *why* they invented so many ways of going about this has long been an enigma. The typological obsessions of early prehistorians culminated in Bordes's production between the 1950s and 1960s of a 'definitive' Neanderthal tool catalogue. It had over 60 categories, with the edge number, location and shape of tools being counted as different types. Comparing Le Moustier to his excavations at Combe Grenal – comprising 50 layers through a sequence 13m (43ft) deep – Bordes picked out repeated patterns in relative amounts of tool types, and on this basis proposed that Neanderthals had lived in five main Mousterian sub-cultures, one of which included Quina.[*]

Although extraordinarily influential as a heuristic tool for recording assemblages, Bordes's explanations didn't account for technology as a dynamic process. Ethnography combined with newfangled computer analysis led to the realisation that a lot of the tool variability simply reflected function: Neanderthals doing different things, in different places. Observations within hunter–gatherer communities showed that retouching of flakes was often about resharpening them, rather than setting out to create a particular edge.[†]

This meant that Bordes's tool categories were actually more like points on a spectrum as Neanderthals rejuvenated the edges of their tools. Furthermore, when the methods of core flaking were considered rather than just shapes of retouched tools, things got much messier. For example, a succession of 10 layers from late MIS 5 to early MIS 4

---

[*] The others were Ferrassie (big scrapers and Levallois), Denticulate (serrated and notched tools), a biface tradition and one 'catch-all' category he called 'Typical'.
[†] One 1970s study reported that historically, Aboriginal knappers paid as much attention to the overall appearance of their scrapers as Westerners do to their pencil sharpeners.

at Combe Grenal look quite dissimilar based on Bordes's tool categories, but in terms of how the flakes were made in the first place, they're virtually all Levallois.

The other problem with the typologies is the relatively recent recognition of two things. First, many of the assemblages Bordes analysed came from quite thick layers. Almost everywhere, recent geologically based analysis indicates that such deposits contain many separate phases. Le Moustier is a case in point: four originally defined layers turned out to contain at least *20* sedimentary levels, and when the lithic assemblages are examined at this finer resolution, there are definite technological differences. What Bordes saw as a single biface-dominated phase in fact begins mostly as Levallois. This might sound like an academically arcane issue, but it matters because for many years prehistorians built large-scale behavioural models based on connecting particular types of Mousterian with the climate or the animals being hunted.

Today it's obvious those models don't stand up to scrutiny, even if variation in technologies and the amount of retouching between different sites and phases is real. And this is where the second recent shift in understanding is important. Very often, early excavators kept little besides the retouched tools. They threw away many cores and virtually all the small flakes. Yet as researchers paid more attention to technology, experimenting with their own knapping and attempting refitting, they realised that the practice of discarding supposed 'waste' artefacts from archaeological assemblages meant the loss of vast amounts of information on *how* Neanderthals made things.[*] Once again Le Moustier is a case in point. Un-excavated areas show the layers here are so rich there's sometimes more archaeology than dirt, yet considering the gigantic amounts of sediment removed by Hauser and Peyrony, the amount of lithics in the old collections is paltry.[†]

Rather than writing everything off, however, the current generation of researchers have returned more than a century later to 'excavate the

---

[*] The majority of stuff that comes off any core is under 2cm (0.8in.) long, yet can still be technologically distinctive.

[†] More careful excavations in the 1980s produced artefact densities about 30 times greater than those from the early twentieth century.

excavation'. Their field seasons are ambassadors for twenty-first-century methods. Over several weeks the depth might increase by less than a hand's span, because everything is kept. Pieces over 2cm (0.8in.) have their precise 3D coordinates plotted using lasers, while anything smaller is located to within a 50cm (20in.) grid square. Truly minuscule splinters are recovered through wet sieving.

This modern 'total collection' policy, combined with meticulous examination of technological details, has led to a new appreciation for the complex interplay between Neanderthals and stone, at both individual and assemblage level. Formalised as the theory of 'techno-economics', it explains to a large degree why they chose to knap in particular ways and varied the intensity of retouching. Consistent patterns exist across virtually every site, confirming that not only did they preferentially choose high-quality stone when expecting artefacts to require resharpening due to long-term use, but for the same reason it's the largest flakes that tend to be retouched. And when Neanderthals transported objects between sites, they moved those made from the best rock farthest, never bothering to carry poor-quality stone to richer areas. This implies not only constant weighing up and decision making, but an extraordinary knowledge of the geology across wide regions.

Change and Time

Yet that's not the end of the story. While – as a rule – available stone resources remained largely unchanged through time, the techno-complexes were not static. One enduring legacy of Bordes's typology is that the different assemblage types do seem to show a chronological pattern when compared stratigraphically between numerous sites. Across south-west France Neanderthals were making a lot of Levallois-rich assemblages during MIS 5, but far less so as time went on. Instead, in MIS 4 the Quina techno-complex appears, but is then itself succeeded by growing amounts of Discoid technology as well as some assemblages with numerous bifaces. Remarkably, this sequence has persisted over the past 30 years, although some nuances have been added: there's definite overlap in some places; for example, Quina lasts into MIS 3, when many Discoid assemblages have already appeared.

And the very youngest layers – only preserved at some sites, but always *above* the late Discoid – see a reappearance of Levallois with large scrapers.

This flowering of technological diversity from about 150 ka onwards didn't happen in the same way everywhere. Neanderthals in other regions were doing different things at different times. While the caves of south-west France got the lion's share of research historically, spreads of lithics across the plains of northern France were harder to understand. That changed over the past few decades, with accurate dating methods uncovering a different technological sequence. Though the precise reasons why are still the subject of debate, there seems to be a much clearer correlation with the regions' more severe climatic and environmental fluctuations compared to the south of France.

As the climate began to cool after the Eemian peak around 123 ka, Neanderthals there favoured simpler flaking methods than typical Levallois. Then, during a cold dip around 110 to 109 ka, that florescence of blades mentioned earlier takes place. Conditions see-sawed over the next 20,000 years, boreal forest repeatedly growing and disappearing, but overall it steadily grew colder. Blades persisted but in more diverse forms, and Neanderthals explored new Levallois methods as well as invented a streamlined approach to making points.

The end of MIS 5 witnessed rapid, dramatic, repeated cycles from warm to cold. Neanderthals are most clearly visible when the forests rebounded, and once again they're making a range of lithics. But as it got colder and massive aridity led to vast steppe taking over, bifaces become important for the first time in hundreds of thousands of years. In the space of 10 generations, full-on glacial conditions kicked in at the beginning of MIS 4, and though Neanderthals making large Levallois flakes pop up during brief thawings, eventually northern France seems to have been abandoned.

While Neanderthals re-materialised almost immediately with rising temperatures at the transition between MIS 4 and MIS 3, culturally they now looked very similar to those from south-west France. Levallois points and true blades vanish for good, but is this because they were no longer useful, having been developed in earlier forest-steppe environments, or were they the work of unique Neanderthal cultures that went extinct during MIS 4? It's intriguing that these northern French populations during MIS 5 display much more obvious technological exclusivity, which could support the cultural interpretation. Blades, bifaces and points all occur with Discoid and Levallois, but never with each other, intimating that these were either fulfilling very specific functions at particular places, or reflecting cultural traditions.

Beyond northern France, one of the strongest cases for true Neanderthal lithic cultures exists at a much larger scale. A distinct 'biface divide' splits Europe down the middle, with the west as a Mousterian world, where bifaces followed ancient traditions. Broadly symmetric, they have sharp perimeters flaked all – or most – of the way round. In contrast, Neanderthals in central-eastern Europe developed a very different way of doing bifaces. Known collectively as the 'Keilmessergrüppen', they're defined by asymmetry, with one bifacial sharp edge opposite a natural or artificially blunted margin.

Mousterian- and Keilmesser-making Neanderthal knappers lived at the same time, used both Levallois and Discoid for flake production and hunted similar species. Nonetheless, they held totally different ideas on what a biface was, from how it should be made to resharpening methods. Clearly there was a cultural border of some sort, but unpicking whether it was to do with populations who never came into contact, or something more subtle, remains a significant challenge.

And enigmas also exist about the main techno-complexes. Discoid and Levallois were apparently widely known technologies, yet wherever we find the former, it's almost always the only method being used.* Did some Neanderthal groups traditionally learn only to knap this way, or did they do different things elsewhere? Our inability to truly track groups from site to site means we need to look for other clues. The possibility that particular techno-complexes were adaptations to certain environments is appealing, but for neither Levallois nor Discoid does it really hold up. Moreover, there's very little evidence from use-wear that the tasks they were used for were any different.

There is one techno-complex that is quite restricted geographically, shows a strong climatic correlation and is *never* found with any other core technology: the Quina. This could therefore represent a particular

---

* When reliably unmixed assemblages are considered.

way of Neanderthal life, and we'll explore this further in Chapter 10. But what's notable in terms of thinking about technological diversity as being potentially related to different cultures across the continent is that the Quina appears in southern France right around the time blade technology *disappears* in the north, and later fades out just as bifaces increase in importance. For the last 40,000 years of their existence, Neanderthals were manifestly experiencing massive upheavals in climate and potentially also population disruption, but rather than failing to adapt, their archaeological record is brimming with evidence of innovation and cultural evolution.

Weaving It Together

Though Neanderthals' lithics were collected for over a century, it took decades for them to be systematically studied. Thanks to advances both in thinking and analytical methods, today we have unprecedented insights into what they were doing with stone, and why. Researchers zoom in and out from continental scales to single refitting sequences, while some of the greatest modern insights have come from the humblest objects: the stony dross they threw away by the tonne. A crucial lesson has been that, while Neanderthals undoubtedly had cultural norms, they were also innovative individuals. It took *someone* each time to devise and refine different techniques, whether adapting to an unfamiliar type of rock, or the invention in many times and places of blades.

The enduring myth that Neanderthal technology was stuck in some kind of cognitive mire, bogged down by minds unable to innovate, is false. These were neither unsophisticated nor fixed and unvarying people. Theirs was a dynamic dance with stone, waltzing to different rhythms that merged external factors with ideas, choices, foibles. Elemental boundaries – what the geology made possible or prevented – surely imposed constraints, but thanks to technical mastery and a laser focus on what they wanted, creative responses were possible. As routinely as breathing, Neanderthals paid attention to their rock: selecting the choicest types, playing with novel ways to fragment, shifting concepts and skills as needed.

Their technologies were also centred on quality and efficiency. Even if Levallois began the Neanderthals' Middle Palaeolithic revolution, a bouquet of other techno-complexes and particular methods blossomed. Time as well as stone was managed, whether via sustainable technologies that could be carried and resharpened many times, or disposable flakes made for the moment. In addition, over time they developed more complex ways of obtaining what they wanted. The icing on the cake was ramification, with second generations of lithics splitting material up in more complicated and specialised ways than ever before. Nothing they did was thoughtless, and while they followed structures, flexibility was there too in the shifting between categories, such as flake-to-core or waste-to-tool.

All these features underpinned how Neanderthals' very world was growing. They were untethering themselves from geology, and in the process opening up the landscape to ever farther exploration. By increasingly separating where and when they made, used and resharpened artefacts, their activities and minds stretched through space and time. More lengthy activities and movements imply expanded memory and planning. With the Middle Palaeolithic, we're witnessing, if not the birth, then the maturing of minds with the ability to fly into the future and imagine what would happen days or even seasons in advance.

Put this all together, and you perceive Neanderthals as hominins at the top of their game. Their lithic technologies seem to have buffered them from all but the toughest climatic and environmental challenges, potentially even stimulating new inventions. From 150 ka onwards there is a strong impression that they evolved ever-more creative solutions as their geographical range expanded.

Thinking of Neanderthals as experimenters pushing boundaries may be unfamiliar, but this new view is grounded in the archaeology. Excitingly, it's also there in materials much rarer than stone: the true scale of their nearly vanished organic technological world is just now being appreciated.

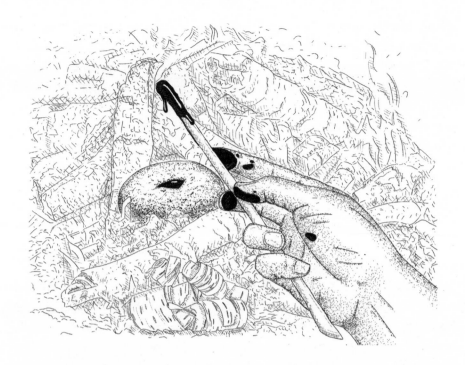

CHAPTER SEVEN

Material World

Inside the body, the bone
Under the skin, the blood
Over the skin, the fur
Across the fur, the hand
From the hand, the fire
Before the fire, the wood
Out of the wood, the tar
Held by the tar, the stone
Scraped by the stone, the red
Beneath the red, the shell
Inside the shell, the secret.

Over 99 per cent of all artefacts from the Middle Palaeolithic are stone. Unlike organic things it cannot rot. Objects made from living things, whether animal or plant, are extraordinarily rare in comparison; teeth and bone survive better than wood, but not always. Yet such substances make up the vast majority of hunter-gatherer technologies, making it likely that there's a whole 'ghost' realm of Neanderthal artefacts we're missing. Sometimes we see their shadows: use-wear on lithics from many sites matches wood or plants. And very occasionally, precious objects survived the millennia, testifying to the existence of lost multitudes.

Until recently these were meagre clues, but the past three decades have seen a boom in discoveries. Today wood, bone and shell are central to new perspectives on Neanderthals as technical artisans in substances other than stone. What it tells us about their behaviour has in many cases been revelatory.

Let's begin with wood. Few Neanderthals lived in ice-blasted wastes, and while not always abundant, trees would have been part of everyday life. Interglacial populations walked beneath spreading beech boughs, watched the blossoming gold of autumn larches and very likely possessed as much tree lore as knowledge of stone. As early as 1911 when a spear tip was found in cliffs by the bustling seaside resort of Clacton-on-Sea, Britain, prehistorians had suspected that pre-*H. sapiens* hominins fashioned wooden objects. It wasn't until later that its impressive age of 500 to 450 ka was realised, and shortly after an entire spear appeared at Lehringen, Germany. It's huge – nearly 2.5m (8ft) long – quite thick, and probably for thrusting. More importantly, it came from the Eemian, and was therefore the work of Neanderthals. Like a hunter's calling card it lay snapped into pieces beneath a huge elephant skeleton, but it hadn't been well recorded. It wasn't until 1995 that truly sensational evidence emerged for the finesse of their wood technology.

A damp November day that year saw researchers from around Europe travelling to the enormous brown coal mine at Schöningen, Germany. They were drawn by dramatic claims about a dream Neanderthal site with hearths, masses of butchered animals and wooden weapons. Lithics and beautifully preserved animal and plant remains from ancient lake deposits had been found three years earlier, leading to preservation of a 4,000m² (4,800yd²) block of the mine's sediment. As colossal machinery loomed over the archaeologists dwarfed by the surrounding pit, the post-apocalyptic scene was a far cry from the ancient world being revealed. From fine sediments within the 40m (130ft)-deep stratigraphy, enigmatic small wooden objects had appeared, one of which was nearly 1m (1.1yd) long with carved, pointed ends. Totally unique then, it was in fact a harbinger.

When the invited archaeologists arrived a couple of years later, scepticism was transformed to wonder. Schöningen turned out to contain one of the most important archaeological finds of the twentieth century. From the dense black and grey silts of Locality 13 II-4 came proof that the apparently outlandish claims about Neanderthal weapons were true. Alongside dozens of butchered horses lay scattered the instruments of their demise: elegant, finely tapered wooden spears.

Dating between 337 and 300 ka, the Spear Horizon is a vaguely diagonal spread covering about 50m² (60yd²) along the shore of an ancient lake. Just one layer among many at Schöningen, it alone contains over 15,000 finds. Mostly bone, there are however numerous pieces of wood including in a relatively small area, eight fragmented spears. The most complete was only snapped in two places and was nestled among remains of one of the 50 horses butchered there.

The spears blew apart notions of primitive Neanderthal woodworking abilities: they're far from pointed sticks. Finely crafted from thin spruce and a single Scots pine, their tips are all at the stump end: the hardest part.* The shafts were systematically carved off-centre for increased strength, a trick also seen some 200,000 years later at Lehringen. Weighted towards the tips, just like javelins they were probably engineered to fly. One 2.5m (8ft) spear stood out as significantly longer, hinting at a multiple-weapon system. Experiments show that the shorter throwing spears easily range to 30m (33yd), and the long lance would allow precision killing while avoiding close contact with frantic prey. The Lehringen spear is similarly lengthy, with a thicker base.

Excavation of the jaw-dropping Schöningen deposits has been continuous since 1995, with the Spear Horizon now one of over

---

* Alaskan Yupik and Athabaskan Indigenous communities will source spruce from driftwood, and stumps are seen as prime woodworking material because of their strength and resistance.

20 studied areas. It's possible to track Neanderthals through space along the lake shore and through time, since archaeology is found in under and overlying layers too. Eventually the number of spears may even increase, since other worked fragments have been identified among huge spreads of splinters.

But weapons aren't the only wooden objects now known to have been made by Neanderthals. New finds in 2018 from Southern Europe confirmed the scope of their skill. Multiple worked sticks with single pointed ends came from two open-air locales: Aranbaltza in northern Spain dating around 90 ka, and Poggetti Vecchi in Italy aged about 200 ka. Much shorter than the Schöningen spears, their length, damage patterns and use-wear strongly suggest that they were digging tools, though probably also useful for prodding, poking and aiding walking. Despite seeming less exciting than spears, these artefacts were in fact crafted with just as much attention.

This extended to raw material. One of the two Aranbaltza sticks was yew, well known as extremely hard-wearing yet flexible wood. Famously used for the feared medieval English longbows, it was also the choice of both the Clacton-on-Sea and Lehringen spear makers. Meanwhile at Poggetti Vecchi, all 40 worked fragments of wood – representing at least 6 tools based on the numbers of handles – were of boxwood: even tougher and denser than yew, this grows very long, straight branches. Hours and hours would have been needed to carve such tough wood, but the choice was intentional: among many traditional societies the hardest available species are selected for digging sticks precisely because they're durable. In contrast, softwoods may have been used for the Schöningen spears because no suitable hardwood was around.

The technical know-how goes beyond selecting species. Just like the spears, CT scanning revealed that the Aranbaltza yew tool was made from off-centred heartwood. The digging sticks at both sites were carefully worked with tips ground smooth, while tiny charred traces show that fire was used to help remove bark and outer wood. There's even recycling evidence: the Aranbaltza stick looks like it was chopped off a longer object – perhaps a spear – while some of the Poggetti Vecchi sticks may be discarded, worn-down tools.

The latter, which were found by chance during construction of a swimming pool, were in fact jumbled among a mass of animal bones, mostly straight-tusked elephant. Without butchery marks it's impossible to prove a connection or be sure what role they played – notably, one had a sharp tip and two others mysterious notches – but otherwise explaining their presence is difficult.

The objects mentioned so far were used out in the landscape, and clearly sometimes linked with hunting. Did Neanderthals have other kinds of wooden tools? In north-east Spain is an enormous rockshelter known as Abric Romaní, which has produced some of the most important data on Neanderthals in the past three decades. When excavations began in 1909 there was little indication that an astonishing archive lay concealed beneath its attractive yet ordinary travertine overhang. In fact, it's the same calcium carbonate waters which formed the overhang that make this place special: layers of flowstone were repeatedly deposited across the floor of the rockshelter. Each covered the detritus from Neanderthal occupations, conserving it in breathtaking detail.

Once would have been amazing enough, but this happened at least 27 times across 12 layers, spanning 40,000 years.[*] Hundreds of hearths, many tens of thousands of lithics and bones are preserved alongside perishable things like leaves, pinecones and carbonised wood.[†] Other wooden objects rotted but left impressions behind cast in flowstone: Middle Palaeolithic equivalents of the Pompeii bodies.

Abric Romaní holds a totally unique record of the stuff left by Neanderthals each time they moved on, including a hundred or so wooden objects across multiple layers. Mostly fuel for fires, some are worked artefacts. At least one resembles the Aranbaltza stick, but others are very different. Two carbonised, gently curved objects from a layer aged between 50 and 45 ka look very much like wooden platters, somewhere between the diameter of dinner and side plates. Another was flat, but with a long protrusion at one end, possible providing a handle.

The most remarkable find came in 2011 from a level dated around 56 ka. While yet to be published, press releases showed a large

---

[*] A borehole has found that the deposits go down far beyond 20m (65ft) and extend back to 100 ka.

[†] Burned without air, like charcoal.

cleaver-shaped tool complete with flat blade and handle; exactly as you might find in the kitchen of a keen chef. It presumably served to slice soft things, and stunningly shows the role of wood technology in everyday Neanderthal domestic life.

Joinings

Neanderthals were indisputably carpenters, but they also pioneered composite tools. The technological echelon represented by assembling multiple parts into one object gives greater control, more shock absorption and saves time and energy since it allows modular repairs. Composite tools mostly have an 'active' stone part – the business end – and a handle, or haft. Some Neanderthals may have used simple wedged handles, but in other cases use-wear points to bindings: probably sinew and tendons or even plant fibres. And amazingly, some ancient adhesives have survived. Suspicious black residues on Middle Palaeolithic artefacts from Syrian sites turned out to be 50,000-year-old bitumen, a ready-to-stick natural asphalt. Neanderthals elsewhere also spotted its usefulness, shown by residues found in 2012 at Gura Cheii-Râșnov Cave, Romania. This finding makes chemical identification of bitumen/oil shale in dental calculus at El Sidrón especially intriguing;* moreover, as the same individual had severe tooth chipping, he may have used his mouth to manufacture or repair hafted tools.

Even more complex demonstrations of Neanderthal hafting technology exist. In the 1970s archaeologists excavating another German brown coal mine at Königsaue found two small black lumps from a lakeside occupation, dating around 85 to 74 ka. One had certainly been part of a composite tool: three surfaces bore imprints of a lithic tool, a wooden surface and the unmistakable whorls from a partial Neanderthal fingerprint. It was only in 2001 that chemical analysis identified unique biomarkers from birch trees; specifically, tar derived by cooking the bark in low-oxygen conditions.

---

* Natural bitumen sources aren't especially common in Europe, but there is an oil shale deposit within 20km (12mi.) of El Sidrón.

Today at least two other examples of Neanderthal-made birch tar are known, widely spread in time and space. One was dredged from under the North Sea, before being picked up on an artificial Netherlands beach. A sizeable lump of birch tar still half-covering a flint flake, it was directly dated to around 50 ka, and quite astonishingly comes from the same submarine zone that produced the Zeeland Ridges Neanderthal skull fragment. Over a decade earlier an almost identical object, together with another smeared lithic, had been found in river gravels at Campitello, Italy. They're far older, taking back this technology to the early Middle Palaeolithic around 300 to 200 ka.

As-yet-unidentified lumps and potential hafting substances exist at a number of sites, but the most spectacular – and surprising – case was published just in 2019. Very small flakes from Fossellone Cave and Sant'Agostino Cave in Latium, Italy, have residues from pine or conifer resin. Both are roughly the same age, around 55 to 45 ka, but in technological terms one is far more complex, because it's mixed with beeswax. Pine resin is less shock-resistant than birch tar, but experiments show that adding wax makes it nearly as good.

Dripping from bark and scenting forest air in warm regions, pine resin is easily noticed; simply leaning on a trunk for balance can leave you covered in the sticky stuff. But what about beeswax? As Chapter 8 will discuss, Neanderthals may well have had an appetite for honey that perhaps led to their investigation of wax as a substance. Its inclusion in a hafting adhesive recipe is testament to their concern for material quality, and an ability to experiment and innovate.

But just how common was hafting in Neanderthal life? Hafting polishes are found on many artefacts that aren't weapon tips, from retouched tools to ordinary flakes. Impressively, almost half of sampled artefacts at the northern French open-air site of Biache-Saint-Vaast had hafting micro-polish, while work on Syrian bitumen sites in some cases suggests that around a third had residues. The many small flake and bladelet systems would make sense as hafted objects, and then there's also conifer resin – possibly heated – from at least one individual's calculus at El Sidrón. It may be that, at least for some Neanderthals, composite tools were far more quotidian than the rarity of their archaeological traces would lead us to believe.

The scarcity of preserved plant remains long hampered our understanding of their use by Neanderthals, but the role of other more tenacious organic materials has also only recently been appreciated. This includes shell, the bio-mineral properties of which are in some ways similar to stone. Very rare Lower Palaeolithic shell artefacts exist, but from 120 ka onwards Neanderthals developed true shell-based technologies. So far 13 sites – all in Greece and Italy – have produced several hundred worked shell tools. At Cavallo Cave, southern Italy, in addition to some fragmented pieces across various layers covering some 10,000 years, the richest level is Eemian with over 120 retouched shell parts. And in the same region and period, Neanderthals at Moscerini Cave also produced shell tools. Along with unretouched fragments, 170 certain tools came from a very small excavated zone.[*]

Once again, shell tools show how Neanderthals were choosy about their materials. Smooth clams (*Callista chione*) were especially favoured, being a decent, palm-filling size, and with striking glossy surfaces. Similar species, though, were hardly used, while mussels at Moscerini may have been eaten but were never knapped. And even though smooth clams are edible, there isn't always evidence that they were food waste. At some sites they were collected from beaches, but at Moscerini almost a quarter look as if they were ferreted out fresh from the sands.

Why use shells? They offer naturally ergonomic blanks, but being blunt they require retouching to create an edge. And though they dull faster than stone, shell tools are auto-renewable: Neanderthals could freshen up edges simply by briefly using the tool on a hard material. Some species are also surprisingly robust, and use-wear confirms that they were cutting meat and skins, in addition to scraping wood.

Most come from Quina-like assemblages, and it seems that Neanderthals noticed the resemblance between such flakes and the long, curved edge and shorter back of shells. They certainly were able to apply the same technical know-how: the unusually strong force

---

[*] Dug in 1949, this site is now buried under construction debris from the coastal highway between Rome and Naples.

needed for Quina retouching is also essential in shell knapping; even the gesture is similar. What all the shell sites have in common is local scarcity of high-quality rock, with Neanderthals forced to use poorer stuff, including very small beach pebbles. Clear economising was going on, with the shells being broken up to gain small, immediately useable pieces, each of which could then also be resharpened. In other cases, however, shell was sourced from up to 15km (10mi.) away, which starts to look more like a preference.

One riddle remains: why are shell-knapping sites not found outside the central Mediterranean? Smooth clams today are abundant farther west and right around the Atlantic coast, though perhaps less so in glacial periods. Nevertheless, the absence of shell tools in Iberia is odd, given that Neanderthals were certainly foraging on seashores, as we'll see in the next chapter. Perhaps some shell assemblages are in those now-sunken sites, where denizens of dark sea caves may be sliding over ancient knapped remains of their relatives.

From Stone to Bone

Neanderthals of course had access to a third, far more abundant organic resource for tool making. Animal bodies connected two great pillars of their lives: technology and subsistence, both in the hunt and afterwards as raw materials. For decades the orthodoxy claimed that antler, ivory or bone tools were virtually absent from the Middle Palaeolithic, such that they became markers for the emergence of 'modern' *H. sapiens* behaviour. Thanks to advances in analysis combined with shifting expectations, today's prehistorians have uncovered a very different picture. The Neanderthals emerged from a Lower Palaeolithic world where animal remains were already entangled with the production of lithics, but they both amplified and transformed these early traditions.

Their forebears well understood the utility of bone or antler hammers for the shallow flaking techniques needed to make bifaces. But the dawn of the Middle Palaeolithic saw a shift where entire bones or big hunks began to be replaced by smaller shards, especially from limb bone shafts. Mostly used for final shaping, retouching and resharpening of tools, in some sites they're extraordinarily numerous, yet absent in others. Growing research into these objects – generically labelled as 'retouchers', although this could also imply resharpening – has unlocked fascinating details about Neanderthal technical skill and traditions.

Outstanding evidence of the ascent of bone tools comes from the Spear Horizon at Schöningen. From an area of just 50m² (60yd²) there were 15 massive bone hammers, some of which bore damage showing they'd also been used to smash other bones for their marrow (many of which had previously themselves been used to resharpen lithics). Hammering and retouching require specific levels of force and knowledge of how to heft and where to aim, so although early Neanderthals at Schöningen were obviously using large bones for different tasks, they weren't yet preferring shards for particular activities.

Nonetheless, there is clear selection in the *kinds* of bones. At Schöningen horse carcasses were available in abundance, and their lower limb bones – known as metapodials – were definitely preferred for tools, especially multi-use objects. Metapodials aren't very meaty, but their thick, strong, flat shafts are perfect for spreading percussive energy.

Within 100,000 years across the early Neanderthal world, bone technology becomes more refined. Metapodials are still favoured, but rather than whole bones, it's almost always shards being used. Research at Les Pradelles, France, a sinkhole formed by a collapsed cave, exemplifies how systematic Neanderthals had become. Recent reinvestigation of immensely rich Quina layers dating between 80 and 50 ka has so far identified some *700* retouchers. Two-thirds come from just one assemblage, where they're twice as common as lithics.

As hunters, Neanderthals were intimately familiar with the physical properties of bone, and variation between species. At Les Pradelles most of the butchered animals were reindeer, yet Neanderthals here preferred bigger beasts for their retouchers: horse and bison bone tools are twice as abundant as would be expected. A similar predilection for larger species is seen in many other sites, for example where roe deer was mostly hunted, less common moose or aurochs bones were selected.

Beyond species, the body part was also considered. Rear limbs seem to be the equivalent of flint for lithic artefacts: Neanderthals typically preferred them, but could be flexible if necessary. Back at Les Pradelles for example, there are exceptions to the metapodial retoucher 'rule' since occasionally jaws, shoulder blades, hip bones, ribs and even toes were used. In other contexts, extremities rather than shafts of big limb bones were chosen, harking back to the early Middle Palaeolithic. But elsewhere it was very different body parts: horn cores, horse teeth and even at the vast cave of Kůlna, Czech Republic, mammoth ivory. Carnivores weren't scorned either: among sporadic sabretooth cat remains from Schöningen an upper forelimb in the Spear Horizon was used both as a hammer and retoucher.*

Pickiness about size even extended to the bone shards themselves. A decent length was needed to achieve the right 'flick-of-the-wrist' retouching gesture, and so Neanderthals consistently chose pieces over 5cm (2in.) long. The average length of retouchers at Les Pradelles is nearly double that of unused bone fragments. Neanderthals had exactly in mind what they wanted, and wouldn't settle for any old splinter. In at least some sites they were taking out the good bits *during* butchery, rather than sifting through random piles of waste, which makes sense as fresh bone is stronger and more elastic. New refitting work shows this at micro-scale. At Scladina, Belgium, while smashing a butchered bear femur for marrow, a Neanderthal carefully picked out only the four longest shards to use as retouchers.

Despite being excellent judges of quality, however, Neanderthals sometimes made unexpected retoucher choices. Like many sites, the Spear Horizon at Schöningen is a palimpsest of occupations, so the fact that about 75 per cent of all the bone tools are from the horses' *left* side is remarkable. It may be something to do with how grip and gesture work better for right-handers using left-side bones, but what's important is that this was an intentional choice.

We can even get a hint that Neanderthals had particular tasks in mind for some retouchers. Incredibly detailed analysis of the location, shape and nature of damage reveals that larger and thicker ones are more heavily used. As the battering changed the bone surface from

---

* Already weathered, this was likely picked up rather than from a hunted animal.

flat to concave, knappers shifted to different zones, sometimes as many as five times. The surface of retouchers was often scraped prior to and during use, and at Le Rozel, northern France, scraping is more common on retouchers used multiple times, suggesting Neanderthals took more care preparing tools that they expected to use for longer.

One remaining puzzle exists. While individual retouchers were more intensively used in some techno-complexes, especially Quina, there's no obvious explanation for why some sites have huge numbers but others hardly any. Local quality or availability of stone doesn't seem related, and nor does the number of retouched lithics, age of the site, its function or the animals hunted. The answer may be that these retouchers were tied to contexts and dynamics that are archaeologically elusive, such as the site's place within a wider occupation cycle, or the social make-up of group members.

Retouchers are by far the most commonly found sort of organic tools, but Neanderthals also used bone in other ways. Consistent crushing damage on the ends of some limb bone shafts suggests that they were used perhaps for indirect lithic knapping, where a piece of bone acts as 'middleman' between core and hammerstone, focusing the force. Other bones with smoothed or polished surfaces have clearly been used to rub materials in different ways. Sometimes this is visible on retouchers, but mostly plain bone shaft fragments. At Schöningen Neanderthals were intensively using sharp bone tool tips on medium-hard materials, possibly even as 'knives' to cut tough muscle fibres. Other pieces there, including a fragment of ivory, had been slowly worn smooth and even bevelled from rubbing on softer stuff, probably animal hides.

Detailed analysis on larger assemblages elsewhere has managed to sort different bone tools according to the type of damage and direction they were moved in. At Combe Grenal some have scratches parallel to the long axis and wide, polished ends, while shorter fragments with pointed ends display very different, sideways wear patterns. Exactly what any of these were used for is unknown, but Neanderthals were obviously selecting bone tools by task.

Just like shell, bone was also occasionally knapped and shaped. This practice has Lower Palaeolithic roots, though how often

Neanderthals did it might be masked because bone is so often damaged by carnivores or other natural processes. Knapped mammoth tusks, however, do tend to stand out. Nineteenth-century excavations at Barme Grande – one of several rich coastal caves on the Italian–French border – uncovered still-connected remains of a butchered young mammoth, whose tusks seem to have been split and flaked. At Axlor, northern Spain, bone fragments appear to have been retouched into scraper and 'chisel' edges, with use-wear indicating they were used for hide working. In other contexts, Neanderthals even knapped bone retouchers and other tools to alter their outline, strong evidence for desired *forms* as well as properties.

Did Neanderthals also make weapons from bone? Maybe. Some of the best candidates come from Salzgitter-Lebenstedt, Germany, a rich open-air reindeer kill site dating to at least 55 to 45 ka. There are more than 20 shaped bone artefacts here, including mammoth ribs flattened into points around 0.5m (1.5ft) long.* The most remarkable – and totally unique – object is a wedge-shaped point made from reindeer antler. Its conical tip and chamfered base recall nothing so much as the business end of a hafted weapon. Only 6cm (2.3in.) long, it would work as a light spear point, or just possibly, a dart or arrowhead. The latter technologies are almost universally assumed to have been invented by early *H. sapiens*, but they are hinted at by lithics in one other Neanderthal site (which we'll explore in Chapter 15). Otherwise, Salzgitter remains the only candidate for Middle Palaeolithic bone weapons.

Tools to Minds

The range of organic and lithic technology we now know Neanderthals employed has ballooned, but what does it tell us about them? Leagues beyond the most complex primate, bird or other animal toolmakers, their expertise even compared to more ancient hominin ancestors is striking. While Levallois was long viewed as a cognitive pinnacle for Neanderthals, similar levels of sophistication are involved in the

---

* Near the original site in the town of Salzgitter there is a residential street named '*Mammutring*', or Mammoth Crescent.

other lithic techno-complexes, even if they functioned rather differently. All demanded excellent control of knapping mechanics and an ability to sustain a multi-stage project requiring foresight.

Over the past three decades the clear trend is for technological lines between Neanderthals and early *H. sapiens* to grow fuzzier: even if not always abundant, supposedly 'modern' things like bone tools were far from absent. The only lithic techniques seen in the contemporary African Middle Stone Age but not visible in the Neanderthals' Middle Palaeolithic are controlled heating to improve stone properties, and the 'pressure notching' technique for creating serrated weapon tips.[*] In fact, these are far from widespread and cognitively aren't radically different to other things Neanderthals were doing.

Given the obvious craftsmanship in the Schöningen spears and other wooden objects, we must surely now be thinking of Neanderthals as carpenters. Such objects actually mean greater investment of time and energy than any biface or even a Levallois core. And although bone tools are generally less heavily worked, they show that Neanderthals' concern for selection and quality extended across all raw materials. They were among the first to fully recognise that animal bodies offered more than food. Carcasses were increasingly treated like 'bone quarries', and using retouchers both drew on and deepened knowledge of animal anatomy. Retouchers not only enabled more specialised lithic tools, but also massively boosted resharpening and therefore the useable life of artefacts.

The production of hafting substances takes things to another level. Birch tar drops can form fortuitously from bark in campfires, but to gain useable amounts Neanderthals needed to maintain careful control of the fire's temperature for extended periods. Moreover, the chemical purity of the North Sea tar supports the idea that by 50 ka, Neanderthals had significantly finessed their technique. Add to this the way Neanderthals obviously aimed to improve the natural quality of pine resin by adding beeswax, and you have cognitive complexity equivalent to the plant gum and mineral hafting recipes known from early *H. sapiens* sites in southern Africa.

---

[*] Pressure flaking involves a kind of retouch where knappers effectively squeeze flakes off using focused compression, rather than percussion.

Composite tools in themselves also imply impressive mental capacity to plan, design and anticipate. They bring together multiple episodes of material sourcing and manufacture for each constituent part, even before their assembly. And renovation is built into the very nature of composite artefacts: worn-out stone edges were replaceable, but handles were likely much longer-lived, and far-travelled. The adhesives were themselves probably carried around: a second piece at Königsaue had been rolled and folded carefully by a Neanderthal, but potentially manufactured elsewhere.

The more we look, the more we see that many of the things Neanderthals made moved considerable distances. Tree ring data shows that the Schöningen spruce spears certainly weren't carved by the lake, but felled in summer at higher altitude (probably in the nearby Harz Mountains). One spear tip even showed signs of repair to damage likely sustained in another hunt, perhaps for beasts other than horses. So far there's no hard data that bone tools were transported, but considering Neanderthals would have needed to resharpen lithics while on the move, they probably were. There's an intriguing hint from the rockshelter of El Salt, Spain, where bone retouchers seem to have already been old when used. And anomalous species might also reflect this: a giant deer retoucher from Moula-Guercy in south-east France had obviously been sourced from a butchered and smashed bone, but it's the *only* representative of this animal in the layer concerned.

If many things were carried, can we imagine Neanderthals having their own possessions? Items that took significant time to make like digging sticks – or that fitted a particular body like spears – were likely recognised as belonging to the individuals who produced them. If this is the case, did most Neanderthals develop the skills needed for a wide range of objects and materials, such as those in a composite tool? Or could these artefacts embodying multiple levels of know-how have been communal projects? We may well be looking at the emergence of craft specialists in at least some of the technical realms of knapping, wood carving, adhesive production and other activities like hunting or hide working.

Perhaps one artisan has already been found. Chemical analysis of the El Sidrón 1 adult male found traces of bitumen in his calculus.

The only likely explanation is using his mouth in composite tool manufacture or repair, boosted by the presence of intense chipping on his teeth and plant residues. This is also a reminder of the lacunas in the archaeological record, as without the chemical analysis, we would have no idea that bitumen was probably being used from Iberia through Eastern Europe, and as far as the Near East.

It seems unlikely that individual Neanderthals developed their sophisticated technologies without social learning contexts, and reasonably elaborate communication. Modern knappers may sometimes self-teach, but that typically requires some kind of textbook, if not video tutorials. Tool-using primates learn largely by observing and copying, but the range of skills and great accomplishment across diverse materials demonstrated by Neanderthals strongly implies some sort of teaching, matching the fact that *directed* instruction is common to all living humans. Combined teaching by showing as well as telling is the most effective, and young Neanderthals likely learned not in a formal way, but by cultural and bodily immersion. They would have heard how a cobble with good structure calls out when struck; felt with their body the right angle and force to hit a core just so.

Youngsters bashing stones together must have created echoing choruses around sites as their hesitant rhythm wove through the more confident strikes of elders. This context of inter-generational learning is vital for maintenance of cultural traditions, which for Neanderthals are discernible beyond the obvious lithic techno-complexes. The central Mediterranean shell-knapping tradition or birch tar technology extending across tens of millennia and three regions in Europe are other examples. At a micro-scale, Schöningen also implies cultural traditions. Time and time again the hunters returned to exactly the same part of the lakeshore, chose the same tree species for their almost identically made spears and used particular parts of the horses they killed as tools.

While minds create things, things also create minds in a manner that extends far beyond the individual or even the generation, and can transform whole species. For Neanderthals, new experiences or encounters opened up fresh ways of thinking about the world. It's not a stretch to suggest that their technological innovations probably impacted other aspects of their lives. Composite tools are a case in point: the inherent process of *joining together* must have reinforced concepts of connectedness and collaboration, crucial for hunting and social networks. And since composite tools are made up of materials connecting different places and times, these objects had a unique capacity to act as potent mnemonics, expanding the vistas of memory and imagination.

Birch tar itself suggests other interesting ideas: to comprehend that bark would transform to sticky, pungent black liquid essentially means grasping that matter could transmute. Rather than being destroyed by fire, it would be utterly remade. Alchemy is a loaded term, yet Neanderthals certainly had concepts not far off this. As tar was cooked, cooled and solidified, then was reheated and softened once again, so cycles of change were witnessed and understood. We see similar radical physical transformations from ores to liquids and finally solids in substances called metals, which wouldn't be invented until many tens of millennia later.

When we regard Neanderthals on their own terms, we perceive that they were fundamentally experts as well as experimenters. Deep knowledge underlay discerning material choices and targeted planning. As connoisseurs of the crafts they were immersed in every day, Neanderthals combined convention with adaptability. They invented new ways to take things apart and join them together. The impacts of this spiralled outwards, and underpinned the way Neanderthals lived increasingly complicated lives and expanded their activity in the landscape.

This mental fluidity allowed them to adapt to whatever situation arose, creating novel forms of production, use and rejuvenation, while engaging with and exploring the world and everything in it in richer ways than ever before. But to do all this, they had to eat.

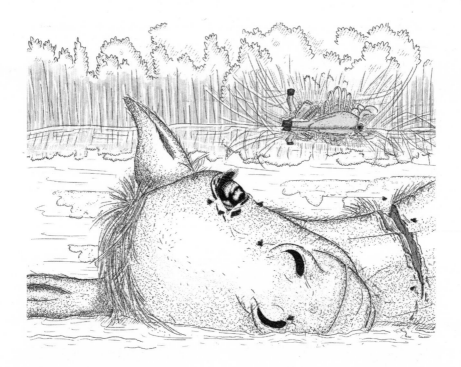

CHAPTER EIGHT

Eat and Live

Like an unsliced belly, a fat silence lies over the lake. Dawn barely filters through autumn-yellowed reeds, but already the frogs are so loud that, at first, the people don't hear the horses. Instead, the ground under their feet tells them stamping hooves approach. The first thirst-quench is the best killing time, and everyone knows it. Reaching the lake edge, the huge horses slow: ears forward, eyes scanning wide. They learn fast and the people often change where and how they meet them: today crouching in brush, covered in scent-masking mud. Nervous snorts and horse-scent heralds the herd's arrival; flaring nostrils detect no danger. Muzzles lower, whiskers break the surface of the water. The horses' minds are cooled by the first flowing draught, throats relax – then the reeds explode – noise, danger – screaming beasts and people – too many behind – forwards into the dragging water – spears gripped so hands ache – and the bite, the bite of the wood in the neck …

A dark, stilled eye stares up from the lake, reflecting clouds and tree filigree above. Sluiced with blood, the floating bulk is too heavy to drag from the churned shallows, where it fell and broke the spears. So they begin to unmake it there in the lake. Swiftness is vital: high sun will be on them soon, pushing meat-smell out into furred snouts twitching deeper in the woods. They heave the skin off, peeling the animal from the flesh. Hooves and tail are left on: the hide's first use will be to carry the meat away. The taking-apart places are found and cut, releasing legs, then rump, head, neck, and lastly, ribs. When every part is up on the clean sands, slicing and ripping begins in earnest. Watering mouths and lusting tongues are slaked by pink, warm marrow; there's too much to carry anyway. The flints' tearing slaughter song is joined by staccato rhythm as scraped-clean bones hammer and resharpen tool edges: the horse unmakes itself. When fat-heavy slabs – tacky with blood – are shouldered, the head, organs and tendons slung in the hide, the people begin walking.

Warm sunlight breaks over the shore, shrinking the pooled shadows inside frantic hoofprints, illuminating puffs of wiry hair and blood-stained sand. Silence congeals around crushed reed stems floating in the lake. Settling into silty darkness below, snapped spears stab through time.

Tonight, as 350,000 years ago, some bellies will feel deliciously full while others ache in vain. Subsistence is survival, and it's always loomed large in Neanderthal research. But as any anthropologist – and restaurant critic – will tell you, food isn't simply nutrition. What, and how, we consume is woven into other fundamental aspects of life, from technology to culture. So working out what was on the Neanderthal menu offers many avenues for exploring their lives. It's no surprise that a species living across such an enormous range of environments and climates ate many types of food, but our increasingly colourful picture of Neanderthal cuisine is also connected to more powerful investigative methods. Today, claims that 'Neanderthals couldn't eat X, therefore extinction' have been replaced by much more subtle comparisons between us and them.

Even if food is more than fuel, biology still provides a baseline for thinking about diet. With more massive bodies and bones sculpted by intensive lifestyles, just how much energy did Neanderthals need? Moving heavily muscled and shorter legs is harder work, but so are unconscious, essential functions like beating a bigger heart. And as brains are greedy organs, even slightly larger ones cost more calories.

In total, we're talking a whopping 3,500 to 5,000kcal *every day*. That's over twice today's typical adult guidelines, and even beyond what world-class athletes burn through. Yet it multiplies at the extremes: female Neanderthals grew bigger, heavier babies who needed more milk. Weanlings ate more food and needed carrying; that cost fell on parents and perhaps others.

Living in tough and cold landscapes also pushes energy needs even further. Travelling through boreal forests filled with deep snow is especially draining, and some hunter-gatherers from these regions have been recorded eating *gigantic* amounts of game: over 3kg (6.5lb) per day, giving around 5,500kcal. Since Neanderthals likely needed about 5 to 10 per cent more energy on average, those from harsh environments and lacking highly insulating clothes must have needed up to 7,000kcal per day.

What does that look like? It's the calorie equivalent of a blow-out Christmas Day binge – a fry-up at breakfast, roast dinner and champagne, cheese platter, plus leftovers and trifle for dinner – *each and every day*. To feed a group of 10 Neanderthals for a week, you're looking at 300,000kcal. Three reindeer every seven days would hit that target, but it's almost 50 per cent more than managed by typical wolf packs. And since human nutritional requirements aren't the same as wolves or hyaenas, living largely on lean meat quickly leads to starvation. To get enough vital micronutrients – fats, vitamins, minerals – you need fat, brain, tongue, eyes and marrow, implying double the number of animals. So while one massive rhino carcass gives a million kcal, it's not enough to stay healthy.

Bony Jigsaws

It's clear Neanderthals chowed down on an awful lot of food. But working out exactly what has been complicated. For a long time, researchers had nothing to go on except animal – or faunal – bones, but simply counting frequencies will be misleading. The trouble once again is taphonomy. Faunal remains can end up in archaeological layers via processes that have nothing to do with hominins: random deaths, floods or as carnivore prey. Early prehistorians didn't realise this or overlooked it, and things only started to change in the second half of the twentieth century, when claims Neanderthals were actually systematic scavengers emerged. This was connected to theories that 'true' hunting only became common with later *H. sapiens*, and predictably led to notions Neanderthals were too stupid to kill big beasts.

Yet the 'meat thieves' idea was on shaky ground. Some hunter-gatherers do employ 'power scavenging', but *exclusively* surviving this way is extraordinarily hard, because carcasses with much remaining meat are a rare and massively in-demand resource. Hyaenas are uber-scavengers, typically arriving at a fresh carcass within just 30 minutes day or night, and can crack open all but the most resilient bones to suck them free of marrow. For anything beyond the meanest scraps, Neanderthals would've had to drive them off early and repeatedly.

The final blow to ideas that Neanderthals eked out an existence on gristly tatters came from the archaeology. During the late 1980s and 1990s it increasingly became obvious that there was very little direct evidence of scavenging, versus an awful lot for hunting. Predator gnawing was almost always *above* cut marks, and this pattern of primary access to carcasses is there early: despite wolves and sabretooth cats lurking at Schöningen, they had to wait. It was carnivores squabbling over Neanderthals' refuse, not the other way round.

Today's researchers examine every bone for telltale surface alterations. Cracks, abrasions or stains reveal whether they lay on land surfaces or were quickly buried, while creatures also leave calling cards: tooth and beak marks, even acid erosion from stomach juices. It's often possibly to identify the species responsible, whether hyaena or hawk. Once naturally accumulated remains have been excluded, researchers then look for unmistakably hominin signatures: burning, cut marks left by lithics or distinctive battering and fractures of fresh bone. Nineteenth-century prehistorians' jaws would drop at the range of powerful lab kit now routinely used. High-powered optical microscopes pick out slice and chop marks, and electron beams trace the cross-sections of scratches at the nano-scale.[*]

---

[*] U-shaped scratches are natural, whereas v-shaped examples point to lithic tools.

Emerging biochemical techniques such as ZooMS[*] can even identify the species of otherwise uncertain fragments.

Individual bones are important, but it's assemblage-level damage frequencies that disprove scavenging. If fewer than 10 per cent of faunal remains have carnivore gnawing, it's fairly safe to say most of the bones in a site came from Neanderthal food waste. Meticulous studies using such methods give us more detailed insights into their subsistence than ever before.

Zooming into Fumane Cave, north-west Italy, we can see how this works. Known since the 1880s, its rich archaeology was only revealed in the 1960s when work on the adjacent single-track lane cut through sealing landslide deposits. The Fumane sediments are made up of numerous Neanderthal layers including Level A9, which formed over around 1,000 years between 47.5 and 45 ka. It covers an area about the size of a school classroom, yet is just 15 to 20cm (6 to 8in.) thick. As well as some 50 hearths and lithic scatters, A9 produced over *100,000* pieces of bone. Careful analysis unpicked their story, showing that just 0.1 per cent of bones[†] had damage from carnivores or rodents, while butchery rates were at least 15 per cent. Of the 18 species – from cave lion to marmots – the most common were herbivores, and the bigger animals like red deer and bison had the most butchery damage.

Confirming big game hunting is one thing, but much can be missed without systematic analysis of *all* bones and teeth. Just like the lithic manufacturing waste, many bone fragments were once routinely dumped during excavations, leading to inaccurate behavioural models linking tool types with particular ways of hunting. Recent re-excavation at Combe Grenal revealed that massive amounts of smaller animal remains had been thrown away during the work decades earlier by Bordes, leading to misleading species and body part

---

[*] ZooMS or Zooarchaeology by Mass Spectrometry is a rapid collagen identification technique that can determine the type of animal from even tiny, otherwise unclassifiable bone fragments.

[†] As with many intensely occupied Neanderthal sites, most of the bones were tiny (92 per cent under 2cm (0.8in.)), and only just over 1,200 could be matched to a species.

percentages.* It also hammered a final nail in the coffin for the scavenging hypothesis, since the supposedly high proportion of teeth at Combe Grenal – taken as evidence that Neanderthals could only get heads rather than fleshy carcass parts – shrunk from 80 per cent to just 2 per cent of faunal remains. Uncritical use of data from most sites excavated some time ago must obviously today be regarded with caution, but as well as re-analyses, fresh discoveries over the past three decades have produced a transformation in our understanding. So what did Neanderthal cuisine look like, and what does it imply about their wider behaviour?

Hunting the Truth

Let's get into the meat straight away. How exactly were Neanderthals killing all the beasts found in Fumane and hundreds of other sites? Spears are clearly one weapon, and in addition to those from Schöningen and Lehringen, in 2018 the Eemian site of Neumark-Nord 2 produced compelling evidence for their use. Among more than 100 massive fallow deer – mostly prime-aged stags – two largely complete skeletons aren't just smoking guns, but come with bullet holes, Neanderthal-style. The hip bone of one and neck of the other bore deep, tapered punctures that can only have been caused by spears. Rather than being thrown, both wounds match experimental damage from being rammed in with a lunging action. On autumn days 120,000 years ago, Neanderthals at Neumark stalked their quarry through dense hornbeam forest to the lakeshore, where the canopy opened up to the sky and the deer had nowhere to run.

The contrast with Schöningen, some 200,000 years older, is obvious. Those spears always seemed to contradict claims that Neanderthals' shoulder joint anatomy prevented effective throwing.† They seem too long for short stabbing spears or even despatching

---

* The new fieldwork recovered *23 times* more bones and teeth because extremely fine sieves were used, conserving fragments down to 1.6mm (0.06in.).
† This anatomical theory spawned rather creative interpretations for Neanderthal spears including snow probes, despite the interglacial contexts and even direct association with elephant bones at Lehringen.

lances, and moreover are weighted like javelins, which only makes sense if they were intended to be thrown. However, we need not think in binary terms, and it's quite possible such objects were 'dual-wield' weapons.

All spears found so far are wooden, but experiments show that slicing stone-tipped weapons have many advantages. The wounds they inflict bleed rapidly, sapping prey strength and reducing aggressive encounters. And there's evidence Neanderthals did use this kind of weapon, with probable impact damage on Levallois points from a number of sites, while at Umm et Tlel, Syria, the tip of a lithic point was found still embedded in a wild ass's backbone.

However they were used in the past, spears neatly skewer scavenger theories. Furthermore, Neanderthals were quite comfortable taking on *enormous* beasts. The extinct horses at Schöningen weighed well over 500kg (1,100lb); nearly twice the size of the species immortalised in Upper Palaeolithic art. But what about the true giants – elephants and mammoths? Their sheer bulk and speed is incredible, but hunter-gatherers do successfully hunt elephants without guns. It's difficult to prove to a legal standard that Neanderthals also took down these behemoths, but circumstantial evidence is powerful. Carcass sites with butchery waste place them at the scene, and at Lehringen the weapon is even present. What is notable is that, aside from rare kill-site skeletons, mammoth and elephant aren't usually dominant in Neanderthal sites. With creatures that huge, it seems that, rather than moving whole bodies or heavy limbs, Neanderthals were mostly transporting soft parts away from carcasses. And the amount of consumption overall could be an underestimate, since the thickness of elephant flesh can prevent cut marks showing up on bones.

This might explain why no traces were found on at least 11 mammoths at Lynford Quarry, Britain. A few parts from other species – horse, reindeer and rhinoceros – were butchered, but the presence of some 50 bifaces and thousands of pieces of knapping debris points to more intensive activity best explained by butchery and possibly hunting of the mammoths.

Other sites are far less ambiguous. Some 100,000 years before the mammoths died and were butchered on the edge of the Channel Plain at Lynford, Neanderthals were also living farther south on what's now the island of Jersey. Within deep deposits of the ravine-cave La Cotte de St Brelade, on today's coast, two mammoth-rich phases stand out. Unlike the densely fragmented faunal remains in other layers, these so-called 'bone heaps' contain abundant, mostly complete and blatantly butchered bones from at least 18 mammoths in total, together with some woolly rhinoceros. Interpreted for decades as a mass kill site, recent reinvestigations have questioned whether the cliff topography was suited to driving herds over.

Instead, the bone 'heaps' may in fact be taphonomic phenomena caused by unusual preservation when Neanderthals abandoned the site at the onset of frigid phases in MIS 6. Rather than being trampled into splinters, the butchered parts they left became buried in talcum-fine loess: crushed rock dust blown from advancing glaciers hundreds of kilometres away.*

Submarine mapping in the past few years in front of La Cotte has revealed a network of parallel gorges, so instead of a jump site, Neanderthals may well have funnelled animals up towards the ravine's cul-de-sac. Hips or shoulder blades aren't the richest body parts and are unlikely to have been dragged far, so kills must have either been very close by or actually inside the ravine. We're probably not looking at the slaughter of entire herds, but even a single cornered mammoth would be extremely dangerous and must have required team work to despatch. La Cotte has one more oddity: some of the skulls were stacked against the stone walls, ribs placed vertically, and in one unique instance stabbed diagonally down through a skull. It's possible that Neanderthals were exploiting these animals' bodies not just for food, but as a means of structuring space.

The best evidence for some element of specialised mammoth hunting comes from Spy, in Belgium. A fair number of bones exist within the nineteenth-century collections, but most unusually, three-quarters were youngsters, even very young calves. It's probably not down to other predators, as while hyaenas sometimes target young elephants and were definitely using the cave, there's not much

---

* Loess blanketed much of Europe many times during various glacial periods, and was often used by early brick-making industries.

gnawing. Plus, neither hyaenas nor other carnivores routinely transport massive bones like elephant skulls. The fact that young mammoth teeth were found at Spy strongly suggests that entire skulls had been present, making Neanderthal involvement more likely. However, like elephants, we can expect mammoth herds to have been very protective of their young: they're even tough for other predators to hunt because they remain so close to the herd. The high frequency at Spy is therefore truly striking and implies targeted hunting by Neanderthals. The motivation for this probably lies in the fact that baby mammoths represented a rich resource, with about 1kg (2.2lb)'s worth of concentrated fatty oil goodness in their brains; as well as being more nutritious, they may well have tasted better too.[*]

Spy may be unusual because of the age of the mammoths, but bio-geochemical analysis on the Neanderthal remains there gives elemental proof for consumption that matches other sites. Carbon and nitrogen isotopes from hominin bones can give an idea of where they fitted into local ecosystem food chains, and on the whole Neanderthals resemble carnivores like wolves or hyaena[†] with high nitrogen. Isotopes can also give an idea of predator diet niches – who was eating what – and remarkably, some Neanderthals, including those at Spy, appear to have been getting between 20 and 50 per cent of their animal protein from mammoth. This supports the idea that the skeletal remains we find are only the tip of the tusk. Much of the time, Neanderthals were probably selecting only meat, fat and marrow to transport from kills. An interesting point to consider is that decades of mammoth myopia has somewhat sidelined the fact that Neanderthals certainly took on other colossal, dangerous beasts, including those massive horses, various rhinoceros species, aurochs (fearsome ancestors of most cows, 1.8m (6ft) or more at the shoulder), water buffalo and giant camel.[‡] But so far, there's no clear evidence

---

[*] Humans love fatty foods, and tests on frozen Siberian mammoth calves show that their flesh absorbed fatty acids from their mothers' milk.

[†] Carnivores are higher up the food chain, and therefore accumulated greater levels of the nitrogen isotope.

[‡] *Camelus moreli*, 3m (10ft) tall at the shoulder, found at Hummal, Syria.

for hunting hippopotamus, which – perhaps surprisingly – are even deadlier than elephants.*

Rather than focusing on big game, however, what defines Neanderthal hunters is their refinement of a way of life going back more than a million years. They went after almost all sizeable prey in their local range, adapting to big species and medium-sized game. The diversity of habitats and behaviours represented by animals like ibex, gazelle, wild ass, boar or chamois means Neanderthals must have mastered many specialised hunting strategies. But that doesn't mean they slaughtered indiscriminately. Just like furred predators, they flexibly shifted between being generalists versus targeting particular species, but almost always picking the meatiest or fattest animals.

Canny use of landscape features combined with knowledge of animal behaviour must have underlain situations like the 50-odd horses at the Schöningen lakeshore, which were killed in multiple phases over somewhere between several hundred years to just a few decades. Neanderthals returned again and again, probably because driving small herds into the water was a great way to slow down otherwise fast, dangerous prey. Elsewhere, hunters were homing in on opportunities offered by seasonal migrations or breeding herds. A particularly dramatic site like this is Mauran, on the edge of the French Pyrenees, where there are probably remains from several *thousand* hunted bison.† There are just a handful of bones from other species, but the bison are mostly females and younger calves, suggesting that Neanderthals could have been targeting them during summer when they shifted from plains to uplands. Most interesting, although this context might have involved herd drives, the butchery itself at Mauran was definitely selective.

---

* Hippos are extremely aggressive and kill more people than elephants; their remains are found in some interglacial Neanderthal sites, but it's not always clear if they were hunted.
† The fully excavated area at Mauran is just 25m² (30yd²) and contains remains from nearly 140 bison; extrapolating across the entire extent of the site – more like a hectare – means the total number must be far larger.

At other places, solitary-living species were also repeatedly targeted. Rhinoceros aren't sociable, and catching them needs careful stalking or ambushes at predictable locations. In interglacial forests, places with rock-licks or even mineral salt-rich water would have been good bets. Judging by striking amounts of butchered rhino carcasses at Taubach, a late Eemian tufa* site in Germany, Neanderthals do appear to have used some lakes and travertine sites in this way.

This pattern of selectivity extends into how they used their prey. Body parts were understood and valued differently, and pop culture images of cavemen with giant roasting haunches are well off the mark. Instead of lean meat, the fattiest and most marrow-rich parts were prized to balance high protein intake and as a richer energy source. This means offal was certainly relished: brains are around 60 per cent fat, and grey matter is also full of particular lipids – long-chain polyunsaturated fatty acids – vital for health and foetal development. By reading the pattern of cut marks across the skeleton, we see time and again that brains as well as other juicy parts like eyeballs, tongues and viscera were favoured by Neanderthals.†

Schöningen shows how this worked: one horse contains far in excess of 200,000kcal, but they're very lean. So Neanderthals skilfully skinned and took them apart, slicing off flesh from the meaty haunches and withers, rather than taking them away. They paid more attention to getting marrow from the lower limbs as well as tongues and internal organs. But this practice was tailored by species: smaller game saw more intensive carcass processing. It's echoed at many other sites, especially farther down the butchery chain from kill locales. At places like Fumane, where mostly marrow–rich joints from carcasses hunted elsewhere were brought back, almost every limb bone was methodically reduced to shattered shaft fragments.

---

* Tufa is limestone formed from calcium carbonate-saturated groundwaters overlying limestone bedrock; also known as travertine, but the latter is often associated with hot springs.
† Neanderthals weren't the first to enjoy viscera: Boxgrove, Britain, shows that hominins around 500 ka were skinning animal heads and removing the soft parts.

All the Small Things?

Big beasts are certainly the cliché for Neanderthal meals, but smaller furred or feathered critters* were eaten far more often than used to be imagined. For many decades it was assumed *H. sapiens* were more efficient, inventive hunters, and Neanderthals just weren't up to the task since small game often need different strategies and gear like traps or nets. This notion spun out into extinction theories, since if Neanderthals were stuck in a big game rut, they weren't therefore accessing the wider protein in a given landscape, and nor could they fall back on those species during big game shortages. But were they really 'killed by rabbits', or rather the lack of them, as some headlines claimed?

Casting a more careful eye over the archaeology itself makes things look rather different. At around the same time that the Schöningen horses were dying, earlier Neanderthals were eating rabbits at Terra Amata, southern France. It's now just one of nearly 50 sites across Europe and Western Asia where exploitation of small game has been identified. Almost half involve rabbit or hare, despite them being supposedly hard to catch, and nearly the same number of sites include birds. Les Canalettes, a rockshelter high on the Causse du Larzac plateau, south-east France, is an interesting case. In Layer 4, dating around 70 to 80 ka, despite there being very few cut marks researchers showed that nearly 70 per cent of all identified bones were from butchered rabbit. This was thanks to distinctive breakage patterns: after a couple of incisions, Neanderthals could tug the skins off and, especially after cooking, simply pull the carcass apart. Incredibly rare, direct evidence comes from eastwards, at the Abri du Maras rockshelter in the Ardèche gorge. Natural mineral films covering lithics have in some cases preserved rabbit or hare fur strands, alongside a few butchered bones.

Rabbits elsewhere — and even the occasional marmot — were probably eaten like this too. But bigger rodents that take a bit more butchering were certainly also eaten. Beavers' fatty tails would have been succulent treats, and their architectural creations were part of Neanderthals' landscapes: at the northern French Eemian site of

---

* Small game is roughly defined as animals below 10kg (22lb).

Waziers, as well as butchered beaver bones the remains of a lodge were preserved.

Multi-stranded analysis has increasingly boosted evidence for Neanderthal use of small game. At Payre rockshelter on the west bank of the Rhône River, they were butchering birds and fish from MIS 6 onwards. Tiny feather fragments were on a tool used for butchering meat, while 'greasy' polish from cutting up fish, residues of piscine muscles and even scales were found on other tools. Fragile fish bones are actually incredibly rare in archaeological sites, perhaps explaining their absence at Payre, but freshwater fishing by Neanderthals wasn't an anomaly in that region. Fish scales have been identified on lithics at Abri du Maras, and unlike Payre, their significance is backed up by some 150 bones from sizeable perch and chub. In the absence of carnivore damage, Neanderthal fishers would seem to be responsible.

Exceptional cases like that tip the balance in other sites with fish bones. Extremely fine sieving during excavation of Walou Cave in Belgium recovered over 300 freshwater fish bones and scales. None had predator damage, and what's more, they're most common in the richest archaeological layers. Right next to a river, Neanderthals here may well have been taking fish by their front door. But how? No certain hooks or harpoons are known, however spearing or using stone traps in rivers are possible. Bears simply wait by a likely spot and smack the fish out, but subtler approaches work too: fish tickling or 'guddling' works well for shady bank-dwelling species.

So much for terrestrial and watery realms; what about birds? There's now evidence for hominins eating birds more than a million years ago, and by the time of the Neanderthals it's at so many sites that hunting must have been involved. At some places it's not common, sometimes just single cases of butchered dove, swan or duck bones, while back at Abri du Maras there are a few feather fragments from birds of prey and probably duck under the mineral films. Elsewhere Neanderthals were regularly eating birds, over long periods of time. Three of the Gibraltar sites have a few butchered rock doves in various layers, while at Fumane Cave black grouse – a classic game bird – was common. More surprising there are the large amounts of chough, a small cliff-dwelling member of the crow family. Chough in fact seems favoured by Neanderthals at many sites, including Cova Negra, Spain,

Figure 6 *What Neanderthals ate was as diverse as the landscapes they lived in.*

where Neanderthals stayed for short periods during a cool phase before 120 ka. Mostly hunting deer, wild goat and tahr (a kind of mountain sheep), they also targeted a range of birds. Along with rabbit, butchered birds are found in five levels, but the richest is Layer 3b with over 100 bones from 12 species. Unlike Fumane, though, it's *all* medium or small birds: partridge and rock dove but also kestrel, owl, chough, jay, magpie and the colourful roller. Despite being relatively scrawny they were comprehensively sliced and nibbled; oddly for the crow family species, only the wings were present.

Catching birds has long been believed an advanced hunting technique, so how did Neanderthals do it? Many species would have been living alongside them, soaring over cliffs above caves, but woodcock, jay and roller at Cova Negra must've come from nearby woodlands. Special throwing sticks might work in wetland contexts – there are candidates at Schöningen – and though Neanderthals collected sinews and tendons and may have had plant-based cord, we have no evidence of nets. Similarly, nobody has ever found preserved darts or bows, but the small bone point from Salzgitter and the tiny Levallois points or even bladelets from a number of other locales must have been hafted and could have formed part of small projectiles.* However, birds don't need to be caught on the wing. As with fish tickling, their natural instincts can be exploited: some species will 'freeze' on their nests, while today choughs in Alpine ski resorts are remarkably interested in human detritus, providing ambush opportunities. All that sticky birch tar and bitumen could also have been used for perch traps. As well as meat, birds of course mean eggs: handy pre-packed protein, fat and vitamin-rich snacks, they're found at Schöningen and must have sometimes been greedily slurped down.

We don't know if Neanderthals ate reptile eggs, but they certainly gorged on tortoises. Butchered remains are among the bones at Cova Negra, and also just seven hours' walk away (though rather older at 350 to 120 ka) at Bolomor Cave, another site rich in small game. Rabbit and diverse kinds of birds from swan to ptarmigan and members of the crow family were eaten, plus at least 20 tortoises. The Bolomor Neanderthals even had a favourite preparation: roasted upside-down to weaken the shell and soften the meat, then hammer open, tear off the limbs and slice out the innards. Tortoises in fact offer one of the best cases for a regional Neanderthal cuisine, as they were consumed at a number of sites across their native warm Mediterranean and Near Eastern regions. In some places they're almost a staple: more than 5,700 remains representing at least 80 tortoises come from multiple levels at Oliveira Cave, Portugal, sometimes accounting for over half of identifiable bones. Intriguingly, wherever tortoises were eaten they were often cooked inverted in the Bolomor fashion, though the shell-hammering technique changed over time. And strikingly, there's no

---

* There are actually no definite bird-hunting weapons from earlier Upper Palaeolithic contexts either.

evidence for Neanderthals eating European pond turtle, even though they were found in Northern Europe during the Eemian.*

Hunting small game, however, seems to have involved choice. Abric Romaní is in a similarly rich landscape to Bolomor and Cova Negra, but none of the rabbits and birds from its long sequence show any trace of hominin involvement. And at relatively nearby Teixoneres Cave, Neanderthals were occasionally hunting rabbit, but not the birds. It's also noticeable that during cooler periods, unlike some Upper Palaeolithic cultures, Neanderthals don't seem to have gone for Arctic hare. Perhaps big mammal herds provided more than enough food, and finding hares that live in the open wasn't worth the effort.

Thinking about economics brings up another sort of subsistence that's relatively easy, yet until recently regarded as unlikely: seafood. Imagining Neanderthals sitting on a beach slurping down mussels is somehow even more incongruous than catching river fish. But beach combing, rock-pooling or even deep wading can all offer rich rewards for little energetic cost. Slow and laborious to collect, in nutritional terms molluscs and other coastal foods are gold: bursting with vital long-chain omega-3 acids. Plenty of non-marine animals take advantage, from bears prying shells open with their claws to Asian macaques smashing shellfish or crabs,† and recent findings prove that Neanderthals were also doing so at least as early as our own species.

As Chapter 5 discussed, most beaches where Neanderthal feet got sandy are currently underwater following sea levels rising at the end of the last glacial period. Nonetheless, some sites on today's coast would have been close to interglacial shores and, depending on the submarine topography, within a few kilometres even when the ocean fell. Among them is Bajondillo, a rockshelter in the southern Spanish city of Torremolinos. Here in layers dating between 170 and 140 ka there's well over 1,000 broken-up mollusc remains, almost all mussels.

---

* The travertine site of Gánovce, Czech Republic, not only produced a shell cast from a pond tortoise, but also a Neanderthal brain cast plus imprints of feathers and rhino skin.

† Nineteenth-century accounts of macaques eating crabs exist, but it was only confirmed after the 2004 tsunami, and subsequent excavations that proved it had a considerable history.

Direct heat forces these creatures to open their shells, and since many were charred only on the exterior, it seems Neanderthals knew this trick. Most interestingly, over several millennia even as climate cooled, mussel eating continued. Shells only disappear when the coastline was around 8km (5mi.) away, implying that seafood continued to have importance even though other hunted game changed.

In fact the numbers of sites with some evidence of seafood eating is large: over 15 in Iberia and elsewhere around the Mediterranean. In the richest layer at El Cuco, close to the Atlantic coast in northern Spain, Neanderthals collected nearly 800 limpets plus the odd sea urchin: a delicacy in many coastal cultures today. And on the Portuguese Atlantic coast at Figueira Brava, there's striking diversity in marine creatures being eaten. The overall number of shellfish here are smaller, but this may be because they were being processed intensively: within the sequence, distinct sub-layers of shells and fragments are visible, more abundant than the sediment. On top of that are the remains of well over 40 crabs and a variety of fish species that can be foraged from rock pools or in the shallows.

There are no seafood sites known from along more northern Atlantic coasts. As sea levels fell, La Cotte de St Brelade in Jersey would have been reasonably close to the sea, but only for short periods, and Neanderthals seemed focused on big game. Meanwhile at Le Rozel, a mid-late MIS 5 dune site on the north French coast, wrasse, mussel and thorny oyster are present, yet Neanderthals were apparently not eating them. Le Rozel does have rare walrus remains, a reminder of larger marine creatures Neanderthals could have sighted in the Channel, though they lack butchery marks.

Elsewhere they did sometimes consume big beasts of the ocean. Dolphin, seal and large fish bones with cut marks come from a few Iberian sites, which might represent washed-up carcasses or perhaps stranded individuals speared in the shallows. We can wonder what Neanderthals made of such creatures, with bodies both different and familiar to the terrestrial prey they knew intimately.

Perhaps the most overlooked of animal groups that possibly contributed to Neanderthals' diets are insects. Regarded as common-sense nutritious fare outside Western cultures, they can be traditional bush tucker or urban street food. Eurasia doesn't have many massive, fat grubs and larvae, but like ours, Neanderthals' summer days would have hummed with the background sound of bees. Hunter-gatherer groups – as well as chimpanzees – are well known to risk stings for the great calorific prize of honey, and if Neanderthals had the chance to try it, their ability to taste sweet things means they'd likely have enjoyed it. As we saw in the last chapter, hafting adhesive made from mixed beeswax and pine resin strongly suggests that at least Italian Neanderthals were well aware of other resources found in nests.

In terms of actual insect crunching, however, we shouldn't rule out those closest to hand: parasites. Ticks and lice might be nibbled while grooming hair, and beyond Neanderthals' own bodies are passengers on their prey. Many of the large mammals hunted by Neanderthals would have had warble fly larvae under their pelts. These ravenous grubs – known as 'wolves' – can be over 2cm (0.7in.) long, and after being laid as eggs in the host's legs travel up through body muscles, reaching even the windpipe. All the tunnelling causes a noticeable jelly-like substance to form in the meat, and the warbles create lumps under the skin that leave holes; but on the upside, the hungry grubs are edible. A range of reindeer-hunting Indigenous North American cultures including the Dogrib, Chipewyan and Inuit have regarded them as delicacies, comparable to berries, and since tiny Upper Palaeolithic carvings of the warble bodies exist, they were definitely around in the Pleistocene. If Neanderthals were adaptable enough to gather things like limpets, there's no reason they wouldn't have gobbled up these choice snacks too.

Fanged Ones

Neanderthals ate all parts of creatures great and small, and increasingly it's apparent that their catholic tastes also extended to carnivores. While to some an unexpected choice, cuisine is a matter of perspective. Not so long ago offal was everyday fare in most Western cultures, but today is largely demoted to unidentifiable stuffings or pet food.[*]

---

[*] Exceptions exist: in one suburban Bordeaux pizzeria, you can find the 'Farmer' pizza, garnished with gizzards and tripe.

And the notion that carnivores are unpalatable isn't universal: dogs and cats are eaten in some cultures, while in others, bears – technically omnivores but capable predators – are regarded as tasty. Many of the hundreds of Indigenous societies of North America have traditionally eaten predators like cougar, wolf, and black, brown and polar bears. They were sometimes fallback foods for hard times, but elsewhere normal parts of the diet, and depending on the season, for some cultures bears might be the main source of meat and fat.

Plenty of Neanderthal sites exist with the odd cut-marked carnivore bone such as wolf, fox or dhole (a now mostly Asiatic wild dog). Even butchery of bigger, more dangerous predators – like a lion at Gran Dolina between 350 and 250 ka, hyaena around 120 ka at Maltravieso or a leopard at Torrejones Cave after 100 ka, all in Spain – probably represent chance encounters useful for food and furs. But with bears, it looks like something else was going on. Neanderthals hunted them more than any other predators, encountering three sorts: the familiar Eurasian brown bear and Deninger's bear, which after 130 ka probably evolved into the cave bear. Even the brown bears tended to be bigger than those today, but cave bears were *huge* – about 600kg (1,320lb) – and when standing would have towered over Neanderthals. They also, as implied by the name, preferred using subterranean dens rather than digging holes.*

Wherever bears slumbered, hibernation offered a relatively safe hunting opportunity, a fact known to lions and leopards as their bones are sometimes found far underground alongside bear remains. But Neanderthals too were stealthy hunters in the dark. Among the more than 20 bear butchery sites in Europe are caves in the foothills of the Italian Alps including Rio Secco, studied since 2002. Two layers there dating around 48 to 43 ka reveal that Neanderthals slaughtered at least 30 winter-dreaming bears. As they processed entire carcasses, they focused especially on the fatty chest and limbs as well as marrow and tongues. The bear's own ribs were used to

---

* 'Bear earth' is a deposit found in some caves that shows the remarkable persistence of hibernation dens: some deaths over winter left not only entire skeletons but phosphorus-rich soil from their decayed carcasses.

resharpen the slicing tools, and burning hints at cooking right there in the den.

Other sites show that Neanderthals knew bears' habits well enough to track them even to high-altitude dens like Generosa Cave, also in the Alpine foothills; at around 1,500m (5,000ft) it probably required waiting until spring to ambush groggy and weak emerging bears. Spring den hunts may also be indicated by the presence of cubs, such as at Rio Secco. Fumane Cave, westwards along the Alps, shows us the consumers at the other end of bear hunts. The latest layers between 43.6 and 43.2 ka reveal Neanderthals bringing back choice cuts from bears hunted elsewhere. Some were burned, and tooth marks as well as marrow smashing even on toe bones shows the gusto with which they were consumed.

Systematic hunting was also happening at Taubach, where Neanderthals not only ambushed rhino but at least 50 bears, probably more.* Carnivores and bears can be just as partial to mineral licks as herbivores, and so the Taubach carbonate-rich seeps and pools flowing down to the Ilm River were probably not only easily discoverable thanks to networks of game trails but also reliable locales for ambushing bear. They were certainly intensively butchered, even with defleshed paws and tongues removed, and once again burning may indicate that some of the bounty was cooked nearby.

Whichever species, the lessons from Neanderthal carnivore and bear hunting are several. At least some of it – especially the bears – wasn't scavenging but targeted, even specialised killing. This speaks to serious guts on the part of the hunters, plus collaboration and probably planning. Den hunting was clearly happening, but we should be open-minded about other options like traps, whether deadfalls or pits. At first glance that might seem rather complicated, but we've got abundant evidence Neanderthals worked on multi-stage projects involving wood, and as we'll see in later chapters, they also created complex constructions.

---

* Taubach was discovered in the late nineteenth century during quarrying of the travertine; this is only a sample, and the scale of the original collections was vastly greater.

Shoots to Roots

While Neanderthals' reputation as primarily big game fans is looking ropey, they're never going to be vegan mascots. Nonetheless, it's plants that underlie the most dramatic U-turn in our understanding of their diet. Preserved Pleistocene vegetation is extraordinarily rare and this, together with visions of barren arctic tundra, led to assumptions that plants weren't eaten; or in such tiny amounts that they are undetectable. Initially, stable isotope analysis appeared to back this up. The first Neanderthal analysed came from Les Pradelles, south-west France, and looked virtually indistinguishable from wolves or hyaenas. As more samples accumulated together with other evidence for active hunting, Neanderthals morphed from scavengers cringing around carcasses to macho killers, with little room for plants.[*]

Yet even researchers knew this couldn't be true. Meat is an amazing protein hit, laced with fatty acids and easily absorbed micronutrients, but neither we nor the Neanderthals are capable of surviving long term on 100 per cent fleshy fare. Strict carnivory actually starves the body, terminating in protein poisoning, and for those pregnant or breastfeeding – likely the majority of Neanderthal women at any given time – it's lethal. Alongside meat and fat, Neanderthal survival depended on plants, and therefore the isotopic data wasn't telling the whole story.[†] Sampling matters: there are fewer than 25 analysed individuals and because of issues around preservation, they're younger than 100 ka and come from colder climates. Neanderthals who lived in warmer, plant-rich times and places are missing from the picture. Yet even if we did have such samples, carbon and nitrogen stable isotopes only reflect protein, not carbs. Meat effectively swamps any plant protein in these methods, meaning that even if half of a Neanderthal's protein was vegetable in origin, isotopically they'd still look more like hyaenas than horses.

If Neanderthals nibbling roots and shoots feels unlikely, remember that other archaeological evidence points to their being herbaceous

---

[*] Research bias towards hunting rather than gathering in part existed because the latter was deemed domestic and so less exciting (as well as probably female-associated).

[†] Plants are the best sources of folic acid and vitamin C, among other things.

connoisseurs. If they were well aware of plants' material properties, whether for tools, hafting adhesive or other uses, why shouldn't this knowledge have also been nutritional? On top of that, the existence of digging sticks is strong supporting evidence. So exactly what plants might they have chowed down on? The options were immense. Europe today has over 1,000 edible species, though they've mostly fallen off our cultural radar. In northern latitudes there are slimmer pickings, yet Indigenous tundra-living societies have long known at least 20 to 40 species that are good to eat, many of which would have grown farther south during colder climates. They include fireweed (or rosebay willowherb), sorrels/docks, berries, fungi, roots and tubers, seaweeds and even some lichen.* Even if only 1 per cent of Neanderthals' diet during glacials was vegetable-based, over a year it adds up.

Interglacial Neanderthals striding through lush forest, meadows or wetland had even more choice. For nearly 25 years Near Eastern sites have pointed to this, most obviously thanks to water filtering of ashes from Kebara Cave, Israel. This recovered thousands of charred remains from nearly 50 plant species, a good number edible. Combined with other warmer sites like Amud and Gibraltar, the range of vegetable scraps in Neanderthals' hearths is impressive: nuts (acorn, pistachio, walnut, hazel and pine), fruits (palm, fig, date, wild olive and grape), tubers (wild radish, bulbous barley and nut grass) and seeds (grasses, peas and lentils). Even Eemian Northern Europe was full of options: at Neumark-Nord and Rabutz in Germany, charred hazelnuts, acorns, lime tree seeds and stones from sloe and cornel berries indicate that these may have been eaten.†

The past three decades have seen the image of Neanderthals as the first Atkins Diet devotees really unravel with direct proof of plant consumption coming from many parts of their own bodies, beginning when researchers peered more closely at their teeth. Eating produces

---

* Essentially every plant group has edible members, bar most mosses, liverworts and slime moulds.

† Cornel trees are today native to Southern Europe; their fruit is about the size of grapes and interestingly the wood is massively dense yet elastic, making it so perfect for spears that in early Greek poetry the tree's name could be taken to mean 'spear'.

wear patterns that can be linked to hardness of foods, and it's possible to distinguish between long-term abrasion versus micro-wear: a gauzy overlay of scratches and pits from the last few days or weeks. 3D scanning, modelling and statistical analyses compare the diversity and directionality of scratches and pits with experimental samples, and unlike isotopes, samples can include Neanderthals living in a mix of environments. And in general, as you might predict, individuals from colder and therefore less heavily vegetated contexts do show more meaty-looking wear. The Spy Neanderthals in Belgium even had wear similar to recent hunter-gatherers from Tierra del Fuego, who are known to have had a very meat-heavy diet. Yet contrary to clichés of ice-bound lives, even cool-climate Neanderthals don't show wear patterns as extreme as recent Arctic peoples including the Sadlermiut Inuit, many of whom ate dried and frozen meat and cracked bones open with their teeth.

In contrast, teeth from Neanderthals like the Tabun 1 woman who lived in warmer and more lush landscapes had wear from chewing tough, abrasive stuff extremely likely to be plants. Krapina is especially interesting: it was probably occupied just before the intensive warming of the Eemian, meaning full forests hadn't yet developed. Yet remarkably, the micro-wear from these Neanderthals most closely matches later prehistoric *agricultural* peoples who ate a lot of fibrous plants. And a clear picture of individual variability exists: woven through large-scale climate-connected patterns, Neanderthals from the same site and even the same layer aren't always identical, showing that not everyone was eating the same stuff.

And you can zoom in further to other kinds of oral evidence, including the dental calculus many Neanderthals suffered from. A 'biofilm' made up of mineralised spit, squished food detritus and the remains of bacteria that feasted on it, calculus is essentially a microscopic archaeological deposit of what was eaten. Stringent analysis protocols make it possible to rule out contamination, whether from ancient sediment or starches coming from researchers' sandwiches. Combined with residues on lithics, the new picture of Neanderthal meals is phenomenal.

Out of some 40 sampled individuals so far, the prize winner for most diverse grot in their teeth is Shanidar 3, who we met earlier having been stabbed in the chest. Before this individual's death, date palms, plants from the pea/lentil/vetch family and unidentified roots/tubers were all eaten. The latter also occur as residue samples on lithics from the same level, and artefacts from many other sites seem to roughly match the evidence from charred remains. They include seeds, nuts, leafy plants/fruit, pea family plants, unidentified roots/tubers, fungi and grasses. Grasses are especially intriguing because collecting and processing the seeds is very time-consuming. One or two cases might be dismissed as by-products of collection for something like bedding, but more points to targeting for food.[*] This is supported by the fact that seed starches from wild grass relatives of barley or wheat are also in some calculus samples including Shanidar 3.

Once again, diverse plant evidence in calculus isn't just from the Near East. All the way up in north-western Europe, the climate at Spy around 100 ka was definitely cooling, yet both Neanderthal adults had traces of grasses and, startlingly, waterlily root starches. These certainly imply that Neanderthals were *actively* foraging for plant foods, and were quite comfortable wading about in water. A truly twenty-first-century development in calculus analysis has been using DNA to try and identify foods, though as a technique it's still teething. Among the scores of bacterial or viral matches, a few intriguing results came back. The individual from Spy with meaty-looking wear on her teeth had calculus DNA that matched rhino as well as wild sheep; seeing as sheep aren't really represented in the faunal assemblage, could this be food eaten before arriving at Spy?

Perhaps the biggest surprise was thrown up by El Sidrón. For the sample in question, dental wear pointed to a mixed diet but the calculus had no large mammal DNA *at all*;[†] instead, matches for pine, mushroom and moss appeared. This triggered 'vegetarian Neanderthals!' headlines, but the reality is complex. The mushroom

---

[*] There have been suggestions that Neanderthals maybe ate 'chyme' – the half-digested vegetable mush in herbivore stomachs consumed by some hunter-gatherer cultures such as the Cree, Inuit, Chipewyan and Kutchin – but much of that would be grass, which doesn't match plant residues in dental calculus.
[†] Even DNA from the sediments found only hominin matches, rather than any faunal genetic material.

species is widely consumed outside Europe and North America, but the pine is trickier. Some northern hunter-gatherer cultures do use inner pine bark as an early spring food, but the DNA-identified species is native to East Asia, which doesn't make much sense for an Iberian Neanderthal. And the moss is tiny, has no food history anywhere and is used in biotechnological applications, raising contamination as a possibility.* On the other hand, it's recently been found to contain complex carbohydrates, so would have had some nutritional value. Perhaps this is a case where Neanderthals knew something we didn't.

Certainly *someone* at the El Salt rockshelter in Spain was eating plants, since hearth sediments have furnished us with the first Neanderthal faeces samples. Aside from providing abundant comedy headlines ('What the crap?', 'Poop scoop', etc.), biochemical research showed that among mostly animal-based compounds, there was indisputable plant matter, probably from roots or tubers.

Many of the new plant-detecting methods described here are still developing, and there's uncertainty in the details. But amid the harvest of data, one thing we can believe is that we're only seeing a fraction of the plants Neanderthals were really eating. That's especially true for cooked foods, which break down faster. And coming back full circle to the isotopes, boundaries are being pushed here too. The most recent studies homing in on amino acids still show animal proteins as dominant, but are now coming up with evidence of plants too. For the Spy Neanderthals, up to a fifth of protein was potentially coming from non-animal sources. Given that local environments at the time weren't that rich in vegetation, that implies some really focused foraging, and perhaps considerable processing by cooking.

Homo gastronomus

So 'meat and veg' were, in some times and places, a fair description of what Neanderthals consumed ... but how? Were there bubbling stews alongside fat-dripping roasts, or was their kitchen-craft mostly raw?

---

* *Physcomitrella patens* or 'spreading Earth-moss' has exceptional properties for genetic research and new medical applications such as genetically engineered cancer drug production.

Certainly that's possible for some food, but cooking not only makes things edible, it also improves nutritional value and often aids digestion, whether meat or plant. While we'll explore debates over Neanderthal control of fire in the next chapter, there's good evidence for some level of meat cooking. Faunal remains that combine burning damage from different temperatures most likely indicates roasting on a fire, since bone would be burned more than areas covered by flesh.* And some cooking would be largely invisible, for example if Neanderthals were roasting fillets or organs, rather than meat still on the bone.

Roasting megafauna is an old caveman cliché but it's actually inefficient and fuel-hungry. Stews are better, cooking flesh while providing marrow-rich broths to sup. Then there's grease: study after study show that Neanderthals were consistently organising hunting around marrow and fat: the animals targeted, the extent of butchery and what was brought back. Moreover, the spongy, grease-rich ends of long bones are nearly always missing. Carnivores love these too, but in assemblages with no sign of their presence, we're looking instead at Neanderthals methodically processing them. This would be either boiling to render the grease,† or pulverising them into an oily bone paste. All this matches hunter-gatherer cuisines in cooler climates, which are often similarly focused on extracting juicy fats.

What about plants? Grass seeds need soaking or charring and grinding, and though acorns are highly nutritious they also require soaking to remove bitter tannins.‡ Calculus studies back this up: some at El Sidrón contained heat-cracked starches, and 40 per cent of the starches from the Shanidar 3 individual appeared to have been boiled. In general, when the type of plant is identifiable, it's often those with hard seeds that benefit from cooking. Such complex practices put Neanderthals surprisingly close to the way hunter-gatherers prepared wild plant foods

---

* It's technically possible that meaty joints thrown into fires as trash could also develop this kind of burning without it being anything to do with cooking, but it seems unlikely Neanderthals would have hauled heavy joints back to caves only to discard them still covered in flesh.
† The bone shaft fragments left over from marrow smashing can even be used for broth-making.
‡ Small pits scooped from sand next to a river is one method for acorns.

at the roots of modern agriculture, but how did a culture with no ceramic or metal containers do this? It's possible to boil liquid by putting red-hot cobbles into any vessel, but heat-cracked stones are very rare in Neanderthal sites. However, there's more than one way to stew a mammoth. You can simply hold the container over a fire as long as the liquid level remains high. It could be a large skull, a naturally hollow stone or even a bark box, but one of the most obvious natural 'pots' would be the stomach or skin of the animal you've just killed.

Thinking about cooking raises the question of where the 'kitchen' was. Some eating might have taken place immediately after the hunt, especially of less easily transported things like blood or organs, and burning at some big kill sites like Mauran might reflect meals during the time it took to butcher multiple bison. But as we'll explore in Chapter 10, there's abundant evidence that Neanderthals moved significant amounts of their prey elsewhere for further butchery and consumption. Even if we assume meat was shared with others waiting at places where game was transported, the hundreds or thousands of kilograms of flesh at places like Schöningen do seem rather too much to have used before spoiling. Given the risk and energy outlay from tackling big beasts and transporting the heavy joints, it would make sense if Neanderthals had some way to store or preserve their surplus.

Doing so requires particular skills, knowledge and advance planning, whichever method was used, which is partly why it's been relatively little discussed. But there's also the lack of direct archaeological evidence. Unlike later prehistoric sites we don't find any big pits, so Neanderthal preservation methods must have left little trace. Freezing is one possibility: some Arctic cultures like the Inuit do this with fish, which can be eaten like ice lollies. During glacials this may have happened by default and would actually have helped preserve vitamin C. Yet much of the time Neanderthals lived in less harsh conditions, so stored foods must have taken other forms.

Another option is smoking. Calculus from two individuals at El Sidrón contained chemical markers for woodsmoke showing that some Neanderthals were living alongside smouldering fires, but there may be other explanations. And preserving this way is perhaps more likely outside caves: another way to view the burning at Mauran is as involved with conserving meat and marrow. But the simplest way to store meat (and the hardest to detect archaeologically) is by making jerky – it requires only drying, and can be either kept or further processed by mashing up with fat and marrow into something like pemmican. As Chapter 9 will show, at some exceptionally preserved sites there are microscopic traces of pulverised bone fragments and cooked fats around hearths that might reflect this kind of processing.

Storing plant-based foods would require quite similar processes to storing animal products. Autumn berries could complete pemmican-type recipes, while leaves, grains or roots could be dried or ground. These would all have been more likely to happen at living places rather than while out and about gathering. A few sites do have enigmatic stone blocks with abraded surfaces, and at La Quina some lithics have cracked grass starches that could relate to grinding or dry heating preparation.

A whole other range of options exist with fermentation. By storing in low-oxygen conditions, meat, fat, fish or plants can all be allowed to go some way along the path to putrefaction yet still be edible. In a way this is like predigesting food, and it's especially useful for things such as brains, which are prone to going bad. Unlike cooking, this also preserves some key nutrients, especially vitamin C.*

A multitude of fermentation recipes exist today, some simple and others complicated. Game hanging represents the early stages, although sometimes mould can develop. *Kiviaq* is more complex: a Greenlandic method where hundreds of little auks (tiny seabirds) are sewn inside a fat-smeared seal skin and stored for months until soft and green-tinged. Vegetable fermentation is also common, whether sauerkraut, kimchee or fermented tofu.

In many ethnographic accounts fermented foods aren't emergency fallbacks or delicacies, but built into the normal diet. Can we believe that Neanderthals did this? The practice was already established 9,000 years ago among Scandinavian postglacial hunter-fisher-gatherers, and depending on the method it's not cognitively very taxing; but getting fermentation wrong can be lethal. There are regular botulism deaths from eating badly kept seal or using the wrong birds for *kiviaq*.

---

* As proteins break down into amino and fatty acids they become more readily available to the body.

Neanderthals already had multi-phased carcass-processing systems, so adding another delayed stage might not have been hard. Keeping food underwater is one simple method, raising intriguing possibilities for places where we know animal carcasses were submerged. This includes horses at Schöningen or the mammoth at Lynford. Whatever the method, fermentation takes so long that if Neanderthals used it, it may well have been something they left and only returned to after travelling elsewhere. Future teasing out of evidence for fermentation may come from isotopic studies, since such foods tend to become nitrogen-enriched.

There's another aspect to fermented foods: taste. Traditional fermented foods with strong flavours and scents – sometimes acknowledged in names like 'stinkfish' – are nonetheless often anticipated with relish. Even without spending weeks ripening inside a marine mammal, seabirds are something of an acquired taste, but they're not far off conceptually from cheeses like Roquefort or Stilton, and apparently have a similar moreish effect. Did Neanderthals salivate at the thought of 'marrow cheese' as they crouched eyeing up bison herds, or smack their lips while guzzling lightly rotted reindeer? Living humans experience at least five tastes: sweet, sour, bitter, salty, savoury (umami) and potentially another that interestingly seems to detect calcium and fats.

Taste, though, doesn't just signal what's good. Bitter sensations especially warn of danger, and genetics confirms that Neanderthals could detect one such compound. Known as PTC,[*] it's found in some plants and is safe only when consumed in small amounts. Intriguingly, this Neanderthal mutation is different to that in many people today and was accompanied by another partially blocking PTC signals. This may mean Neanderthals had a higher tolerance for such flavours, and combined with genetic evidence for a wider array of bitter and sour taste perceptions, sampling unfamiliar plants or fermented meats might have been safer for them. The duo of taste and smell combine to produce what we experience as flavour, so it's even possible that Neanderthals lived in a richer culinary world than ours.

---

[*] Phenylthiocarbamide.

Live to Eat, Eat to Live

What and how we eat is deeply cultural. Even apes don't simply forage what's around, but follow what they grew up with. We use lithics to classify Neanderthal cultures, but food-based traditions were probably also part of their diversity. As hunters of massive prey, they certainly collaborated, but unlike wolves or hyaenas, after killing together they shared the spoils. Aside from mothers feeding offspring, chimpanzee hunters are far less altruistic, bartering scraps for social favour, including sex. That kind of reluctant cooperation isn't what Neanderthals did. They carried forward their collective action from kill through methodical butchery to moving the richest parts onwards, sometimes delaying feasting over three or more stages of processing.

Salzgitter-Lebenstedt shows this in practice. At least 44 and potentially twice that many reindeer were killed here in autumn, likely during different hunts as herds moved down from summer grazing in the Harz Mountains. Animals of all ages were skinned and filleted but only the fattest – prime-aged males ready for breeding – were more intensively butchered. Neanderthals were after the richest parts of these chosen animals, going for marrow, fat and organs, but less so lean meat. Such an obviously selective pattern couldn't emerge in a selfish free-for-all, but from groups with common purpose. The same thing is mirrored hundreds of times over in other Neanderthal sites.

Zooming in on the butchery patterns themselves reveals a systematic approach, far from a disorderly scrum. There were probably just a few – or even only one – skilled cutter per carcass, who knew where to slice so that joints would open, where to strike so that bones would shatter. Highly practised Neanderthals, like modern butchers, leave more orderly, shallower and fewer marks, so assessing number and placement may even highlight different levels of skill. It's interesting that there are much higher cut-mark rates for reindeer limbs at the Pech de l'Azé IV rockshelter than the kill site of Jonzac. Since the sites are both Quina technology and similar ages, this difference could be explained by the presence of seasoned hunters and perhaps other skilled adults at Jonzac. The 'messier' pattern at Pech de l'Azé IV may well reflect the less refined slicing of others, including youngsters learning the craft.

Could Neanderthals have had specialised roles in subsistence? Sharing resources promotes the division of tasks, even if most individuals were multi-skilled. This may be visible in spatial patterns; for example, at Schöningen marrow smashing was happening away from the butchered horse bodies.

We'll explore task areas more in Chapter 10, but who actually hunted is another question. Women on average may have been more vulnerable during pregnancy and when caring for helpless infants, so in evolutionary terms risking rare offspring wouldn't be a great strategy. Certainly in many hunter-gatherer societies mostly male hunters deal with large game, sometimes leaving for days.

But this isn't universal. Though societies with lioness-like women as the *primary* hunters are vanishingly rare, it's not unusual to see them join hunts, take part in kills and dominate the primary butchery. And wherever they are, women and children frequently team up for local hunting of smaller game.* Perhaps most surprising, in some hunter-gatherer cultures tiny family groups including women and babies will head off seasonally into the land, supporting themselves for weeks at a time.

Fundamentally, whatever kind of game was hunted or foods foraged, it happened within a social context. Neanderthals probably snacked together on what they found, children learning by observing what was portioned off to transport elsewhere. Tooth scratches show youngsters gradually making more grown-up gestures with eating tools, as little hands grew and managed better. In fact, children as foragers may have contributed a not-insubstantial amount of food, and catching small creatures would have given them the opportunity to practise butchery.

This may well explain why small critters like rabbit, which can be ripped apart – especially after cooking – still have cut marks, but most striking are the tiny birds at Cova Negra. Songbirds in themselves aren't necessarily odd fare, with many eaten as traditional foods or gastronomic delicacies today, with French ortolan among the most famous.† In some hunter-gatherer cultures songbirds are fallback foods, though in

---

* Sometimes it's women whose tracking skills are the most renowned.
† Banned in the 1990s, some chefs have claimed that eating them whole after being pickled alive in Armagnac is an ecstatic – albeit prickly – experience.

others they're regularly eaten, and it's often children doing the hunting. The species like swallow or blackbird being carefully carved, cooked and eaten at Cova Negra were butchered following the same methods as larger birds. Minuscule legs were sliced, and the greasiest bones nibbled and pierced to suck out marrow. Rather than markers of starvation in an otherwise game-rich site, this would make more sense as a way for small fingers to learn how to cut meat, membranes and tendons.

Perhaps within Neanderthal groups there were other kinds of subsistence that followed particular social groupings, but what about hunting specialisation more generally? In places like Cova Negra where the amount of bird butchery overall is very high and resembles specialised avian hunting in some later *H. sapiens* sites, it's a possibility. Certainly in other places like Jonzac or Mauran there is a dramatic focus on one species over very long time periods, but this is balanced by the impression that Neanderthals simply took the best of whatever local ecosystem they were in. Single-species assemblages reflect sweet spots where combinations of climate, topography and animal behaviour made catching them much more likely. Considering the centennial or even millennial scales involved, it's hard to know if there was a persistent tradition at work, but specific knowledge would have been maintained over at least several generations.

Specialisation is one aspect of behaviour that, over time, has been used to characterise Neanderthals as less productive or capable in subsistence terms than early *H. sapiens*. Yet this notion too has unravelled in the past two decades. For example, some early South African *H. sapiens* populations gathered marine molluscs so intensively that millions built up in huge piles, and shrinking shell sizes over time implies longer occupations and *over*-exploitation. In contrast, Neanderthals generally seem to have 'browsed' shellfish. Even places like Bajondillo, El Cuco and Figueira Brava in Portugal, where seafood was important, were nodes within wider site networks across many environments.

This generalist adaptation led to claims that Neanderthals were inefficient since they didn't fully extract available resources. But recent research shows in some cases that they did over-exploit creatures just as easy to catch as molluscs. Over multiple layers at Oliveira Cave, tortoises become significantly smaller, likely reflecting over-hunting,* which may even have led to their Iberian extinction. In early *H. sapiens* this kind of intensive use would be taken as evidence of booming populations, so perhaps the same is true for Neanderthals at that time and place: more babies meant more mouths to feed.

Better understanding of early *H. sapiens* people has also shown how overarching theories about our 'success' don't neatly stack up. Isotope analysis in 2019 showed that Neanderthals and early Upper Palaeolithic communities from several sites in Belgium were virtually identical in their heavy reliance on meat, probably mammoth and reindeer. Even with technological differences, it doesn't look as if Neanderthal hunters here were less efficient.

In the two decades since Schöningen's spears pierced through the heart of scavenging theories, how we think about Neanderthal hunting has been transformed. Even beyond such exquisite rarities, the literal weight of evidence – millions of animal bones from hundreds and hundreds of sites – amply demonstrates hunting prowess for even the most gigantic beasts. Smaller animals and even plants have become uncontroversial parts of millennia-old meals as researchers' analytical techniques have matured beyond imagination.

All this has fed back into wider theories about Neanderthal bodies, cognition and social lives. Following from the micro- to the macro-scale how tools scored through fur and flesh, to the finesse of selecting the choicest parts to carry and eat, their self-confidence at the intersection between lithics and animal bodies is palpable. It's almost as if they could have done the job blindfolded.

Nonetheless, Neanderthals' diet sometimes still ends up being framed unflatteringly in opposition to *H. sapiens*. Niggling assumptions persist that they must have been doing something fundamentally 'wrong' to explain why they disappeared, even though the archaeological reality leaves claims about our superiority somewhat hanging. Even if not at all sites or in such large degree, as a species they're far closer than used to be believed to the 'broad spectrum' diet

---

* Female tortoises take a decade to become sexually mature, and so just like twenty-first-century over-fishing, adults were being killed before they could breed.

that supposedly underlay early *H. sapiens*' success. Instead of trying to pick holes in what Neanderthals ate, we could instead ask why early *H. sapiens* became more specialised despite it being riskier. Perhaps relying heavily on shellfish or small game might not have been a choice, if they were outcompeted by Neanderthals for the best-quality foods: large mammals.

What Neanderthals *didn't* eat may also be manifesting other things. Plants or animals enjoyed by some hunter-gatherer societies can be ignored or even shunned by others. Just like smell, taste is part of the ancient core of hominin brains, woven deep into memories and identity. The scent of certain foods may well have evoked concepts of seasons or connections to particular places for Neanderthals. Perhaps those living in Wales would have been surprised – even turned their noses up – at what others of their own kind from Palestine consumed with relish.

The biggest question is how it all ties together. In terms of their diet, Neanderthals weren't stuck in an evolutionary rut. Just like lithic technology, over time we see increasing proliferation and fragmentation: their range of foods multiplied, and animals' bodies were taken apart ever-more carefully and thoroughly. The next two chapters explore how these patterns, from single hearths to entire landscapes, reveal them to be hominins who were moving things around at greater scales than ever before, and building new connections between themselves and the world.

CHAPTER NINE

Chez Neanderthal

'Go soon' the wind whispers. 'It's time' roar the deer, their breath pluming in the cold air. 'Leave now' says the dawn frost, burning its message into the grass. The people listen. Shadows inside the rockshelter grow colder, hard skin forms on puddles. Their hunting trips shrink, taking only what's close by, knowing they will soon leave. A last deer gives itself up, is taken in to be un-made right there next to the fires. Cranes flap overhead like scraps of summer chasing the sun; half the deer is eaten. Soon drips from the roof edge will harden into icy fingers. A morning of excitement: a dangling fluffy catch, tail ringed like the path of a stone through water. They turn its wide claw-pads over to know it, snag fang tips with their thumbs, stroke long whiskers: cat. An animal more dreamed than seen, it is the last message, the last meal. With a final fire, they suck its fat into their cheeks, roll and pack its deep-furred pelt. Then the people haul themselves up, begin walking. Their noise fades to silence; the rockshelter exhales.

Time spins up and begins to defocus. Thrushes come to scratch at ashes, cocking their heads. Rabbits blur inwards and outwards as a falcon swoops down; it stays to winter under the overhang. Below its blade-sharp gaze, the wildcat's body settles into the earth. Membranes and guts ripple with the maggot-dance beneath. Blanketed by last year's leaves, only furry gnawers bother the summer-dried bones. As trees flare from green to yellow to orange, the people return. But, just as the old ones foretold, the floor has become too wet for fires, and the rockshelter will welcome no more dwellers. They depart, leaving roots and mosses to embroider the edges of bone heaps. The years flow past as limestone-filtered waters seep and rise around the wildcat skeleton, the deer, the wood, the ashes, the stones. The sleeping time has come.

Place is solid, but time is the wisp in archaeologists' fingers, disappearing just as we try to grasp it. The incredible details now amassed about Neanderthals are so far beyond the dreams of pioneer prehistorians, they verge on science fiction. Yet fully reconstructing the rich tapestry of their lives – seeing not just the threads, but the weave – is formidably hard. Phenomena like lithic techno-complexes are visible at geological timescales, yet require explanations operating at human ones. And for a long time researchers foundered in the quicksand of deep time, the connections between things out of reach without knowing how old they were. The advent of direct dating methods was transformational, but just as important has been better understanding of how time creates sites.

A vital baseline in trying to understand what any particular site means for Neanderthal behaviour is both what they did and how long they spent there. However, even with super-accurate dating methods applicable to minuscule samples, in the vast majority of locales it's not possible to distinguish occupation phases shorter than a millennium, never mind a century. That's because an archaeological layer can swallow up surprising amounts of time: sediments separating different deposits of artefacts can erode or slip away, leaving a mixed mass of objects. In this way, a hand's breadth thickness may be a palimpsest of a thousand summers.

Remarkably, measuring the true number of occupations represented by particular layers has now become possible, in some circumstances. Pioneered at Mandrin Cave, south-east France, an ingenious method known as fuliginochronology* is opening a window onto just how crammed full of time some Neanderthal sites are. Close inspection of odd black smudges within lumps of carbonate – a mineral deposit growing on walls and ceilings – revealed nano-scale stripes.

Essentially tiny stratigraphies written in soot, they formed when the fires of Neanderthals in residence 'smoked' the roof and walls, leaving thin soot films. If nobody was there, plain carbonate sealed them over, then the cycle repeated and layers built up. Just like barcodes, they're unique, allowing pattern matching between different chunks within and between layers.

This sooty archive provides the only known means of counting the minimum number of times Neanderthals stayed during reasonably thick archaeological layers, and the results are startling. One 50cm (20in.)-thick level at Mandrin Cave covers at least eight periods of occupation, already quite a large number. But the level below – of about equal thickness – represents up to *80* occupations. It's a stark warning that the appearance of layers can be deceptive, and a reminder that more than 99 per cent of all the assemblages archaeologists study aren't from single occupations but represent patterns of behaviour over at least one if not many generations. 'Time-averaged' assemblages like that are far from useless, but to understand them better we need to discern the minutiae of Neanderthal life.

To make the ideal archaeological site, you need 'high-definition' conditions that prevent lithics and bones from different occupations blurring together. Fine sediment that accumulates rapidly but gently and remains un-eroded is perfect. Such locales are precious not only because they represent brief phases in time, but also because, if undisturbed, they preserve spatial patterns revealing what Neanderthals were doing in different parts of the site.

The 'Rosetta Stone' for deciphering the wider archaeological record of the Neanderthals has always been finding *single* episodes of action or presence, ideally lasting just a few days, or even minutes. Individual refitted groups of lithics of course represent very short timescales, but finding a whole layer at the same temporal resolution

---

* This combines the Latin for 'soot' and the Greek for 'chronology'.

is almost unheard of. Yet thanks to twenty-first-century excavation methods, we know they do exist.

To study these kinds of places, modern technology as well as infinite patience is vital. Lasers record object positions in 3D, providing data to digitally reconstruct the vertical or horizontal spreads of artefacts. Details such as lithic clusters around hearths or micro-layers invisible during digging emerge on screen. A key approach is to look for 'special' things: unusual stones or rare animal species that stand out like a UV glow against masses of other fragments. Combined with refitting and microscopic sediment analysis, today we are as close as we may ever get to being able to 'watch' Neanderthals going about their daily lives.

Chapter 7 explored the wooden artefacts of Abric Romaní, but the exceptional conditions of preservation there also show in stunning detail how Neanderthals used space in-between periods when the rockshelter was too wet to live in. After they left, fresh travertine flows entombed the entire abandoned living surface, preserving everything just where it lay. The timescale for each archaeological layer is certainly more than a few days, but is likely to represent decades rather than many centuries. Excavations have now reached Level R, some 60,000 years old, and it'll take many years for this material to be studied.[*] But analysis of younger levels from M to P, dug up to a decade ago, have already produced amazing results.

Level O, dating around 55 to 54 ka, contains one of the most high-definition records of Neanderthal life in the world. From sediments covering the 270m^2 (320yd^2) surface of the rockshelter, but under 1m (3ft) thick, some 40,000 objects were excavated. Digital analysis of their positions clearly shows at least three major phases, although each is probably made up of more than one occupation. Magnetic data from hearth sediments indicates that everything formed within just a few centuries, so each phase is likely to represent Neanderthals coming and going over a few generations at most.

---

[*] Given the depth of un-excavated deposits, there's another 40,000 years of Neanderthal archaeology beneath: a further century's work for multiple generations of researchers.

The middle phase, Ob, is the richest, and contained a completely unique discovery: a whole, butchered wildcat. Through every level at Abric Romaní Neanderthals normally smashed up bones so systematically that picking out individual creatures is almost impossible. To find a mostly complete skeleton is extraordinarily unusual, and points to a very brief snapshot in time just before the site was abandoned and another travertine layer built up. The last Neanderthals to live there for many centuries caught and skinned the cat, probably cooking it on a nearby burned area. Bellies full, they were already thinking ahead: missing toe tips and tail bones show that they took the thick, striped pelt onwards with them.

Where the Hearth Is

The wildcat is an astounding find, representing no more than a morning or afternoon more than 50 millennia ago. Even in high-definition places like Abric Romaní, the typical layers are still tricky: how is it possible to separate overlapping vestiges of activity from within the dense spreads of artefacts and bone fragments? The answer is to begin at the heart and spiral outwards: a black carbon molecule, an incandescent ember, an ashy aureole, a lattice of branches. A circle of eyes glowing in the dark. Hearths are archaeological touchstones. They lie at the centre, where the warp of time and weft of space connect. Like beacons shining through the fog of millennia and confusing haze of data, they offer anchor points precisely because they were also the cores for Neanderthal life.

Fire is one of the most potent symbols in the grand story of human evolution. It gives light and heat, but so much more, protecting against predators, cooking food and transforming other substances. It even extends social life by banishing the darkness. Just as our houses have been built around fires throughout history, hearths also structured Neanderthal existence: an instantly recognisable centring of space. We don't need to imagine them sitting face to face; we can literally see it in the way artefacts encircle the ashes and charcoal.

Charcoal fragments, burned artefacts and heated sediments are found across hundreds of sites, but exploring what they mean is challenging. Fireplaces are fugitive artefacts, their structures as fragile as the ash inside them. They can be smeared or obliterated through erosion, trampling or distortion by overlying layers. Experimental studies show that sometimes heated sediments survive where other fire evidence does not, and micro-layers from different burning phases are especially strong evidence of hominin action.

A key locale that has developed our knowledge of Neanderthal pyro-technology is El Salt, Alicante. This rockshelter is about 350km (220mi.) south of Abric Romaní and, as well as being broadly contemporary in age, has hosted a similar decades-long project. As excavation began to show black circles that looked like hearths, researchers wanted to try and better understand what these features were. Creating experimental hearths right by the limestone cliffs allowed them to closely match conditions to those the Neanderthals of El Salt experienced.

They found that fresh hearths typically show a triple-layer structure: heat-reddened soils at the base, then a black layer, topped by ashes from the burned fuel. In many archaeological sites the ash layer has disappeared due to natural processes, which was often the case at El Salt. The black layers they'd excavated were therefore charred remnants of whatever had been on the floor *beneath* the fire, confirmed by microscopic analysis as essentially weeds and leaf litter. Therefore any artefacts in those black layers are more likely to be from *older* occupations.

Despite advances in identifying hearths, Neanderthal pyrotechnical skills remain hotly contested. Nobody today denies they used fire (as hominins have for over a million years), and hearths undeniably become more common through the course of the Middle Palaeolithic. From about 120 ka fire was obviously part of everyday life. But the question of whether Neanderthals simply scavenged it or could *produce* it is, perhaps surprisingly, still debated.

The problem is that some sites exist with relatively rich archaeology but the presence of fire is barely detectable. What's more, the most often cited cases are Roc de Marsal and Pech de l'Azé IV, south-west France, where Quina layers really stick out as lacking hearths or even much charcoal, despite their presence in earlier periods. Given that the layers date to the MIS 4 glacial, the apparent lack of fire right when it was coldest is puzzling.

Is it possible that Neanderthals could use fire once they had it, but had forgotten or never knew how to make it? Theories that they simply toughed it out with thicker clothing and raw diets not only seem unlikely, they're also contradicted by the presence of *some* charcoal and burned lithics and bone, albeit far rarer than in other layers at the same sites. Perhaps Neanderthals were just scavenging fire from natural blazes? However, in high-latitude tundra environments like those when the hearth-less layers accumulated at Pech de l'Azé IV and Roc de Marsal, lightning strikes are very uncommon. If wildfires were extremely intermittent, then Neanderthals would need to be exceptionally good at conserving embers; in which case they could just as easily have daisy-chained flames from one hearth to another.

There's another explanation for all this. Maybe Neanderthals were perfectly able to make fire at will, but simply changed how and where they used it according to different ways of living. If during the Quina phases they tended to make hearths just outside caves, there would be no trace inside except for a bit of charcoal and a few burned remains: exactly what's found.

Palaeo Pyrotechnics

Whether or not fire-on-demand was universal among all Neanderthals, birch tar technology strongly argues this was the case for many populations from at least 300 ka. Precisely what their blaze-starting skills consisted of is less certain, but being curious, inventive and surrounded by knapping, at least *some* Neanderthals must have noticed how struck flint gets hot and creates natural sparks. For a long time archaeologists found virtually no special 'strike-a-light' tools, but it now appears that they were simply being economical. In some sites, up to 75 per cent of bifaces show battering in the centre of one or both faces. Microscopic study and experiments show that they formed when either another flint or potentially iron pyrite – both known to produce sparks – struck at an angle and then dragged along the biface's long axis.

More complex methods might also have been practised. Just as you speed up your barbeque by using chemical firelighters, new research suggests that Neanderthals may have done something similar. Manganese dioxide is a deep-black-coloured mineral naturally found in small concentrations at many sites. But some locales have exceptionally large amounts, for example up to 1kg (2.2lb)'s worth across several layers at Pech de l'Azé I.

When examined closely, many of the hundreds of small pieces here and in other sites had been rubbed. Furthermore, occasional finds of limestone blocks with remnants of black powder point to Neanderthals sometimes grinding manganese. One explanation comes from the fact that this mineral has pigment properties that we'll explore later. However, it's also a great fire accelerant, especially when powdered, making wood catch alight faster and burn more efficiently. As yet there's no direct evidence that Neanderthals did use manganese this way, but it's an intriguing possibility.

Once alight, Neanderthals were interested in carefully managing their fires. The majority appear like simple campfires: flat, circular deposits of charred material and ash, functioning well enough without surrounding stones. But sometimes Neanderthals invested in constructing hearth features. At Abric Romaní and Roca dels Bous, another rockshelter not far away, they chose to kindle fires within natural hollows, improving the heat retention. In some cases they deepened these first, and most impressively in Level O at Abric Romaní Neanderthals were controlling airflow into pit hearths by digging small trenches. It's probably no coincidence that in well-preserved sites like this, there's also more evidence that stone blocks or cobbles were placed by fires, probably helping to avoid draughts or direct heat.

Given what we know about Neanderthals' material choices for artefacts, it's not surprising that they also took care in the fuels they used. Wood is by far the most common, and similar to how they hunted, largely they took whatever was around. Being very abundant, pine is therefore the most commonly burned wood, but sometimes it seems to have been selected despite other species being available. For example, in Level J of Abric Romaní, among over 1,000 identified charcoal fragments, all but one were pine, the exception being birch.

El Salt has a more diverse pattern, potentially linked to its local environment, which was milder and more varied. In addition to pine and juniper there's maple, evergreen oak and even yew. But interestingly, the species vary between different hearths. This is particularly visible in Unit 10, where among mostly pine-fuelled hearths, maple was much less common or widespread, and even rarer oak and boxwood seem to have been burned only in a couple of fires. Concentrations like this probably reflect Neanderthals staying for short periods, but it's difficult to know if the species choice was deliberate.

The length of time Neanderthals stayed in any place would have affected how they used fuel. The Abric del Pastor rockshelter in south-east Iberia is over 800m (2,600ft) up in the mountains, and pine wasn't as abundant in the cool, dry climate. Most of the remains in hearths are juniper and terebinth (or mastic wood),[*] but they're not ideal fuels. Slow-growing juniper has very twiggy, tough branches and produces little dead wood, meaning that collecting it is difficult. Terebinth is also less suitable than pine, and would have made a heady smoke. It's likely that Neanderthals only switched to burning these species after using up the limited local stocks of pine, suggesting that at least sometimes they were living at Abric del Pastor for more than a couple of nights.

What Neanderthals do seem to have been picky about is the *type* of wood they burned. Where studies have been done, microscopic features as well as the small size of twigs and branches point to naturally fallen or dead wood being the most common fuel, rather than freshly felled green wood.[†] It's easier to collect, burns better – especially if resin-rich, like pine – and especially suits cooking. Forests in their natural state generally have abundant dead wood, and by walking just a kilometre, Neanderthals could find enough to fuel multiple small fires for up to six months.

This means that unless they were staying for significant periods of time, cutting fresh wood made little sense and daily foraging would

---

[*] Terebinth interestingly is edible, and can also produce a sticky gum.
[†] In Level M at Abric Romaní firewood averages 1 to 3cm (0.4 to 1.2in.) wide, and mostly under 25cm (10in.) long.

be enough; exactly what's seen in hunter-gatherers who don't stockpile. On the other hand, some high-latitude Indigenous communities including Athabaskan and Yupiit peoples in Alaska and Itelmen in Kamchatka, Russia, will drag entire standing dead or recently fallen trees back if they encounter them while out and about.* At Abric Romaní, Neanderthals did bring some larger branches to hearths, and there may even be evidence of a woodpile of sorts in front of the rockshelter in Level M. As we'll see in the next chapter, Levels N and Oa contain entire tree trunks, but whether they were hauled in for fuel or something else is less clear.

A most unexpected fuel for Neanderthal hearths is coal. Already mentioned in Chapter 8 as having excellent evidence of rabbit hunting, the Les Canalettes rockshelter is also unusual because Neanderthals there were burning brown coal. It wasn't a case of glacial wood shortages, since the coal is most abundant in hearths during phases when warm-loving species like elm, maple and walnut grew locally. There's another possibility: Neanderthals were *intentionally* experimenting with fossil fuels. Brown coal isn't easy to spark, but once alight it burns low, hot and even; adding 500g (1.1lb) to wood embers will dramatically prolong a fire's life.

How would they discover the coal in the first place? Perhaps simply by paying attention to what washed up at the edges of rivers, where they sourced a lot of the stone they knapped. The closest brown coal deposits to Les Canalettes are about 10 to 15km (6 to 10mi.) north, at the confluence of two deep river gorges, and it's highly likely that Neanderthals came across eroded nodules.† Being ultimately of a woody origin, it would have combined familiar yet strange properties, tempting curious investigation and realisation of its usefulness. Most fascinating, brown coal was burned through the multiple occupations at Les Canalettes, representing at least many centuries, if not millennia. Either Neanderthals repeatedly discovered it or there was a long-lived tradition; whichever the case, the peaty aroma of its smoke would have marked their stays there.

---

* Some Alaskan communities harvest trees by debarking and returning after several years when the tree will be dead and ready-dried.
† Deposits here were mined since medieval times.

A different, abundant material that has a similar improving effect for fires is bone. Although hard to get alight and quick to burn out, adding it will double the life of a wood-fuelled fire, potentially a huge advantage in open tundra. Plenty of sites and hearths do contain burned bone, of course, but identifying whether Neanderthals used it as fuel is tricky. Sometimes what seems to be a bone fire turns out with detailed analysis to have a lot of wood too, preserved only as microscopic remnants or chemical traces.

On the other hand, Neanderthals would have lived with one eye constantly on their fires, and since – as we'll return to later – they sometimes burned butchery waste, it's likely they noticed that bone could make the flames last longer.

Creating a blaze required skill, but Neanderthals also had to maintain the flames, especially as there's more than one kind of fire. Across many hunter-gatherer societies there's a spectrum in how fire is used: big open-air blazes for protection, pit fires for roasting, small cooking fires, hide-smoking fires, sleeping hearths for warmth, even anti-insect 'smudge' fires. Neanderthal sites show a strikingly close match to the variety in ethnographic data. The smallest fires are just 20 to 30cm (8 to 12in.) across and often seem temporary, potentially kindled for a single day or even one task. They're visible as individual circles of charcoal and ash in some places like Abric Romaní, but in sites without high-definition preservation, they're blurred by time into thick charcoal and ash layers speckled by tiny burned bones and lithics.

Yet Neanderthals also made larger, more permanent hearths. At Abric Romaní, sizeable 'combustion zones' 1m (1.1yd) or more across have been excavated, and being surrounded by masses of lithic and faunal debris were obviously the focus for activities over many days or even weeks.

It's not just about size either. When tiny bone fragments from inside hearths are examined, their varied colour and condition reflect burning at quite dissimilar intensities. Some hearths only smouldered below 300°C, while others blazed at over 750°C. Some of the lower-temperature examples look a lot like ethnographic sleeping fires: small and located close to back walls, their heat reflecting off stone to keep slumbering Neanderthals cosy.

It's even possible to see how individual fires were sometimes used in diverse ways. Microscopic analysis showed that one shallow pit hearth at Abric Romaní had sometimes burned freely, but at other times appears to have been banked down, leading to lower oxygen concentrations.*

Occasionally, there are hints about the kinds of tasks certain hearths were involved in. Back in Unit 10 at El Salt, quite a lot of the maple wood being burned was rotten. Fully decaying timber – beyond dead wood – is a poor choice of fuel, unless you're after a lot of smoke. Supporting the possibility that Neanderthals had deliberately chosen it are over 200 fragments of maple seeds, most likely from fresh branches. Although technically edible, burning leafy branches would certainly have boosted the hearth's smokiness: perfect for curing hides.

So far we've only thought about Neanderthals using fire at particular locales, but there's another possibility. Many hunter-gatherers use fire as a different kind of tool out in the landscape, sometimes to communicate, other times on a larger scale. Mimicking wildfires, they can drive animals and even manage the environment, since burning opens up vegetation, forcing new growth that's a magnet for herbivores.

There just *might* be hints of this among Eemian forest-dwelling Neanderthals, precisely when we might expect to see this kind of behaviour. In the Neumark-Nord sediments, just as Neanderthal lithics become visible, there's a sharp spike in charcoal particles: 10 times background levels. The pollen shows more sun-loving species too, like blackthorn and hazel; something was opening up the forest. What's difficult to tell is whether this was natural burning creating a landscape attractive to Neanderthals, or if they were the fire starters. It's clear, however, that there's a connection, since the pattern lasts for 2 to 3 millennia, then as their archaeology disappears, so the forest begins closing in again.

---

* Adding to the unusual character of this hearth, there were also sediment grains that didn't match local rock.

Burning Time

Hearths have more to tell. Prehistorians' desire to gain the highest possible definition in assemblages has always come up against the question of layers with many hearths. Through Level O at Abric Romaní, for example, there are 60. Did Neanderthals have more than one active fire at a time, or do they reflect amalgamations of separate phases? This is not an abstract musing, since one of the biggest uncertainties in our knowledge of Neanderthals is the size of groups they lived in, and multiple synchronous hearths implies larger groups.

Stratigraphy is key to finding out: if hearths overlap vertically, they were obviously used at different times. However, if they seem to be at roughly the same level but are spread out horizontally, the problem gets gnarlier. The solution is analysis of the surrounding spreads of objects, looking for refits or other connections between different hearths, especially in both directions. If many things moved back and forth, it's good evidence that those fires were active at the same time.

Over the past decade researchers have done this meticulous work across entire sites, and managed to unpick palimpsests that would otherwise have been extremely hard to discern. Unit 10 at El Salt extends just 35m² (42yd²) and is only 50cm (20in.) deep, but contains more than 80[*] hearths. By comparing the overlaps of hearths and refitting, it's been possible to separate out eight distinct phases each just 1.5cm (0.5in.) or so thick. Combined with the rate of sedimentation, researchers have been able to reconstruct the cycles of history in this one layer to an astonishing degree. Neanderthals were coming to El Salt for just a few generations at most, then abandoning it entirely for several centuries, before returning. Digitally hovering above the site, hearths appear and disappear in a slow heartbeat, pulsing as Neanderthals came and went over 1,000 years.

---

[*] Less than 15 per cent of the full area has been excavated, so this is probably a minimum hearth count.

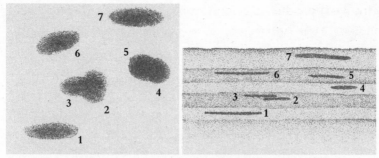

Apparently overlapping hearths are shown in stratigraphic view to be from separate phases.

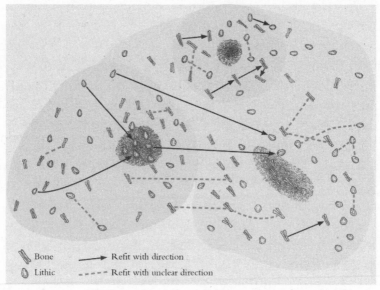

Refitting objects (lithics, bones) can suggest which hearths were in contemporary use.

Figure 7 *Using spatial relationships to untangle site chronologies.*

And where hearths are widely spaced out, making it harder to identify their sequence, refitting artefacts back to individual stone nodules – known as raw material units, or RMUs – means it's still possible to pick out individual knapping phases. By plotting these spatially, researchers could see that the RMUs cluster round particular hearths, and don't seem to connect them. Every hearth really does seem to be from a different occupation.

The dissection went even further. Adding up all the RMUs and other isolated artefacts (tools made elsewhere and left at the site, or small flakes showing an artefact had been imported, resharpened and then removed) produced a maximum possible number of 'events' that happened at each hearth. Being cautious, researchers counted each knapping sequence and single tool as separate events, but that's certainly an overestimate. In reality, a Neanderthal would probably have sat and knapped more than one stone block – creating several RMUs – and perhaps also left behind one or two blunted tools. Moreover, it's likely they didn't travel alone. Put together, this means that even if the average number of events per hearth is over 100, that's still probably no more than a few days' activity by very few Neanderthals.

Some of the El Salt Unit 10 hearths had extremely low numbers of events, probably representing the remains of a single visit. For example, along with only 33 animal bones, one fire was associated with just 43 artefacts, comprising 8 RMU knapping sets, 2 imported tools and 11 flakes that must have come off cores which were then carried onwards. But the highest definition yet achieved for *any* Neanderthal site comes from another rockshelter less than 5km (3mi.) south-west. Since 2005, excavations at Abric del Pastor have uncovered the entire 60m^2 (70yd^2) surface, so we can be sure no hearths are unaccounted for. In Level IV, which is only 70cm (27in.) thick, at least four sub-levels contain more than one occupation phase. Crucially, some phases have minuscule assemblages clustered around a single fire: IVc-1 even has more animal bone fragments (95) than lithics (22), which form just 6 RMUs. This particular phase looks exactly like the ephemeral trace of a few Neanderthals staying one night, doing a bit of knapping, eating, then moving on.

It's unlikely that we'll ever push past this level of definition in time. But beyond the frankly mind-blowing ability to 'see' a single evening from more than 90,000 years ago, the full analysis from Abric del Pastor provides an answer to the conundrum of whether multiple hearths mean one or more visits. When studied in detail using RMUs and 3D mapping, *every* level at Abric del Pastor with multiple fires turned out to consist of individual phases, each with one hearth. This strongly suggests that it was only ever visited by very small groups of Neanderthals.

But there's an exception. The lowest level yet reported at the site, IVd-1, contains four phases, each with a hearth. One has far more artefacts, RMUs and refitting sequences than the others, which in theory might be due to an unusually long occupation. However, the numerous lithics aren't matched by more animal bones or butchery. This makes it especially likely that there were instead more Neanderthals sitting round the fire, even if it was just for a night or so.

Home Design

Hearths are both chronological and spatial anchors in our understanding of how Neanderthals used their sites. They were the nuclei around which life happened, with everyday tasks made manifest as artefacts circling like electron halos. In evolutionary terms, hearths represent an important threshold, revealing the emergence of consistent patterns in how space itself was managed.

Undoubtedly some of this was based on practical considerations: small or awkward spaces would limit choices, while the woodsmoke in calculus at El Sidrón is a reminder that breathing in fumes was a regular risk. Even without knowledge of respiratory diseases, anyone who's experienced streaming eyes around a campfire understands that avoiding smoke is a good idea. Airflow models elsewhere hint at how the wider living space was used. This kind of analysis of Level N at Abric Romaní suggests that some hearths would have smoked out the rear area where Neanderthals slept, so they were probably only used during daytime.

But beyond this, did Neanderthals have 'fireplaces' as we understand them, remaking hearths in the same spot because of cultural tradition? This is very tricky to confidently identify. At Mandrin Cave, some of the soot 'archives' deposited on the walls came from a series of hearths built in precisely the same place, front and centre of the shelter. But this may still have been chosen for ventilation. In other places like Abric Romaní, Neanderthals were raking out hearths between uses, albeit apparently within a single period of occupation. Equally, variation in burning intensities on hearth stones – shown by colour differences – might still form during one visit if the fire was used for varying purposes.

Better evidence for true, multi-phased hearths comes from thin intervening layers, marking pauses in use that were long enough for natural sediment to cover the old ashes. This very likely reflects Neanderthals returning to Abric Romaní after some time away – probably rather longer than a season – and choosing to light new flames inside a familiar hearth. Back at Abric del Pastor, the hiatus in time may be even longer. One of three hearths in Level IVb directly overlies fire-reddened cobbles from a previous phase, potentially decades older. Whether or not the Neanderthals or their forebears had visited before, the stones were visible and it would have been obvious that this was a place of old ashes and charcoal.

That Neanderthals sat around hearths perhaps generations old is a striking image, redolent of how customs solidify into 'place'. In sites with exceptional preservation, fragile traces also exist showing how they were dividing up space itself. This implies pre-existing notions of 'correct' areas for particular activities, at both an individual and group level.* Big sites that could in theory accommodate larger groups and longer occupations are most likely to be where splitting up of space happened. However, exploring this once again encounters the problem of proving that different zones in a site were used simultaneously.

One of the best candidates is Abric Romaní, but even using the RMU method, though shorter phases can be defined, the hearths within them can't be reliably separated. Moreover, refitting connections between different fires and activity areas are all one-directional and often concentrated in the top of the levels. This means that they're probably objects moved by later Neanderthals recycling old artefacts, and using lithics alone there's nothing to say the 60 hearths from the whole of Level J, for example, aren't from 60 different visits. However, there's little point in recycling old dry fragments of animal bone, and these show a distinct difference in movement compared to other artefacts. While lithics tended to be carried inwards towards the hearths near the back wall, the faunal fragments almost all go sideways or outwards to the area beyond the roofline, as shown by refitting.

---

* Individuals make choices, but for large-scale spatial patterns to emerge, it takes many people.

Neanderthals were obviously using different areas for particular tasks according to the material.

The more ancient Level Ob has been subject to full-spectrum refitting, and – translated from dry scientific description – the results are like a map allowing you to imagine dawn in a Neanderthal home 55,000 years ago, just after its occupants have left. Standing back against the rear wall, look down and your feet are surrounded by smashed-up teeth, jaws and possibly skulls. The stone hammers and anvils that processed deer, horse and aurochs still lie there. Straight in front of you, wisps of smoke rise from a large bonfire, in its ancient sense: fuelled by intensely burned butchery trash, obviously reused many times.

As the sun rises to your right, it picks out a roughly circular mass of thousands of lithic and bone fragments: splintered detritus from final butchery stages, plus more hammers. A greasy smell suggests that cooking happened here too. The ground is compressed: many feet have walked here, tired legs folded to sit, and where children poked the ground with sticks, there are small pieces of bone pressed down into underlying deposits. Look to the rockshelter's margins: the ground is comparatively clear, but things have still been happening, and at the western side there's the still-bloodied skeleton of a cat, dragged spitting from its tree under last night's moon.

Seeing such detail in how Neanderthals separated this space is remarkable, but it's only the beginning. The refitting of Level Ob also unveiled complex connections between different areas, though their meanings are tantalising. Fresh bone and lithic knapping refits moved from the rear of the rockshelter to areas immediately adjacent, and someone carried a freshly shattered aurochs tooth right across the site. Even being conservative, the cumulative patterns point to at least two areas around here being active at the same time.

Tracking by animal species is especially revealing. Aurochs and horse remains seem limited to the inner areas, although in slightly different zones, and micro-wear on their teeth suggests that they weren't being hunted at the same time of year. The aurochs look to have been taken over weeks or months covering an entire season, but in contrast the horses were killed over a very short period, perhaps just a week or so. Since Neanderthals were certainly staying through at least two phases in Level Ob, it's not entirely clear how or if these two hunting patterns intersected. But if hunters were bringing in horses in rapid succession – maybe all at once – this would explain why their bones aren't evenly jumbled with those of the aurochs.

It's highly likely that catching aurochs and horses required varying hunting strategies. Aurochs lived in smaller groups and didn't migrate, whereas horses may well have appeared seasonally in large herds. Could it be the same Neanderthal group doing all this hunting, and just visiting Abric Romaní at different times across the seasons? One hint that this could be the case comes from the common use of the head-smashing corner and bone-burning hearth at the back of the site. Even though the remains clearly built up over time, the fact that all species were processed here is striking; moreover, the shattered horse head parts aren't in *quite* the same spot as fragments of aurochs' skulls, which instead are found with head parts from red deer. It really looks like there were established places within the rockshelter where particular stages of butchery happened; and maybe even expectations of who did it.

Rather than horse or aurochs, the most commonly hunted animals at Abric Romaní were deer, and they have another story to tell. Their remains are found all over the place, sometimes even in small zones without other bones, and come from across all seasons. Despite Neanderthals having comprehensively pulverised nearly all faunal remains in search of marrow, researchers were still able to pinpoint some parts belonging to individual creatures. They realised that within a bone scatter in the outer, east side of the rockshelter, there were parts from a single stag. But unlike the wildcat that had similarly stood out, the deer was strangely lopsided: except some bits of antler and skull, there were only bones from the *right* half of the body.

What was going on? In general, hunted animals at Abric Romaní reflect the typical Neanderthal pattern of bringing back only the richest parts, mostly limbs and some heads. The half-a-deer is therefore very unusual in having other body parts present, never mind located close together. One possibility is that the hunters had a lucky kill close to home, and split the carcass immediately, either leaving one half behind or taking it elsewhere. Alternatively, the carcass was brought back whole, butchered and possibly cooked in the outer area, before the missing left side was taken to the bone-fragmenting 'factories' of the inner hearths where it was rendered archaeologically invisible. Why did they leave the right half of the deer behind? It may be that, like the wildcat, this was a final hunt just before the site was abandoned, and they simply didn't need the meat.

Whatever the real story was, the stag is exceptional proof that Neanderthals living here had a system for butchery, with successive stages happening not just in different places in the landscape, but also within sites. This complicated practice also implies sharing of tasks, and food. And remarkably, there are hints that splitting carcasses was happening for aurochs too. Researchers identified that within the richest inner zone, many of the bones come from the right halves of four aurochs. In contrast, in a different area every aurochs fragment that was identifiable by side came from the left. Deer and aurochs were more likely to be killed individually,* so portioning out between the group – maybe to related sub-units, like families – would make sense.

Fascinatingly though, the horse bodies show no such pattern and instead are widely spread across the whole rear zone. If herds were being hunted virtually simultaneously in seasonal terms (though not necessarily all the same year) then a predictable glut might mean there was more to go round, and entire carcasses were processed by different families.

These are speculations to be sure, but they're entirely plausible scenarios based on everything else Neanderthals did. And while the pattern isn't identical, other places confirm that segmenting of space for carcass processing isn't unique to Abric Romaní, or to large mammals. Meticulous mapping of bird remains at Fumane Cave uncovered that Neanderthals were butchering grouse and chough by tool and hand, probably cooking them, then leaving much of the body parts in a central dump. But something different happened to

---

* Aurochs are associated with woodland and believed to live in family groups rather than large herds.

the wings. Some were removed whole, others were taken apart, skinned and sliced up, probably for tendons and feathers, but *all* of the wing waste was kept separate from the rest of the bird detritus, and placed up against the east wall. This obvious division of tasks in the space of the cave suggests that different individuals were concurrently dealing with each butchery stage, creating their own trash piles. Furthermore, it must have been a practice repeated many times through the timespan of that layer.

Micro to Midden

Reconstructing the minutiae of Neanderthals moving things around and organising space – in essence, dwelling – is nothing short of extraordinary. But with the most advanced archaeological techniques, high-definition sites can go deeper and explore how Neanderthals' habits were worn into the ground itself. Over time, individuals going about their daily business compress floor sediments, compacting them into micro-layers just a few millimetres thick. Their analysis is known as micro-morphology, using resin-consolidated samples sliced super-thin, then lit up under the lens like geological stained glass. The tiny structures within make it possible to anatomise the content of Neanderthal hearths and floors, based on comparisons with later prehistoric sites and experimental projects.

By combining this technique with spatial data about hearths and activity zones, researchers were able to show that across the rockshelter at Abric Romaní, there was as much diversity in floors as seen in Neolithic houses. Trampled layers were common, but not universal across the whole site surface. Areas with richer spreads of artefacts had micro-morphological samples that pointed to more intensive use, and the converse was also true. This means that Neanderthals were using space in the same way over the long term. It was even possible to see how natural architecture such as large stalagmites and built features including alignments of limestone rocks seemed to define boundaries between 'cleaner' and messier areas.

Micro-morphology has also provided proof that, far from being slovenly, Neanderthals were regularly disposing of their rubbish.

At Abric Romaní some samples taken away from hearths nonetheless showed a *mélange* of tiny bone and lithic fragments burned at different temperatures. They were most likely scraped up from inside and around fires, then dumped some distance away. Other rubbish samples were quite distinctive: masses of mostly unburned, crushed bone and animal fats, plus coprolite (fossilised dung) fragments (species unclear). These matched deposits surrounding particular hearths, and likely reflect Neanderthals tidying up especially messy butchery waste and other waste. Most interesting, this cleaning was systematic: some dump areas were multi-layered, clearly having been used repeatedly.

Cave-proud Neanderthals were far from limited to Abric Romaní. Lakonis is a collapsed cave in southern Greece dating between 80 and 40 ka. Its cemented remnants, today perched above the glittering Mediterranean, must be a contender for most picturesque dig site. Micro-morphology here also found dump areas, and in this case it looks like Neanderthals were intentionally burning butchery and food debris. Meanwhile, external ash dumps at El Salt were identified because they contained boxwood, which had only been burned in a couple of hearths.

The most impressive evidence of Neanderthal housekeeping comes from Kebara. Along with a deep sequence of stacked hearths, it's famous for huge middens: trash heaps. Against the back wall was a pile of raked-out ash so large and thick it had to have built up over a long time period. Microscopic study of the apparently bare floor around hearths found masses of tiny bone splinters, confirming butchery had been going on, but the large waste had all been left in a big dump next to the ash heap.

Something else truly unique was going on in the central area at Kebara. Three circular, densely packed accumulations of animal remains were excavated, the richest and largest about 1m (1.1yd) across. It contained over 3,000 bones and thousands more small fragments, all embedded in a yellowish mass of minute bone slivers. Weird brown sediment halos around the bone circles appeared to have been stained by some sort of organic matter.

How did these things form? Over centuries, it seems Neanderthals had been placing butchery and cooking waste in exactly the same spots, yet the pieces in these circular features were smaller than those that went to the rear dump, and came from meatier bits of the skeleton.

The bizarre round features extended down at least 0.5m (1.5ft),[*] but it wasn't possible to tell using the excavation methods of the time whether they were pits or some sort of structures Neanderthals slowly added to over time, with the surrounding layer building up around them. The brown stains also remained a mystery, although decayed offal might be a possibility.

One last question: what did Neanderthals do with waste from their own bodies? The hominin dung in a hearth at El Salt has already been mentioned, and researchers took excavation here to a Lilliputian scale. Using a vacuum hose to filter sediment down to 1mm (0.03in.), the hearth micro-stratigraphy was teased out. The burned-black old occupation surfaces under the fire showed a triple-phased structure, with concentrations of tiny lithics and coprolite biomarkers more abundant at the top and bottom.

It looks like Neanderthals had first made a fire when arriving, then swept the floor and burned the waste, which included old faeces mixed with animal dung and plant material. This feels like a 'deep clean' on moving into a home, but there's also evidence from Abric Romaní that bodily waste was routinely incinerated along with grass and potentially moss: most likely old bedding.

All Manner of Furnishings

The word 'bedding' doesn't imply four-posters, but nobody likes lying on hard rock. Today there's growing evidence that as well as being particular about where they did things, Neanderthals were also concerned with furnishing their home spaces. At the Spanish site of El Esquilleu, hearth-side sediment included lots of whole phytoliths: microscopic mineral pieces of plants, especially grasses, which are made of silica and so resist decay. Some were still connected together and may have been remnants of some kind of thick leaf-based pads. Something similar was also found around and

---

[*] The excavation did not reach the bottom, so how far they went is unknown.

within fires at El Salt, backing up yet more evidence from Abric
Romaní.

Neanderthals also wanted warmth and comfort when sleeping
under the stars. About 20 years ago, construction north of Poitiers,
France, uncovered a miraculously well-preserved campsite at the site
of La Folie. Somewhere between 84 and 72 ka, Neanderthals stayed
by a river, which shortly after flooded and left behind several metres
of fine silts ensuring what lay beneath was protected. The archaeological
layer was just 10cm (4in.) thick, but extended over more than 10m
(11yd), and contained incredible detail. As well as hearths and lithic
scatters, dark stains about a hand's breadth thick turned out to be
decayed plant matter. Considering their thickness and location in a
zone devoid of artefacts, the simplest interpretation is that this was a
sleeping place.

La Folie contained something even more astonishing. Surrounding
all of the archaeological features was a roughly circular series of small,
slanted pits, each ringed by limestone blocks. They contained traces of
organic material and had compacted walls, and were the first clear
case of Neanderthal constructions. By piecing all the evidence
together, it seems that large wooden poles had been rammed into the
ground, then secured with stone blocks. It's even possible to see how
the stones collapsed inwards slightly when the poles were removed (or
rotted).

This was clearly a built living space, providing both shelter –
probably using hides lashed to the poles – and an enclosed 'home from
home'. The area is so large that it wasn't likely to be roofed, but there
was probably one main entrance marked by a gap in the circle, with a
hearth adjacent. Most interestingly, refitting shows that artefacts
moved between different zones inside the structure: even during a
relatively short stay, Neanderthals were dividing up space. Knapping
was happening outside and around the inner edges, while the central
area seems to have been for processing wood, vegetable matter and
skins. And just like in a cave, the bedding was directly opposite the
entrance, up against whatever barrier was used: the farthest point
from danger.

The La Folie discovery wasn't the first claim for Neanderthal-built
structures, but without modern excavation and analysis, there were

many sceptics of things like the stacked mammoth bones at La Cotte de St Brelade. Other recent finds are, however, tipping the trend towards Neanderthal 'furniture', of a sort. Around 70km (40mi.) south of Paris is a field called Les Bossats, near the village of Ormesson. In the 1930s, Upper Palaeolithic artefacts made by *H. sapiens* were tugged loose by ploughing, though not reported until 70 years later, which kicked off excavations. Beneath that layer, archaeologists found that Neanderthals had been there too, somewhere between 53 and 41.5 ka. A fine covering of sediment meant that knapping debris lay virtually where it had fallen, and in the richest area four sizeable sandstone blocks were found. They must have been hauled in from nearby deposits, and were most likely useful surfaces; in other words, camp tables or chairs.

Placed stone blocks or other large objects are known elsewhere. At Abric Romaní, hearths have limestone blocks arranged around them, some acting as anvils for bone processing. One in Level Ob is located in the rear skull-smashing corner and would have stuck up proud from the surrounding deposits, potentially explaining why this area was reused many times.

The travertine at Abric Romaní in other levels also preserved casts of massive pieces of wood. Fuel stocks is one explanation, but given the small branches in hearths, it may well be that they're remains of structures. In Level N there's the entire tree trunk, and in Level Oa a long, thick pole had been carefully trimmed of branches and bark. Both objects are positioned with one end towards the back wall, making feasible supports over which a shelter could have been formed; and revealed in the side of a hearth, there's even a potential posthole: very regular and rectangular.

Aside from bedding, micro-morphology also indicates that Neanderthals didn't always squat above cold stone floors. Micro-layers 0.5 to 2mm (0.02 to 0.08in.) thick at Abric Romaní exactly match mat-covered floors in later prehistoric houses. Moreover, the state of the sediment just beneath points to non-porous material, rather than woven matting: hides are most likely, and there's even minuscule remnants of probable burned skin in some samples.

The mat samples come from next to hearths, including cooking areas, based on the presence of burned fats and bone splinters.

Unexpectedly, some were located on top of butchery areas and even over dump sediments.

Repeated successions of covered floors and disturbed sediment – full of plant remains, charcoal and bone with different degrees of burning – suggest that the ground was cleaned in-between phases when mats were used, perhaps at the start of occupations. And once more there's evidence of long-term habits, since some of the same zones had covered floors in both Levels Ja and Jb, representing at least many decades. While plant bedding would probably have been collected fresh for each stay, hide mats must have been transported and reveal Neanderthals carrying homely elements from place to place.

Claims that Neanderthal use of space was thoughtless or random – equal to that of hyaenas – are now truly obsolete. On the contrary, they were among the first hominins to create complex, intentional divisions of space, with a surprisingly familiar layout. Hearths are the stable cores around which both Neanderthals and archaeologists gravitate. They ignite our collective imaginations, summoning and illuminating the shadowy wraiths around them. Fires are time-travelling artefacts: alight, they stretched through days or even weeks. Cold and buried, they're memorials for vanished bodies moving around them. The order in which they were used marks out when walls glowed or were in shadow, and sometimes, time travellers emerge: branches left smouldering when the site was abandoned, one unburned end sticking out as if the inhabitants had only just left.

Hearths have long been like a single shared word between different tongues: easy to notice and understand, but set into the otherwise confusing noise of thousands of objects. But today, deciphering the wider material record of Neanderthal homes has become possible too. We can perceive long traditions of task areas, fixtures and even soft furnishings. All this returns us to the question of who was at any given place, and how long for. If Neanderthals organised their material world through fragmentation and accumulation, from individual knapping sequences to the contents of entire sites, then it's plausible that their social groups were also dividing and coming together. Yet how this worked is tangled up with underlying patterns of subsistence, technology and mobility. To truly grasp the system of the Neanderthal world, we must now look at a landscape scale.

CHAPTER TEN

Into the Land

Murmuring rouses him. The sun has sunk, leaving only flint-dark shreds of cloud. Now twilight is all around, fire-glow quickly fading into the steppe. Blinking, and stretching on the eldermother's lap, he sits up. Hers and the others' faces are turned to the western horizon, lit by a still-luminous quartz sky above the shelter. This year, by the river where the great churned muds usually were, the deer had not crossed. So they'd waited for many days, until some hunters left to search farther upstream. Stomachs are long past groans, gone into empty holes.

Then he hears it.

'oooOOO!'

The hunt returning, singing of the meat and the FAT. Remembering its melting taste, and the crisp of charred antler skin, saliva springs almost painfully in his mouth. Scrambling up he's surrounded by moving legs, carried on the swell of anticipation as the people make ready. The eldermother calls out, older children run brave into the dark to meet the hunters. He lingers close to the light of the

hearth – fanged ones always follow kills – but his feet dance as the un-made deer
arrive on many shoulders, haloed by puffed breath in the frigid air. No matter the
cold, tonight all will sleep warmed by blood.

For all the spectacular, intimate details we have about individual
sites, Neanderthals were fundamentally nomadic. Their world
was the land, and moving in it was life. Like everything else they
did, this was far from random. Sites were not simply destinations,
but *intersections*, nodes within networks stretching hundreds of
kilometres. The blood-stained, fur-tufted muds at a kill site – an
ephemeral locale – were nevertheless linked to caves or rockshelters
through the animal bodies taken there to be further divided. All
places Neanderthals went were connected by their physical
movements, and the things they carried. Each new-kindled fire was
a glowing pearl in an unbroken string laced across ridges and
threaded through forests.

To begin understanding how interconnectedness was at the heart of
things, we must rise upwards like smoke from the hearths. Central
points to their home spaces, they were nonetheless linked to the wider
world through their constituent materials. Comparing fuels to
reconstructed ecologies, it's possible to work out where firewood was
foraged from at a local landscape scale. For the El Salt fires that burned
around 55 ka, all wood species were coming from within two to three
hours' walk. In this way, tiny smears of charcoal transport us outwards,
to Neanderthals striding through pine scrub and climbing ridges out
of sight of home. Mapping the great entanglements farther out
between other things and places is extraordinarily complicated, but
doing so opens up truly enlightening vistas into their lives.

How We Move

Most living descendants of Neanderthals have forgotten what it is to
truly move, shifting from place to place seasonally, never mind
doing so on foot. Water and stone were to some extent reliable and
static resources, but plants and animals were more mutable and
survival depended on their availability. Among recent

hunter-gatherers, settling in one place for a long time – or only moving across small areas – is rare, because outside the tropics, environments are mostly not rich enough. In higher latitudes, low mobility only becomes possible in particular circumstances with predictable, high-quality foods that are either abundant year-round or can be stored.*

The archaeological record shows that wherever they lived, Neanderthals were focused on hunting large beasts, even if they also took small game, seafood and plants when possible. This means that whether they stalked cool steppe-tundra or warm forests, moving multiple times each year was still necessary. But the diversity of the environments means we should expect the frequency and distance to vary. Based on what we see in recent hunter-gatherers, open, colder settings require people to be highly mobile and move methodically over huge ranges. And even if deciduous forests generally don't usually involve very long-distance movement, it's not easy to stay in any place for long, as finding large animals is harder, and other resources quickly become used up.

But it's not just about food. Mobility is a never-ending waltz danced between subsistence and technology. Finding stone, and choices in how it was knapped, imposed their own demands that motivated movement. Actually mapping out how this happened is fiendishly complicated, however. Even with the resolution in dating available today, it's almost impossible to be sure that any two sites within a given region were in use at the same time by one Neanderthal group. There's no equivalent to the micro-scale methods for determining if hearths within a single layer really were contemporary with each other. Instead, archaeologists must shift perspective, and consider different kinds of questions, based on how repeated choices by individuals and groups coalesced into long-term patterns shared across many sites.

But before delving into the detail, it's important to understand why measuring Neanderthal mobility matters. Just as much as technology,

---

* Salmon runs of the Pacific north-west coast are a well-known case, allowing numerous Indigenous cultures, particularly Coast Salish peoples, to live semi-permanently in large lodges and villages.

it gives a view into how their minds worked, and has fuelled debates about their cognitive capacity and sophistication. If Neanderthals planned activities ahead of time and had a schedule of sorts for where to move, this implies that they could imagine the future, and had enough brainpower to maintain goals over days, weeks or even months.

Therefore the complexity of mobility systems is important, but the *extent* of travel is another key factor. If groups were moving farther, not only was the planning element more impressive, but it also points to larger range sizes. As we'll explore below, territory size has implications for how connected Neanderthal societies were.

It's been obvious for decades that Neanderthals routinely separated activities between different points in the landscape. They didn't set up a new camp each time they reached a rock source or opened up a steaming carcass. Initial knapping locales – where stone was found, tested and prepared – and animal kill sites are identifiable *precisely* because they lack the later-stage lithic products and the richest parts of hunted animals.

Looking first at how butchery was 'fragmented' through the landscape, this pattern is obvious at places like Schöningen, and continues over hundreds of thousands of years. In south-east France, a rocky bluff at Quincieux around 55 ka preserves a record of systematic butchery. For big game, the least fatty or meaty parts are all that remains: horse hips and still-articulated backbones, heavy woolly rhino jaws, mammoth teeth. For smaller species, heads and whole joints are also missing.

Where did the good stuff go? In many places we can make out an intermediate type of site, essentially hunting camps, where Neanderthals further processed either partial carcasses or selected joints. Some hunting camps like Les Pradelles were used over and over, hinting that they were connected to particular kill sites Neanderthals returned to many times, like Schöningen and Quincieux.

The final destination for food – either directly from kill sites or coming via hunting camps – was to what we can think of as 'central locations', or more prosaically, homes. This includes big sites like Abric Romaní, where masses of third-stage processing happened involving a lot of bone smashing and some cooking, plus presumably eating and sleeping. The richness of the archaeology, combined with concurrent use of different parts of the site following quite specific patterns of activity, is very good evidence that Neanderthals were spending more than a day at a time there.

Did Neanderthals have concepts of place that would match our category of 'kill site' versus 'central place'? Clearly they lived according to traditional ways of doing things, the repetition of which created the spatial patterning we can see both within sites and between locales across the landscape. But beyond this, did they really schedule their lives beyond the next meal? Hunter-gatherers the world over are extremely attuned to seasonal changes in resources such as the arrival of herds, and plan to arrive at particular places at the right time. Did Neanderthals have winter and summer camps, or was their existence more itinerant, living as hobo hominins?

Working this out involves thinking about how things interconnected. As the last chapter showed, even where we have high-definition archaeology, there's no evidence anywhere for truly long-term settlement, on the order of many months. Some places like Abric Romaní were used more often, certainly for several days or perhaps longer, and probably by bigger groups. Even large open-air sites like La Folie with clear activity zones obviously weren't lived in for months. And at the other end of the spectrum are locales such as Abric del Pastor, visited very briefly and by no more than a handful of individuals at a time.

It's these short-stay sites that contain hints of seasonal patterns. Another example is Level 3 at Teixoneres Cave, north-east Spain. Between around 51 and 40 ka there are multiple phases when Neanderthals briefly stayed, interspersed with evidence of carnivores moving in. Most interesting, the hunted species show different seasonal patterns. The time of year deer were killed changes through different sub-levels, but horses were always dying in late spring through to early summer.

This pattern is strikingly similar to that at Abric Romaní, which is around 150km (90mi.) south-west. It seems that in this region of north-east Iberia, Neanderthals moved regularly and hunted deer that were available all year round. But horses were only killed over very short periods; the time of year isn't clear at Abric Romaní, but at Teixoneres it was possible to fine-tune down to season thanks to a bird assemblage accumulated by predators. Chough and magpie remains showed a distinctive bone condition that develops *before* they lay eggs, which cannot have happened prior to mid-spring. Moreover, their carcasses would only have been brought in by carnivores if there weren't any hominins about yet. Put together, the horse hunting was very likely late spring/summer, which is precisely when horses come together for breeding and can be distracted and vulnerable.

What's also interesting about Teixoneres is that it has far fewer lithics and hearths than Abric Romaní but more faunal remains, including a higher proportion of horse. Though they mostly arrived already jointed, the animal remains in general look slightly less intensively smashed up than at Abric Romaní, and in another contrast between the sites, a bit of rabbit hunting was also going on at Teixoneres.

Quite clearly, Teixoneres was a different kind of place, with probably smaller groups staying for briefer periods. It might even have sometimes been more like a hunting camp than a place where meat, marrow and fat were ending up. Yet neither is it like the far more ephemerally visited Abric del Pastor and El Salt, which truly look like places very small parties of Neanderthals stopped at for a couple of nights at most.

Level IV at Abric del Pastor is striking, since Neanderthals were apparently mostly interested in hunting and butchering tortoises with a smattering of ibex that was probably also taken locally. But alongside those species were rare deer, horse and aurochs bones, mainly bits of leg and the odd head. It's possible that, just as Neanderthals carried lithics around as a kind of travel toolkit, they also took other resources: those random bones could be remnants of provisions from elsewhere. In a larger assemblage this kind of subtle signal would become invisible. And the lithics match this. All the RMUs at Abric del Pastor tend to look very fragmented: Neanderthals were dumping old tools and taking just a few flakes off cores that they then carried onwards, presumably along with some of the tortoise meat.

These Iberian sites stand for others across the whole Neanderthal world. Some were short stopping places, others homes for longer periods, but all were points on a cycle of movement. For Neanderthals living by a never-ending odyssey, 'Ithaca' was the journey, not the destination.

For sites where the length of occupation phases is measurable, it's something up to a few centuries. Successive generations grew up in these caves and rockshelters following the same routines and traditions, which became physically part of the site itself through hearths, waste dumps and well-trodden floors. But then things changed and nobody visited, sometimes for 1,000 years or more. Either groups relocated to entirely different areas, or the population itself petered out. It's noticeable that the least 'homely' site, Abric del Pastor, appears to have lain empty for the longest spans of time, which might mean that populations moving into the region were unfamiliar with all its nooks and crannies.

Shifting Stone

Individual sites, even comparing across regions, only give us a keyhole view into what Neanderthals did across the landscape. To understand their true scales of mobility, archaeologists need to map *individual* movements. The most obvious way to do this is by tracking their most abundant resource: stone. Finding out where the lithics in any assemblage originally came from illuminates, at least partly, their actual journeys between sources and sites.

Things are never simple, however, and as prehistorians have learned over the decades, geology is tricky. What's casually referred to as 'flint' includes silica-based rocks formed by varied processes, during many ancient aeons. It takes extraordinary effort and time to map thousands of flint sources, and examine and classify them based on structure, microfossils and chemical profiles. On top of that are different taphonomic permutations causing rock to change on its journey before a Neanderthal even picked it up. Stone from the same 'primary' outcrop looks quite different once it's tumbled down from a ridge, rolled along a river, and eroded out of gravels; all 'secondary' sources.

By creating gigantic stone 'libraries' mapping flint and other kinds of rock sources, it's possible to directly compare them with lithic artefacts. This reveals exactly where Neanderthals had been prior to bringing those objects into a particular site. The results show that, just as with animal bodies, they followed broad rules in transporting stone based on assessing quality and distance. There's usually a lot of the closest available stone types, from within about 5 to 10km (3 to 6mi.), even if the quality wasn't great. This would be collected while doing other things like hunting, within a couple of hours' wandering of the site. And while they'd use it, bad rock was never carried into areas with good stuff.

Stone from farther away is almost always present too, and this is what excites archaeologists, because it directly connects individual places to points in much larger landscapes. The more distant the rock type, the fewer artefacts made from it there will be; typically less than 10 per cent will come from far-flung sources over 60km (40mi.) away. The very longest distances – more than 300km (190mi.) – are found for the purest silica rock of all, obsidian, but even decent flint was sometimes carried over 100km (60mi.). Interpreting what this means in terms of movement is, however, fraught with difficulty.

Considering each assemblage in isolation, you might assume that Neanderthals were making special trips to get decent stone. But since other evidence shows they mostly stayed for relatively short periods at any site, that wouldn't make sense in energetic terms.

Instead, what's much more probable is that far-travelled objects were simply the 'survivors' from a selection of tools Neanderthals travelled with during their perambulations between stone source and other sites. Backing this up is the fact that high-quality stone is hardly ever transported directly in 'raw' state to knap. Instead, as we saw in Chapter 6, in virtually all cases artefacts that came from furthest away are products like Levallois flakes, bifaces and tools often obviously resharpened. Similarly, rare cores from distant sources had already seen a lot of knapping before being abandoned.

Remarkably, in high-definition sites like Abric del Pastor it's possible to actually pick out travel gear. In single-hearth phases with tiny lithic assemblages, the few artefacts made from distant stone must be things one of the handful of Neanderthals who'd stayed there decided to leave behind.

While it's impossible to trace the journey of an individual stone block, it's clear that Neanderthals weren't moving 'as the crow flies'

Above: Reconstruction of La Folie, France. The site features a circle of probable postholes and traces of activity within, showing that its Neanderthal inhabitants used the interior space in different ways.

Below: One of the Schöningen spears, including remains of the horses early Neanderthals hunted on this 330,000-year-old lake shore.

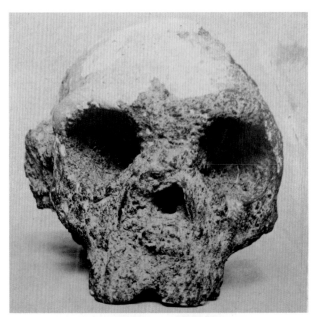

Left: The Forbes woman's skull as it was when it came to George Busk in 1864, still covered with concretions.

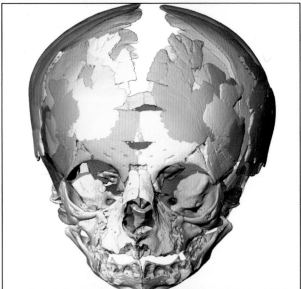

Left: Reconstruction of a Neanderthal newborn baby's skull, based on the Le Moustier 2 and Mezmaiskaya 1 infants.

Left: Boy's tooth from El Sidrón, Spain, showing unusually small scratches from eating using 'flint cutlery'.

Left: The Discoid core from Fumane, as it was found.

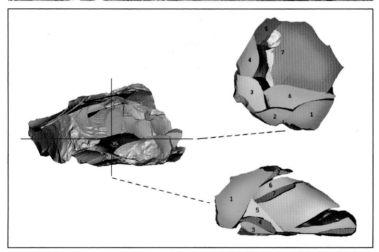

Left: The Discoid core reconstructed in 3D, showing missing flakes taken elsewhere.

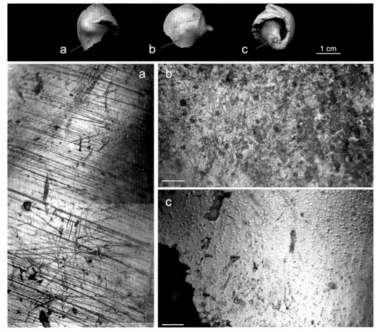

Left: The fossil shell from Fumane, showing traces of abrasion (a), red pigment (b) and polish (c), which point to handling and potentially being strung.

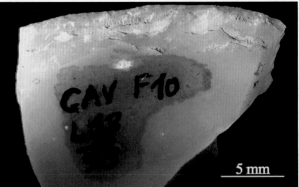

Above: Lissoirs: probably leatherworking tools, made using rib bones selected from large herbivores.

Left: Retouched tool made on shell from Cavallo Cave; it is similar to others from Italy and Greece.

Left: Remarkable travertine cast from Abric Romaní, revealing the shape of a wooden tool.

1 cm

5 mm

a

b

c

d

e

CAV F10

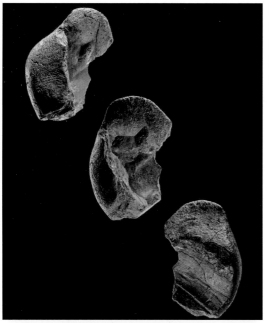

Left: Birch-tar piece from Königsaue, Germany. The imprints of wood (bottom) and stone (middle) implements, and a Neaderthal fingerprint (top), are visible.

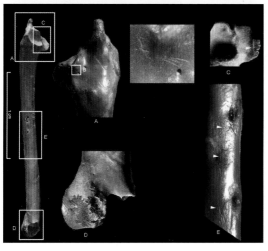

Left: Blackbird's wing bone from Cova Negra, Spain, with miniscule cut marks; was it butchered by a child's fingers?

Below: Neanderthal foot and handprints from Le Rozel, France, possibly from teenagers and children playing on sand dunes 80,000 years ago.

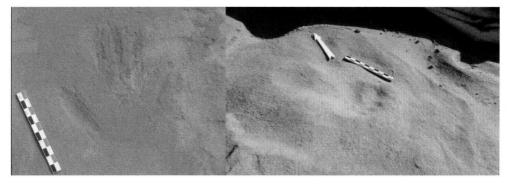

Above: Denny's bone: genetics revealed she was a girl with a Neanderthal mother and Denisovan father.

24.75mm

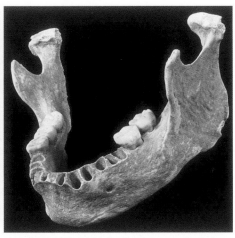

Above: Oase jaw, Romania: a 40,000 year old human with a Neanderthal ancestor just 4–6 generations back.

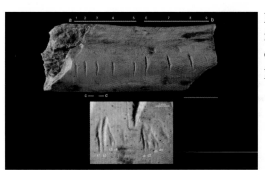

Left: The engraved hyaena bone from Les Pradelles, France, showing equidistant marks that hint at a possible tally system.

Left and Below: Shanidar Z, Iraqi Kurdistan. Found in 2018, this is the first articulated Neanderthal skeleton excavated for decades.

Above: The enigmatic constructions made from stalagmites at Bruniquel, France, around 174,000 years ago.

Above: Samuel Jules Celestine Edwards was a late 19th-century medical student, editor, evangelist and public lecturer.

Left: The skull-bone retoucher from La Quina, France; it was used as a tool despite its unsuitability.

Below: The Amud baby, found lying close to the cave wall with an unusually intact deer jaw on its pelvis.

Below: Frederick Blaschke's 1929 reconstruction for the Field Museum, US; a rather bleak vision of Neanderthal life.

Above left and below: Neanderthal reconstructions by Tom Björklund. In the 21st century Neanderthals have become ancient humans expressing affection, even day-dreaming.

between distant stone source and site. Individual flakes of high-quality, distant rock are sometimes the only things left behind, having been removed off bifaces or tools that were then carried on further. In some cases even cores were carried a considerable way – over 40km (25mi.) – before being knapped, then removed again. We don't know how many places such objects visited before 'dropping out' of circulation, but the discovery of second-generation products made from rocks sourced 100km (60mi.) away suggests that three or more locales might not have been unusual.

It's been suggested that because Neanderthals didn't *routinely* transport high-quality stone in bulk, they weren't well organised. Aside from the fact that they did occasionally do this, what's clear is that it mostly just wasn't necessary. In later periods where we see *H. sapiens* doing this, it's largely because they were so focused on blades, a technology that can be far less adaptable to average or poor stone.[*]

Overall, what we see in Neanderthals matches how recent hunter-gatherers organised their use of stone in the landscape. Personal travel gear is selected based on multiple factors: the expected activities, the amount of travelling and, crucially, what kind of stone would be available en route. This last point is key, because it is yet more evidence of Neanderthals possessing detailed knowledge of geological resources, and thinking ahead. They knew the places with bad rock that were worth bringing good cores to, and conversely, those with nearby, decent stone with which to restock.

And as always, Neanderthals weren't robots. They adapted their approach depending on the geological situation: if necessary, middling-quality stone might be carried some distance, and occasionally even raw blocks were transported quite far directly to sites. Sometimes their previous activity at sites could change things: as Level J in Abric Romaní built up, Neanderthals began reducing the amount of stone blocks they brought in because they were recycling artefacts from older occupations.

*

---

[*] It's even been suggested that moving a lot of raw stone over long distances only became feasible once dogs were domesticated, perhaps by 30 ka.

When we zoom out further and look at techno-complexes, it's clear that the decisions Neanderthals were making about moving stone around varied. In assemblages using technological systems like Levallois or Quina, which were focused on producing flakes that could be carried for some time and resharpened, the amount of artefacts made from distant stone is higher. And in Discoid assemblages, which in technological terms appear to be much more immediate and even disposable, there's very rarely anything that's been brought in from farther than 30km (20mi.) or so. Geology, technology and mobility were all intertwined.

Scrutinising the shifting of stone by Neanderthals has other implications beyond their capacity for planning or managing time and resources. Since the distances their artefacts moved were, for many decades, the only direct measure of mobility, they ended up being used as a proxy for range size. Because at any site almost everything came from within 60km (40mi.), prehistorians proposed that Neanderthals mostly moved around in quite small areas; essentially the size of the English county of Shropshire.

Range size isn't just about land, but about people. If Neanderthals lived in small areas – maybe just across a couple of valleys – then they'd rarely meet other groups. Moreover, without large territories and extended social relations, it was theorised that Neanderthals wouldn't need material expressions of shared cultural values, which can help maintain networks.

The comparison, as so often, was with early *H. sapiens* in the Upper Palaeolithic. While their lithic assemblages also have largely very local stone sources, artefacts that came from beyond 60km (40mi.) are more numerous, and the distances are greater. Predictably, this was presented as reflecting larger territories and stronger social networks. But digging into the data, as well as our assumptions about what stone sourcing means, produces different answers.

First, for people used to walking everywhere, 60km (40mi.) is certainly feasible for a return day trip (for sources located 30km (20mi.) away). That Neanderthals mostly lived within the walkable distance between sunrise and sunset seems very unlikely when ethnographic data is examined. It's actually normal for small numbers of people from hunter-gatherer groups to go off for short trips lasting more than a day. And as we've seen, in Neanderthal sites the type and state of artefacts from those distances doesn't fit the pattern we'd expect if rock was brought straight from the source.

The real proof that Neanderthals cannot have been limited to geographically small ranges comes from the much farther-travelled artefacts sourced beyond 60km (40mi.), and sometimes well over 100km (60mi.). Generally sparse in each assemblage, prehistorians tended to ignore them in large part because working out what they mean is difficult. Yet their rarity doesn't mean they're anomalies, and in fact they were already there in the early Middle Palaeolithic. Fundamentally, they're the best data we have about the true extent of the landscapes Neanderthals moved over.

But what do these distances mean in terms of mobility? Today's best ultra-marathoners can cover 1,000km (620mi.) in a week, and even 'normal' athletes can manage up to 200km (120mi.) in 24 hours.* While Neanderthal bones bear witness to extremely physical lives, and they were probably more efficient striders than us, long-distance running was not a forte. Factoring that in together with shorter legs making them up to 10 per cent slower and the impact of realistic terrain, a day's trek drops well below 100km (60mi.).

Add in the varying travel pace for entire groups including youngsters and those with added burdens, whether hauling stuff or simply from age, and it's obvious that artefacts from beyond 80 to 100km (50 to 60mi.) cannot have been directly sourced. Furthermore, truly immense stone transport distances exist. For example at Mezmaiskaya, not only are there lithics from multiple points across a region extending out around 100km (60mi.), but obsidian was also arriving from 200 to 250km (120 to 150mi.) to the south-east, and flint from some *300km (190mi.)* north-west. That's at the extreme end of the spectrum, but transfers over 100km (60mi.) are known throughout the Neanderthal world.

How can we understand what such vast spans of landscape – more than halfway between London and Le Moustier in the Périgord – mean for how Neanderthals moved around? It mustn't be assumed

---

* Eight minutes is the average time to cover 1km (0.6mi.) at a speed alternating between a slow jog and a brisk walk.

every object in an assemblage was contemporary, but even so, they prove sites were nodes within larger networks taking many days to traverse. And it's unlikely any particular excavated locale will have been at one end of a territory. Therefore, a 300km (190mi.) movement is probably only part of the landscape familiar to the Neanderthal who carried that artefact.

If we also consider that many far-travelled artefacts have clearly been used and resharpened before ending up at a particular site, even starting with a large initial flake, it wouldn't be that long until they were used up. In order to be carried over a number of days from source to the eventual place they were left, this implies two possibilities. Perhaps Neanderthals were hauling spare stone around to replenish their stocks? If so, we should see more telltale waste at medium distances up to about 50km (30mi.). There are resharpening flakes from moved-on bifaces and tools but they're often not from so far, and cores let alone raw blocks being moved over these scales are extremely rare.

The other alternative is that artefacts sourced between 100 and 300km (60 and 190mi.) are material manifestations of Neanderthals covering huge stretches of land while barely stopping and using the tools. Such rapid, extensive schlepping doesn't fit a notion of aimless wandering, but makes sense if Neanderthals were travelling in a targeted way, to known locations. It's possible that, heavily laden, entire groups could have moved this far over the space of a week or so. Particular conditions might push such decisions: a major migration motivator for reindeer is to escape the torturing summer mosquito swarms. But across the breadth of environments Neanderthals lived in, we should expect them to have moved around in diverse ways.

The extremely minimal occupation traces at Abric del Pastor are proof that on occasion, very small numbers of individuals were travelling together, while places like Abric Romaní equally prove that 'home' sites existed where food was arriving having been hunted and sometimes partially prepared elsewhere.

Adding all this up, it may well be that artefacts from extremely distant sources were left by just one or two Neanderthals who'd travelled hard and fast, at a pace beyond less mobile group members, whether heavily pregnant women, the infirm or toddlers. Splitting the group up in this way might be worth it to take advantage of seasonal bounty, and there's an intriguing comparison in some North American Paleoindian contexts.* Very long-distance lithic movements – some exceeding those seen with Neanderthals – are believed to be tools sourced near far-off bison kill sites, brought back along with precious meat and fat by hunting parties.

That we see something similar for Neanderthals leads to the conclusion they too had impressive knowledge of resources across an enormous territory, and probably mentally scheduled when and where to be.

Social Stones

Another possibility exists that would explain some of the lithics from distant stone sources, but it's almost never seriously discussed: exchange. Giving and receiving objects or resources like food is a crucial way humans of all stripes maintain relationships. For hunter-gatherers living in small populations who don't meet often, it's especially important. Despite indications (as we'll see later) that some Neanderthals were genetically isolated, this isn't always the case.

As hominins highly attuned to everything in their familiar home regions, the presence of other groups would surely have been noticed, and encounters taken place. Prehistorians have long framed these as probably antagonistic, but there's no strong reason to believe that. Neanderthals lived in social contexts where food was cooperatively obtained and shared, so the idea of giving and receiving things with close relations who may have moved into other groups – or even strangers – might not have been alien.

All these questions about social networks and mobility come down to how Neanderthal societies were structured. Total population estimates tend to be in tens of thousands or even less. At any point in time there may have been fewer Neanderthals walking about than commuters passing each day through Clapham Junction, London's

---

* This is the term for the oldest North American archaeological traces dating to 18 ka and later but it covers a great many cultures.

busiest train station. Can we say anything about how groups were organised, beyond the fact they sometimes seem to have separated into smaller units?

Studies on recent hunter-gatherers find an average of about 25 people who largely live and travel with each other. Known as a 'band', they're fluid entities: some may stick together, others routinely split up for particular activities. This might mean hunting parties, or even just a couple of adults and associated children heading off to live independently in summer. Temporary rearrangements can happen for many reasons, including imminent births or simply a desire to visit kin.

Actually seeing this kind of thing in the archaeological record for Neanderthals is challenging, but not impossible. Somewhere between 4 to 10 humans can fit around a hearth, which are usually spaced between 1.5 and 2m (1.6 and 2.2yd) apart. So synchronous activity areas, multiple fires and especially rubbish dumps in places like Abric Romaní or La Folie point to between 10 and 20 individuals. In other words, the size of a typical band.

One meticulously excavated site may be the closest we'll ever get to a group photo. Le Rozel, on today's north-west French coast, contains many sandy layers backed up against a cliff and dunes. Astonishingly, a series of levels from around 80 ka preserve hundreds of footprints. In the richest phase, careful size comparisons show that at least 4 and probably over 10 individuals were here. Most fascinating, they're largely adolescents and children as young as 2 years old. With so few adults it's hard to imagine this was a full group, and instead it looks like a pack of youngsters foraging at the beach.

In hunter-gatherer societies, beyond bands there are larger networks that connect communities; often related by blood but also other kinds of kinship, they're termed clans. Bands maintain ties with others within their clan both through random meetings, and via gatherings that, though not necessarily formal, take place around predictable circumstances. Is it possible Neanderthals groups were linked to each other across watersheds or beyond mountains by something like a clan structure? So far no sites have been found where large numbers of contemporary hearths and activity areas would indicate a mass gathering.

But that's not to say situations didn't exist that attracted a lot of Neanderthals to the same place, at the same time. Some seasonal events temporarily bring together even solitary predators: think of grizzly bears lining rivers for the Pacific salmon run. A number of sites point to Neanderthals being present for probably seasonal hunting, with slaughter in profusion: for example at Mauran it's bison, while at Salzgitter-Lebenstedt it's reindeer. With prey gluts competition was reduced, meaning individuals were freer to socialise and perhaps move groups. Equally, encounters with unfamiliar groups would have been less stressful. If exchanges of lithics ever happened for social reasons, this would be a likely setting, with the objects in question – perhaps made of unfamiliar stone – then carried onwards from the slaughtering grounds along with fat, meat and marrow.

That's all speculation of course, but it remains true that despite being very sparsely scattered across the land,* not all Neanderthals were genetically inbred, and so the question is how they maintained DNA diversity. The patterns of lithic movements prove that tough terrain wasn't an obstacle. At least some individuals or whole groups crossed even mighty rivers like the Rhône, and high mountain passes in the Massif Central and Pyrenees. Perhaps some populations that became genetically isolated, like those in Iberia and the Altai, lived in rich environments without large, migratory game herds as an incentive to travel outside their normal range. Fundamentally, there's no solid reason to assume all Neanderthals disliked strangers, especially since, as we'll see in Chapter 14, they were open to intimate relations with other sorts of hominins. If they did happen, rendezvous with strangers on the steppe may have been more like unexpected holiday romances than annual summer camps. But what's interesting too is that over time, very long-distance transfers become more common. Something was changing in the way Neanderthals lived in the landscape after 150 ka, but working out what caused it is one of the hardest remaining problems.

---

* Population density was probably lower than one individual per 1km² (0.4mi.²), based on comparable densities of hunter-gatherers from similar environments.

Rock to Bone

Stone movements had long been the only measure of Neanderthal mobility, but twenty-first-century analytical techniques opened up another route. A growing range of bio-geochemical methods can track the movements of individual Neanderthals using stable isotopes. Mobility isotopes provide information on where an individual had previously lived, although, as with dietary analysis, it only records parts of the life history. Strontium isotopes vary according to bedrock, and get into teeth through food and drinking water.* The data looks more like weather pressure maps, with bands across different regions, than pin-prick locations. But by comparing the values in teeth to the geology where they were found, it's possible to see movement from different geological areas, sometimes with dramatic results, such as the Bronze Age person buried near Stonehenge who'd grown up in the Alps.

The few Neanderthal studies so far haven't thrown up anything like that, but they have demonstrated that entire lives weren't spent in single valleys. The first sample was an adult from Lakonis, Greece, who lived for part of their childhood somewhere up to 20km (12mi.) away. A more distant measure came from an Eemian tooth at the site of Moula-Guercy, France, pointing to a move from geology at least 50km (30mi.) to the south, with matching lithics too.

But there are complications, and these measures may be underestimates. Because tooth enamel can take a year or more to form and everything we know from archaeology says that Neanderthals weren't staying in one place that long, there's potentially some isotopic 'blurring' when crossing between different geological areas. Additionally, if Neanderthals were eating a lot of animals with isotopic signatures from distant regions, this might also confuse things.

To balance out some of these issues, it's possible to combine isotopic data with lithics. Between about 250 and 200 ka, Neanderthals living along the Rhône River returned again and again to the Payre rockshelter. But over time, its place within their wider movements through the landscape shifted. Earlier on, it seems

---

* Or indirectly through mother's milk.

they stayed for longer, hunting prey mostly in the valleys at different times of year. The lithics were largely arriving from within about 30km (20mi.), up on the plateaux and hills. Using oxygen and lead isotopes at the micro-scale, seasonal early childhood movements of a Neanderthal who eventually died at Payre (though was perhaps born elsewhere) can be tracked.

After entering the world in spring, the tooth shows exposure to lead from around 2.5 months old. It's possible that this points to the baby's group having moved for the summer to a region where the natural geology had higher amounts of this element. A subsequent but strong lead peak implies they moved again at least once more in the winter before the baby's first birthday. Nearly a year later, just as the baby hit toddlerhood, another lead phase lasting a bit over a fortnight is followed by medium-to-high levels for seven months. If all these lead fluctuations really do mark movement, then we might be looking at a roughly annual range covering at least two locales. At first the baby must have been carried, and then perhaps toddled after its parents along already familiar paths.

Over time the Payre rockshelter filled up with sediments, and Neanderthals stayed less, mostly in autumn. The lithic assemblages become sparser, more obviously resharpened, and detailed analysis based on over 200 sources show that far-travelled artefacts become both more numerous and diverse. Some of the new sources are more remote – up to 60km (40mi.) away – and reveal that Neanderthals were now more often bringing rock from the river valley gravels, including crossing the Rhône River. However, the diet isotopes suggest that the landscape focus of hunting was reversed, with more prey species pointing to hills and plateaux like red deer and thar (although a preserved fish scale shows some food came from the valleys). Strontium isotopes from some Neanderthal teeth also hint at periods spent up on the plateaux, but another oxygen and lead study on a child's tooth gives more detail.

Forming when its owner was around 3 years old, once more there are distinct lead peaks, but they don't start until around the age of 5. What's more, the season has changed: the first peak comes during early spring, and there's another nearly 18 months later, probably in autumn. The group this youngster lived with seem to have been moving on a different schedule, possibly shifting more often. Moreover, the most distant stone objects aren't heavily resharpened tools but flakes, hinting that the length of movements may well have altered too.

Another kind of isotope – sulphur – allows dietary comparisons between carnivores and prey, and may point to very long-distance movements by some Neanderthals. And most interesting, analysis of fossils from two Belgian sites dating to broadly 40 ka – Spy and Goyet – nevertheless gave quite different results. The Spy Neanderthals were mostly eating species from nearby, primarily mammoth. But at Goyet the sulphur showed Neanderthals closely matching samples from a cave lion and a wolf (or possibly fox). Unlike most other predators there that were hunting local species like horse, woolly rhino and bison, this particular lion and wolf were eating something else. Cave bear is one possibility for the lion, since this was sometimes their favoured prey. But the wolf doesn't match that either, and along with the Neanderthal, looks as if it might have been eating reindeer.

Two scenarios might explain this. The reindeer found at Goyet may have lived farther away – as much as 100km (60mi.) – but were hunted locally during seasonal migrations. Or alternatively, it was the predators, including Neanderthals, who were travelling, and spending enough time far away for their sulphur signal to shift before arriving at Goyet. While distances around 100km (60mi.) would nicely link to remote lithic transfers, unfortunately because this site was an old excavation, the data for artefact sourcing isn't available.

Where We Live

Even if the isotopes haven't yet given abundant evidence for Neanderthals travelling over hundreds of kilometres, when lithics are also considered it's inescapable *something* moved over very considerable distances, and in quite varied ways. For the stone sourcing, there are only two possibilities: Neanderthals themselves were carrying artefacts, or objects were being exchanged between groups from different regions. Either scenario would raise many questions about social organisation, and imply some kind of territorial concept must be part of the answer. Among recent hunter-gatherers, the size of regularly traversed land varies drastically, from 250 to over 20,000km² (100 to 7,700mi.²). That's the difference between a very large city versus a small country. Huge ecological productivity in the tropics means people can get what they need from smaller areas of land, but the ranges for Neanderthals in higher latitudes were more likely towards the upper end.

Is it possible to define a particular ancient territory? A step beyond tracing the distances individual artefacts moved is to examine patterns across multiple assemblages. Recent syntheses across more than 20 sites in south-west France found a decidedly skewed pattern in lithic transport. High-quality flint was being carried sometimes up to 100km (60mi.), largely eastwards along the Lot and Dordogne valleys. But far less was being moved north or south, and equally stone from the north is *only* transferred a few tens of kilometres southwards, with virtually no overlap between the two regions.

The strong impression is of some kind of barrier over which stone wasn't being shifted, but the question is why. There's no obvious geographical boundary, which leaves something to do with either logistics – a point at which lithics tended to be replaced with local stone, rather than transported onwards – or a social factor. What's striking is that, compared to some of the known long-distance movements from Europe and elsewhere, the lithics that were moving up to but not beyond this zone in either direction hadn't come especially far, which might indicate it was about territory rather than economics.

To twenty-first-century Western minds, 'territory' evokes borders, commerce and ownership, but for recent hunter-gatherer societies rather different concepts exist. Conflict isn't unknown, and members of a band's extended social network – the clan – may be more welcome than true strangers. But often, access to land can be fluid. Territorial boundaries may exist without being explicitly defended (although particular places may be restricted), and sometimes bands may freely move over land at least double the size of their claimed territory.

On top of this, it's quite possible that some of the complex mobility patterns seen in Neanderthals might derive from individuals, not groups. Their closest fellow predators in social terms are wolves, where up to a fifth of individuals at some point are wanderers between packs. Adolescents can embark on vast treks over hundreds of kilometres before settling down. One remarkable tracked female, known as Naya, loped all the way from East Germany to Belgium, sometimes covering 30 to 70km (20 to 40mi.) a night. Perhaps some of the longest stone movements were Neanderthal 'Nayas', searching across mountain ranges and rivers for a welcome at a new hearth.

How places and things were connected in the Neanderthal world is perhaps the most testing of puzzles about them. It requires aligning all the jigsaw pieces of technology, subsistence and mobility, as well as climate. Dramatic changes after 150 ka – from full interglacial to deep glacial and everything in-between – were very likely part of the reason for Neanderthals' growing flexibility, as well as specialisation. Everyday experiences began to look quite different depending on where, and when they lived.

One of the ways to explore this is through the techno-complexes. Some unarguably were better suited to different sorts of mobility, and if the need to move was driven largely by finding animals, which itself varied according to climate and environment, then looking for extremes may help us see unique Neanderthal lifestyles.

The sun-drenched Eemian is an obvious place to start. It was the warmest and lushest world Neanderthals experienced, and though sites are rare, they're distinctive. Good evidence for whether forest hunting was different comes from two spear sites: Lehringen and Neumark-Nord. At the former, the spear is thicker than those at Schöningen and looks more like it was intended for thrusting, but it's also still quite long. Perhaps in this case, the prey species made a difference: elephant hunters may have wanted a healthy distance between the bloodied, raging beast and their own bodies. At Neumark-Nord, where two big stags were killed, the spears themselves are missing but their shadows remain in the holes they punched through bone. Both deer seem to have been taken out by close-range, low thrusts, which is definitely far better suited to dense woodland hunting than throwing javelins like those at Schöningen.

What also makes Neumark-Nord stand out compared to the vast majority of Neanderthal kill sites in any time and space is that after scoring two top-condition stags, fat from autumn's plenty, they barely butchered them. Even the most processed individual was simply filleted but not jointed. Moreover, there's no marrow smashing. Why leave rich calories behind? The implication is that they weren't struggling for food. If these forest Neanderthals had become adept at ambush tactics between trunks and brush, then abandoning carcasses might not have been a big loss in energetic terms. More intriguingly, treating carcasses this way indicates they weren't the starting point for staged butchery processing through the landscape. That would make sense if there were fewer hungry bellies waiting elsewhere.

It's quite possible that adapting to the rapid growth of the Eemian forests may have forced Neanderthal society to splinter into much smaller units able to cope with animals that no longer gathered in big herds. Smaller groups living in leaf-dappled light may not have needed to move so often, and the lithics back this up: they're generally not technologically suited to being carried and resharpened for long periods, and tend to be made with very local rocks.

Reindeer People

At the other end of the climate and environment spectrum, the Neanderthals of the Quina techno-complex experienced life very differently. Of all the lithic traditions, Quina is especially distinctive in being notably restricted in time and space. While some early Middle Palaeolithic assemblages have similar tools, the full Quina 'package' doesn't really appear until about 80 ka, during the MIS 4 glacial. Though it continues into MIS 3 after things warmed up, it's largely gone by 50 ka, and primarily found in southern France.

The universal feature of *all* Quina sites here, whatever the climate, is huge amounts of reindeer. The MIS 4 glacial caused a decimation in almost all prey species, apparently reflected in hyaenas basically abandoning south-west France. Such a drastic situation must also have impacted Neanderthals, and in this bitterly cold tundra world, reindeer may have been the only decent-sized game available. Was this extreme situation – some 40,000 years after the golden days of the Eemian – the setting for Neanderthals to adapt by developing the Quina and becoming truly specialised hunters? Most striking is that brief, less ice-blasted phases during MIS 4 were enough in some areas for forest-adapted red and even roe deer to reappear, if only temporarily. But they're never found with Quina assemblages, only Levallois. This suggests that the Quina techno-complex really was specialised, and represented a connection to open tundra south-west France and the large reindeer herds they contained.

Reindeer have always been of particular interest for discussions about how Neanderthals hunted and moved around the landscape, because of their great migrations. Today the distances involved vary, and trying to track exactly how far Pleistocene reindeer travelled is the subject of ongoing research. But what's certain is that they *did* move and gather seasonally. This would have happened during autumn breeding and spring birthing: reindeer fawns from caves in the British Midlands show that there was a calving ground here, but where the animals were coming from isn't known.

In south-west France, the site of Jonzac has provided huge amounts of information about how Quina-making Neanderthals were hunting reindeer. Mobility isotope analyses for the animals suggest that youngsters killed there in autumn/winter had already done a year's migration between geologically different regions. In this case, rather than *following* herds, Neanderthals seem to have been waiting where they knew reindeer would gather, or perhaps intercepting them along a known route between seasonal aggregations.

Jonzac's nearly 1m (3ft)-thick Layer 22 is so stuffed with reindeer it's known as 'the bonebed'. From an area just 3m^2 (3.5yd^2) in extent and 30cm (12in.) thick, over 5,000 bones were excavated. They represent at least 18 individual animals, and whole carcasses were being butchered here. If this cliff-base wasn't the kill site itself, it must have been very close by. Even if not a mass kill, what points most strongly to this having been selective hunting from among large numbers of animals is the fact that some prime parts including entire limbs were left un-butchered. These Neanderthals were not starving.

Not quite 80km (50mi.) north-east from Jonzac is another Quina site, Les Pradelles. Chapter 7 described the huge number of bone retouchers there; mostly made from reindeer bone, they were part of a cycle. Retouchers formed and resharpened the Quina scrapers, which then processed yet more reindeer bodies. What's especially interesting about the best-studied layer, 4a, is that while 98 per cent of all animal remains are reindeer, this probably wasn't the kill site. It's quite clear that Neanderthals had slaughtered and prepped carcasses elsewhere, before selecting mostly limbs and some heads for further butchery at Les Pradelles. Therefore this was a hunting camp, but it must have been located close to a known locale where herds gathered, or a migration route. The archaeology overall points to Neanderthals staying only for short periods of intensive activity, but this happened over and over through some 2m (6ft) of deposits: representing multiple generations and probably *hundreds* of reindeer.

The Pech de l'Azé IV rockshelter represents a different kind of Quina site, where hunting seems to have been during spring. The bone breakage is even more intensive, potentially because there's less body fat after winter and so getting marrow was even more important.

Reindeer often take different routes during their autumn and spring migrations, and while predictable year-to-year during stable climatic conditions, that may have changed from 60 ka with the beginning of MIS 3. The timing of herd movements, routes and locations of gatherings must have been disrupted, pushing Neanderthals to greater flexibility and rapid travelling once the animals were sighted, or perhaps, heard: the first sign may have been the collective noise of clicking hooves moving through tundra herbs, velvety muzzles chewing and bleat-grunts, well before curving antlers appeared over the horizon.

There are small amounts of other species in Quina sites of this age, including bison, horse and woolly rhino at Jonzac. They may have been hunted while groups waited for reindeer to arrive, or even food supplies brought in from elsewhere. They certainly weren't whole carcasses, and often look less heavily processed than the reindeer. Yet in some cases they have more damage from scavengers; perhaps Neanderthals sometimes used hunting camps during other, leaner times of year.

Big surpluses from reindeer probably meant Neanderthals adapted in other ways beyond hunting tactics. A portion of the marrow, fat, filleted meat and other parts* was probably eaten immediately, but it looks as if most was being shifted onwards to other places, along with freshly made lithics that had been knapped to butcher them. At Les Pradelles, for example, there are five times more bone retouchers than the tools they were used to work, and both new Quina flakes and scrapers are 'missing'.

It's also possible that some of the marrow, meat, fat and some lithics were moving over a short distance outside the site, to areas containing hearths but which weren't excavated. As mentioned in Chapter 9, many Quina sites, including Jonzac, Les Pradelles, Pech de l'Azé IV and Roc de Marsal, lack hearths and abundant charcoal. But that doesn't mean fire wasn't being used, as shown by burned flints. If hunting camps were dealing with multiple reindeer carcasses at once, it's quite likely a number of individuals were present, perhaps the whole group. Neanderthals would have needed to spread out their activity areas, just like at 'home' sites like Abric Romaní.

Certainly an awful lot of bone smashing was going on. Not only is everything highly fragmented, but often the fatty bone ends are missing. Either they were being simply crushed, perhaps with marrow, into a kind of bone pâté, or there was more complicated processing. Some kind of heating and rendering into grease would make more sense if all this abundant food was going to be taken onwards, in addition to means for preserving meat and fat. These would likely be happening away from the main butchery area and waste dumps.

Remarkable excavation of a reindeer-hunting site at Anaktuvuk Pass used by Iñupiat people in Alaska gives a hint of what this kind of camp could have looked like. The pass itself has been used for at least 3,500 years, and one area dating back 500 years contains remains from late summer/early autumn hunts. Across some 90m² (107yd²), clear spatial organisation is visible. Without a cave or rockshelter, people stayed in tents, but it was warm enough not to need sleeping hearths. There were fires, however, all within the main carcass-processing and

---

* In fact almost *every* part of reindeer or caribou is consumed by different Indigenous cultures: delicacies include heads, muzzles, mammary glands, foetuses, antler velvet and, for the Iglulik, Netsilik and Copper Inuit, even droppings.

grease-rendering zones; plus separate dumps for animal and lithic waste. While this isn't to say we should expect a mirror image in Neanderthal sites, it's quite possible that the Quina cave or rockshelter inner zones are actually the messy dump zones, and hearths were farther outside.

Whichever time of year, the kind of hunting we see Quina Neanderthals doing implies impressive logistical planning. Beyond timing, having to work at scale to catch and butcher many reindeer may imply that it wasn't just space being separated by activity, but also the group itself. Hunter-gatherers of all kinds have very different habits in terms of who goes on hunts, but tackling herds would always require more hands than tracking and killing single beasts. There are hints at some Quina sites of systematic butchery techniques, which probably means those doing the initial cutting and jointing, as well as trickier things like removing tendons, weren't inexperienced. But marrow smashing was potentially something even youngsters could get stuck into.

During the MIS 4 glacial, when there were hardly any other animals about, large herds would have attracted a lot of attention. Whoever came from individual groups to kill and take apart the reindeer would have faced other hungry carnivores, as shown by cut-marked cave lion and fox bones at Jonzac. Moreover, predictable seasonal hunts in roughly the same place every year may well have been opportunities for another resource vital to Neanderthals: socialising.

Under the Skin

There's one other intriguing aspect to what Quina Neanderthals were doing. It clearly replaces a Levallois-based way of life at the end of MIS 5/very early MIS 4, but why? Levallois is no less suited to butchery, and it's definitely portable. But Quina has two unique features. First, it's more economical in terms of both time and stone, since the cores don't need as much management. And second, Quina flakes may often be smaller, but they're thicker and stronger, and their shape – with one fat natural edge – means they can take far more resharpening, and produce more second generations.

If Neanderthals were having to cope with infrequent but intensive bouts of butchery, there would certainly be a need for resharpening. But something else can explain the distinctiveness of Quina tools and even the particular technique of resharpening: long, curved, steep edges are perfect for hide working.

So far we've mostly mentioned this technology in relation to tooth wear seen in many Neanderthals, but there's plenty of other evidence that they were accomplished hide workers. As a starting point, in many sites the placement of cut marks and even the types of bones suggest animal skins and furs were being carried around. More direct proof, however, are use-wear traces on lithic artefacts. The majority of tools with polish from scraping and rubbing skins would have been used either in combination with holding in the mouth or against a leg or a hard surface like a log to push against. It's even possible sometimes to distinguish lithics that had been used to scrape fresh, moist skins from those that worked dried hide, which marks a second phase of processing.

It's useful to understand quite how complex hide working actually is, both in terms of stages and time. Preparing one deer skin takes about a day, if the weather's good. The first job is to get off all the gore, which needs to be done while skins are still fresh, since drying causes fats to congeal and stick.* This implies that Neanderthals would need to do this very soon after killing, perhaps at nearby hunting camps. After cleaning and drying, skins can potentially be stored for months. Since Neanderthals were habitually undertaking phased butchery and knapping in different places, this was even more likely with hide working since it required a lot of time. And in fact, rolled-up hides exposed to a winter's smoke can even be in better condition.

After this comes softening for pliability, and preserving. Typically it begins with soaking, sometimes with added wood ash, since the alkaline properties help get grease out. Neanderthals might have used rivers, lakes or melted snow/ice as a soaking container, but large animal stomachs or even pits are another possibility. Across the world many traditional methods exist for hide improving, but fats of some

---

* Alternatively, frozen and thawed skins can also be worked.

sort are always involved. Brains and marrow are especially good, as is bear fat, all of which penetrate deeply and help with waterproofing.

Whatever the fatty mixture chosen, after soaking it must be worked into the skin, which is either stretched or in some other way continuously manipulated. This is where using mouths as a means of holding on comes in, while smoking the skin during this phase can help keep it soft (and enable it to resoften later when being used after getting wet).* But smoking is also exceptional for preserving hides, and among northern hemisphere hunter-gatherers, falling-apart rotten wood is almost universally used.

Can Neanderthals truly have used a process as complicated as this? Hide working was already happening well before 300 ka, as suggested by collagen and hair residues from below the Spear Horizon at Schöningen; and polish on a bone tool there can only have got that way from *hours* of use. For Quina sites, use-wear at Jonzac found that a large number of studied artefacts had distinctive polish from working fresh skins, showing that Neanderthals were getting the essential cleaning stage done during initial carcass processing.

One tool, however, had distinctive traces showing it had been used on dried hide. Perhaps it was used to prepare a hide made elsewhere, or was simply brought into Jonzac with previously acquired wear; it was certainly retouched again afterwards.

Beyond specific activities at Jonzac, though, this implies that reindeer hides at hunting camps were, like the meat, fat and marrow, part of a web of materials at different stages of manufacture that Neanderthals were moving around.

Why did Neanderthals need substantial amounts of hides? One possibility hardly ever discussed is that boiling hide together with bones, tendons and hooves is a well-known method for making glue. The released collagen makes an adhesive that's been exploited since medieval times as not only strong, but good for delicate work,† since it's self-sticking, shrinks as it dries and can be remoulded by heating like birch tar. But scraps will do for this, and you don't need to process animal skins to the extent that Neanderthals did.

---

* Hides for footwear in northern latitudes tend to be preferentially smoked.
† Historically favoured in particular for musical instruments.

The obvious answer is of course clothing. Many reconstructions show Neanderthals draped in the barest tattered skins, and it's also been claimed that the relative scarcity of small furred predators means they only wore loose capes. But thermal modelling shows that unless they were super-fat and thickly furred like bears, fitted clothing of some sort was vital. Well-insulated clothing is possible even without a natty wolverine trim,* and Neanderthals could take their pick from reindeer, bison, bear or others. Moreover, eyed needles aren't necessary for tailored clothing, as hides can be pierced with stone, bone or even wooden awls that push thread through instead of pulling.

What about feet? There's no evidence of hard-soled shoes, but softer foot coverings wouldn't leave a trace. Production is relatively simple too: a deer's hind leg provides a ready-made tube to slide onto the foot; if done while wet, it will shrink-wrap as it dries.

What makes the production of clothing especially likely beyond Neanderthals' physiology is the multiple lines of evidence for massive processing to soften and stretch hides, notably seen in their clamping and probably also chewing tooth wear. It matches exactly that seen in hunter-gatherers who do this with skins and also sinews, and what's more is almost identical to early *H. sapiens* samples. It's far less difficult to imagine those populations wearing reindeer parkas, but Neanderthals from open habitats – which are likely to have been colder – actually have even more heavy wear.

So they wore clothes, and probably Quina Neanderthals in particular. But what kind of quantities of hides are we talking? Based on a fairly simple outfit with an upper-body garment and leggings or skirt, each adult would need at least five large hides. To make this from scratch would take somewhere between 20 and 80 solid hours of hide processing depending on species and the particularities of the process. This outfit would need replacing every few years, and then there's children's garments and coverings for infants.† As well as the worn teeth, we can recall how Neanderthals' arms were thickened

---

* Reputed to be extra-good for hoods because it sheds ice.
† Baby wraps would actually make sense as the first garments made of hide, which must go back into the Lower Palaeolithic.

from huge pulling forces: much of it potentially from a lifetime where hide working was a never-ending task.

If that's not convincing, then in addition to the use-wear on lithics from many sites is a recently discovered class of shaped bone tools. Made from rib ends, Neanderthals carved the bone to narrow it and form a standardised, symmetrical round tip. So far there are only five known examples, but they come from two sites: one from Pech de l'Azé, and four from Abri Peyrony, just 35km (22mi.) away, where they're in different layers: one with bifaces, and another Discoid. They're all broken, but a nearly identical, complete object found in 1907 at La Quina may show their original form: it's strongly curved, and the opposite end was left unaltered. These tools are virtual clones of things called 'lissoirs' – smoothers – which are not only found in later *H. sapiens* cultures, but still used today to soften and burnish animal skins. And even the damage on the Neanderthal examples fits this: all are tips snapped off due to intense pressure during use.

Given how discriminating Neanderthals were about the animals and body parts their retouchers came from, it's perhaps not surprising they weren't just selecting ribs for lissoirs, but also by species. Collagen analysis found that all the surviving objects are either bison or aurochs, and strikingly, three are from a layer at Abri Peyrony that's 90 per cent reindeer.

Those sites date to around 50 to 40 ka and show that only a few thousand years after the Quina techno-complex disappears, Neanderthals had extremely sophisticated processing techniques. Whether or not lissoirs were also used by the Quina reindeer hunters isn't certain, but perhaps as well as clothing, they had other motivations for intensive hide working. If we recall the impression that many Quina sites, even places one stage beyond hunting camps, don't look particularly like 'homes' with hearths, then perhaps these Neanderthals were using tents or shelters. We know from La Folie that open-air constructions of some sort were used later, but truly mobile shelters formed from hides would have allowed Quina-making Neanderthals to move flexibly across the tundra and stay warm while ranging over long distances. Even if they obviously returned time and again to particular caves or cliffs for the hunting and initial butchery, the lack of hearths despite there being some burned materials suggests we're missing these parts of sites.

Big tipis of the sort crafted by Indigenous cultures in North America like the Cayuse, Dakota, Blackfeet, Pawnee, Crow and Plains Cree can use 30 to 50 deer skins, but transporting all that was typically aided by horses. It's unlikely that structures on that scale would have been dragged around by Neanderthals, but perhaps smaller tents were used, and there's also the evidence for hide mats at Abric Romaní.

What's interesting about a way of life requiring many skins, whether for clothing or shelters, is that it sometimes results in so-called 'wasteful' hunting, where animals are targeted not for meat but skins. Summer reindeer are pretty lean and warm weather makes meat spoil fast, but their skins are in good condition, whereas by the autumn they're often becoming less so, not helped by holes parasites leave in the hide. Apart from hides and choice cuts like tongues, other parts might be left behind, and there are occasional hints of this in some Neanderthal reindeer sites, Quina or not. At Jonzac there are a few still-articulated limb joints; something similar is known from Abri du Maras in south-east France, and many carcasses at Salzgitter-Lebenstedt also look quite lightly butchered.

Hides can also produce things just as vital to hunter-gatherer life, but often overlooked in their ubiquity today: bags. The average Neanderthal probably had a lot to carry: food, fresh skins, furs, perhaps bedding, not to mention stone. Natural 'bags' might be made from animal stomachs or bladders, but hides are especially useful because they're tough and big. Rolled around objects and bound with sinew or tendon, they could have been slung over shoulders to ease heavy loads. It's perfectly feasible to use a freshly tugged-off skin like this; sites with a lot of foot bones might reflect the limb extremities being left on, like handles on old-fashioned travel bags.

Containers are also connected to food storage, perhaps especially important in the Quina. As noted in Chapter 8, there's no *direct* evidence Neanderthals stored food, but in a glacial setting where most meat was sourced during big seasonal hunts, it may have been necessary. By now it should be clear that Neanderthals partitioned all kinds of activities and materials across the landscape, so taking this a step further by preserving parts for later isn't such a stretch.

Depending on how it was stored, the glut could provide nutrition for weeks or even months, and would have been especially important if winter was coming. Drying meat might have been difficult in soggy weather, but during glacial periods summer and autumn were probably drier. If Quina Neanderthals at hunting camps were staying long enough to get the first stage of hide processing done, there may have been enough time to also begin drying meat, rendering grease or even using hide-smoking fires for preserving edible animal parts. Such multi-tasking would certainly fit the efficiency we see in so many other aspects of Neanderthal behaviour.

What happened to the Quina Neanderthals? If this techno-complex emerged as an adaptation making the most of glacial tundra and its reindeer, how did they cope when things warmed up? As we saw earlier, some later sites include a few horses, aurochs or bison, and at a handful of locales those species actually dominate. This might reflect some Neanderthals trying to adapt to steppe life, but on the other hand in the western Pyrenees there's still a link between cold climate, reindeer and Quina as late as 45 ka. After this, however, the 'package' seems to finally disappear.

Yet when talking about 'the Quina' we encounter the tension in archaeologists' desire to classify lithic techno-complexes versus the variety in what Neanderthals needed and wanted to do. A few sites here and there have lithics that look in all respects to be Quina technology, complete with massive amounts of retouchers, but they're living rather different lives. One case is the Italian cave of De Nadale, up in the Alpine foothills. It contains an MIS 4 assemblage closely resembling the contemporary Quina in south-west France, except here Neanderthals were mostly hunting giant deer and red deer. These species have totally different ecology and behaviour to reindeer, and it may be that Quina-making groups moved into this region and adapted, or it was an independent innovation.

Living Landscapes

How did Neanderthals think about their world? Beyond daily tasks – waking well-rested or groggy, helping children eat, finding flakes within stone, following animals – what did travelling, pausing for an afternoon or spending many days somewhere mean to them? Although it's already featured in this and other chapters, the word 'home' is tricky, automatically bringing to mind twenty-first-century Western concepts of a fixed place. But for people living by moving, home will be many locales: the heart of existence moves too. Intimacy with kin is one constant that would have made a Neanderthal feel 'at home', but their deep familiarity with the wider landscape was probably another kind of relationship.

Neanderthals' thoughts and memories must have included individual, well-known places: the bluff rising beyond the hill; the stand of pines by the ford; the dark opening in the cliff. Oft-repeated or extraordinary experiences are especially powerful in binding memories, and are more likely to be recalled when particular locales were revisited. Landscapes for Neanderthals therefore weren't abstract spaces, but a continuous flow of lived encounters, both new and remembered.

This is a kind of history, which would have been soaked up from childhood through attention and probably direct communication. But *places* are something more, their meaning manifesting through slow accretion over time of custom and memory. And then there are the archaeological deposits themselves, accumulating through centuries and millennia. Histories in mental and material form probably influenced how Neanderthals chose to use particular places, stimulating even more distilled forms of activity and living. In this way, locales like Jonzac might emerge conceptually as something like *Reindeer-Hunt-Cliff.*

Where truly deep deposits were visible, obviously formed over more than a lifetime, something else might have been sparked. These cultural condensates communicating time itself may have had the power to stretch Neanderthal imaginations beyond the 'now'. We know they recognised and interacted with older objects in rockshelters and caves, but did the sight of deep bone beds or clearly ancient hearths spark a sense of those gone before? We can see in some places that decades or centuries old objects stuck up from the ground – physically crossing time – and must have stimulated an impression of permanence.

Moreover, some recycled lithics in caves don't look as if they came from surface scatters. If so, Neanderthals may have been the first archaeologists: unearthing eroded ancient remains of their forebears, turning them around in the light and making them part of the living world once more.

Fingers digging down into soft earth, feeling out sharp edges, is a reminder that Neanderthals' experiences of place and land went beyond the visual. They knew rock by its texture as much as its colour, listened to the song of water not just for flow but temperature, and picked out tree species by the changing sound of wind soughing through their leaves. For Neanderthals, as with many Indigenous cultures, the land itself may have been conceptualised not as some kind of stage upon which to tread, but as *someone* with whom a relationship or communication was possible. One must always walk in company.

It may be speculation to muse on how the land was perceived, but as they strode, crept and ran, Neanderthals inscribed themselves upon it. Hundreds of millennia before anyone wrote the first word, their footprints were a kind of signature. We heard earlier about those at Le Rozel, which in total has over 250 prints through several layers. Some are short step-by-step trails, but mostly they're isolated prints. The smallest feet – those of a toddler – left the lightest trace, and there's even a perfect handprint: fingers spread wide, pressed into the sand, as if raised in greeting through 80,000 years.

The oldest known Neanderthal prints were left more than 250,000 years before those at Le Rozel. On the slopes of the extinct Roccamonfina volcano, southern Italy, they were believed to be the Devil's tracks after being revealed by eighteenth-century landslides. In fact, they were left around 350 ka by three early Neanderthals whose feet sank into cooled and rain-softened pyroclastic ash and mudflows. More than 50 prints show how all moved differently: one zig-zagged down, another took a cautious curving path, slipping and sometimes dropping down a hand for balance, while the third ploughed in a straight line.

All three walkers were under 1.35m (4.4ft) tall, exactly the calculated height of Le Moustier 1, making them young teenagers. It's quite possible that this gang witnessed the eruption that had produced the mixed ash, pumice and rough rock under their feet, and a later ash fall that covered their tracks. Where were they going? Tracing the three trails upslope, all led from a flat ledge speckled for about 50m (55yd) by yet more hominin footprints, as well as those of animals: a Neanderthal routeway.*

Underground footprints also exist. Around the same time as the Le Rozel children were playing on the Atlantic shores, high in the western Carpathians of Romania an older teenage Neanderthal was exploring Vârtop Cave. They gently stepped through strange 'moonmilk' that later hardened into flowstone. Somewhat extraordinarily, this is the only trace Neanderthals were ever in the cave. But elsewhere other youngsters left traces in living sites. Theopetra Cave in Greece contains a handful of diminutive footprints older than 128 ka, one of which was probably aged just 2 to 4 years old, while there are hints another wore thin foot coverings.

That so many of those leaving footprints were youngsters is striking. In many hunter-gatherer societies, children receive little formal teaching from adults, but learn instead by hanging out with their peers. Wandering, exploring, testing and playing grown-up would have happened in living sites, as well as through the surrounding land. It must have opened up exciting adventures as well as risks, and suggests that Neanderthals' experience of the land changed through their lives.

Ancient tracks bewitch us today, but traces of living creatures would have interested Neanderthals too. To survive on the bounty of the earth means constant attention to the land, and muddy trails worn into the ground by the passage of many hooves, or even bent grasses left by silent paw-pads, would be noticed. But more than this, as hominins lacking the claws, teeth or great speed of other carnivores, mastering hunting would mean practice, collaboration and most particularly knowledge and planning. By combining understandings, such as which species drink at dawn, with imagination would allow Neanderthals to predict the myriad ways an animal becomes vulnerable. This wisdom may well have been distilled into the skill

---

*The upper tracks were already exposed in medieval times and bear marks where locals had tried to enlarge the small footprints to accommodate booted adult feet to unknowingly walk in their ancestors' footsteps.

of tracking. For recent hunter-gatherers, whatever the environment or method of hunting, this is crucial. Tracking is far more than following trails. It's about full attention to the entirety of the world, so that traces even of ants become familiar. Practised trackers can identify not only animal species, but sub-species, sex, age or even physical condition. It's also possible with larger creatures to recognise familiar individuals.

Neanderthals evolved with 30,000 generations of hunting heritage already behind them. Tracking may have developed nearly 1.5 million years ago, as footprints from Ileret, Kenya, show early hominins stalking the muddy margins of a lake where both trails and their makers could be watched and ambushed. Pursuit or endurance hunting is strongly connected to tracking, and involves chasing prey to exhaustion whereupon killing is simple. Wolves and hyaenas work this way, and it's especially suited to open environments like steppe.[*]

What makes tracking so important in this kind of hunting is that it opens up the ability to predict animals' behaviour. Even novices can just about manage it, as shown by the Lykov family who fled from religious persecution into the Russian taiga in 1936. They survived undisturbed over 240km (150mi.) from the nearest settlement for 40 years, and a vital supplement to their marginal farming was hunting. But they had no weapons, and the Lykov boys taught themselves to simply chase animals through the forest until they dropped. The Lykovs were rarely successful, sometimes making just one kill a year, whereas Neanderthals had had lifetimes to hone their skills. Even if they preferred ambush to pursuit hunting, tracking must have been part of this.

Tracking has also been suggested as a means by which even more sophisticated hunting evolved, with impacts far beyond subsistence. When trails become hard to follow, a hunter expert in the behaviour of their quarry can predict where its tracks will re-emerge. A frightened, rapidly tiring deer may be more likely to hide than continue running, and knowing this can make the difference between loss or kill. But it's more than knowledge: this skill, known as

---

[*] Hot or very cold climates work best, as animals are more likely to suffer heat exhaustion.

speculative tracking, entails visualising the animal's state of mind. It requires the cognitive ability to understand that others have outlooks and emotions different to one's own, supposedly possessed by just a handful of species other than humans.

Could Neanderthals have done this? Certainly speculative tracking for hunting would have been useful in tricky environments like forests where animals could disappear. But it would also have helped in more open landscapes, for example predicting when and where herds might appear.

Moreover, if Neanderthals' ability as hunters involved tracking and the ability to imagine themselves into other minds, then surely this also involved considering the thoughts of other hominins. Among expert Indigenous trackers of the Kalahari, Southern Africa, the traces of close relations are as recognisable as their faces.* And physical hints of the presence of strangers, whether footprints, lithic scatters, spreads of charcoal or old carcasses, would have been pored over and of extreme interest, provoking excitement.

To live in such a world was to be surrounded by presences of many kinds. Places and people were connected by flows of stone, meat and other materials, but were also physically linked by tracks on trails. These routes were like rivers of memory, immersing Neanderthals from the time they were carried upon a warm chest, to the slow striding elder, recalling the reindeer winters before trees filled the valley.

The past five chapters have been a deep dive into Neanderthal life, from micro-scales at individual locales where generations exist in each scrape of the trowel, to networks of travellers and objects extending hundreds of kilometres. Compared to the millions of years before, Neanderthal existence was a major upgrade to hominin life. They lived in more complex ways than anything before, and perhaps the best way to think of what the Middle Palaeolithic represents is *amplification* and *enhancement*. Whatever the ecosystem, they were top-level hunters and canny foragers. Their lithic technology was more efficient and specialised, and they pioneered new ways of using organic materials. But something else more profound was going on.

---

* Even when wrapped in foot coverings, idiosyncrasies of movement create unique patterns.

Neanderthals were also the first to truly begin stretching their lives through time and space in a way no earthly creature had ever experienced. They disintegrated stone and animals' bodies in more complicated, systematic ways and moved the pieces farther than ever before, and even the shift in retouchers from entire bones to shards mirrors this growing intensity.

And as objects and activities became more specialised and separated through space and time, they cast a net of presence, action and intention across land, and in the memory. Theirs were minds that looked out to the horizon and knew it intimately: how the paths changed in spring, when the fords would fill and perhaps even how many sunrises until the great cliffs came into view around the river bend. One might even claim that the first revolution imbuing land with social meaning belonged to Neanderthals, not *H. sapiens*.

Moreover, all this was happening with ever-more collaborative processes, from hunting to butchery, and in the movement of resources and sharing of food. It's even visible within sites, showing that Neanderthals weren't just supporting their kin, but opening up new ways to relate to each other within their most intimate spaces.

As more things than ever were extracted and moved around, *how* one could do simple things like knap became replete with possibilities. Substances were mixed and things combined. The potential for society itself to diversify expanded: identities could be formed around more than categories such as age. Skills and proficiency across ever-more specialist crafts, from hunting to hafting to hide working, were new ways for Neanderthals to place themselves in the world. Proof they did so is worn into their very bones and teeth as intensifying cycles of embodied expertise.

The variety of Neanderthal lifestyles are like dances emerging out of the rhythms of particular environments, and manifested by techno-complexes and mobility patterns. But the precise tempo and choreography was always unique, a *pas de deux* shared with the creatures they lived by and with. And even as movement across the landscape grew in scale through time, concurrently place developed greater social significance. Choices about what happened and where had a finer grain than ever before, and internal space was divided accordingly. Neanderthals were accumulating the first great archives, the detritus of life forming unconscious monuments: permanence despite transience. Like wells of memory in a vast, shifting world, it was in these places that Neanderthals first kindled human history and geography.

And burning bright at the centre are the fires. Like suns wielding enormous gravitational forces, everything and everyone spiralled around the hot heart of *home*. Fifty thousand years later, Neanderthals' hearths still have weirdly cosmic properties. Amid ancient surfaces densely spangled by myriad artefacts, fireplaces are like archaeological wormholes, bridging the impossible chasms of time separating us from long-vanished dwellers. As researchers encircle hearths, recording and excavating, their presence is like an afterglow of human attention, reanimating empty spaces. Time collapses, and it's almost as if our fingers reaching out might graze the warmth of Neanderthal skin, sitting right there beside us.

Those bodies – once incandescently alive, now dry bone behind glass – weren't simply engines that needed refuelling, or automata for making endless sharp flakes. Just as our days are suffused by social interactions, the kernel of the Neanderthals' world was in their relationships. Physical proximity had been the medium of hominin intimacy for millions of years, measured in touch, glance, guise, but the Neanderthals added new currency: material substances. Composite technology is a time-travelling mind made manifest, and they were surely among the first humans to tell stories. The things they collected, took apart, carried and brought back together were about more than survival. They also mark an amplification in communication, an inexhaustibly rich channel to express connections and meanings beyond the mundane.

CHAPTER ELEVEN

Beautiful Things

The firebrand glow on the walls dies like a fading sunset, as footsteps whisper away into faintness. The smoke lingers, but eventually its particles dissipate into the inky blackness. Winters upon winters pass, the years marked by rumbling heartbeats of hibernating bears. The cave is in stasis, warmed only by their slow breaths. Thickly furred generations wear the ground into bowl-shaped nests, while all around calcite-loaded water trickles, drips, flows. Micro-thick layers rise into leathery drapes, frozen mid-billow; pale fingers stretch upwards over tens of millennia. Sometimes the bears wake early. Confused in the dark, they wander towards a deep chamber. Questing ursine snouts touch hard, cold rock-bones, snapped and lying in long walls. Many bears, many times raise their heads, scenting water, clay, cave; once, long ago, some caught lingering traces of charred bone. But in the blackness, their wide eyes are blind to the strange edifice in front of them.

Yet more years crawl by, before a muffled crashing marks a rockfall sealing the cave. Dusty air billows into the chamber, disturbs dead, flat pools on the floor.

Without illumination, the trembling water contains no shivering reflections of the carefully constructed stalagmite rings. They wait.

To find the strangest Neanderthal site, you must go to the Aveyron valley, a less-visited region of south-west France than the Périgord. Twisting for kilometres through deep gorges, the river passes a hill near the town of Bruniquel that hides a secret. Deep inside a cave something wondrous lies, utterly bizarre and *old*; even for Neanderthals. In 1990, when cavers broke through massive roof-fall debris, they had no idea of what lay more than 300m (330yd) inside. Stalagmites were bestrewn across the floor of a broad chamber, but what at first seemed random resolved into two roughly circular forms. After initially being radiocarbon dated to older than 47 ka, it remained an intriguing anomaly until a new project began in 2013. Then the real revelation came: a large series of dates using the uranium-series method[*] showed unequivocally that this subterranean structure was made over 174,000 years ago. Instantly, Bruniquel became one of the most important Neanderthal sites ever found.

Meticulous study found complexity at every level. Over 400 stalagmites had been snapped off, and from among the broken pieces Neanderthals had selected wide, straight mid-portions, obviously with particular sizes in mind. By aligning their 'speleofacts', they formed two rings on the chamber floor. The largest is more than 6 by 4m (6.6 by 4.4yd), and contains two small speleofact piles, with another two heaps placed externally at either end. A second smaller, but more circular, ring is set to one side.

While the heaps of broken columns recall a Classical ruin, closer inspection shows this isn't a random jumble, but a built construction. Each ring is made from up to four layers, some sections buttressed with vertical pieces, and one zone features five elongated speleofacts stood up side by side. The intricacy goes beyond support and into architecture. Tucked in behind the five 'sentinels' is a doubly poised

---

[*] A radiometric method that measures the decay of a uranium isotope in flowstone and stalagmite.

creation: a flat plate balanced on a cylinder, itself holding up other pieces.

This is already a place that goes beyond unique, into jaw-dropping. But there's other things going on. Burning was identified at multiple points along the rings, and within the small piles. In fact, about a quarter of all the speleofacts have been exposed to fire, and in some cases it seems as if blazes were lit *on top* of the structures. Some fragments of burned bone are also visible, the largest of which – inside one of the piles – may be bear.

Both the speleofact constructions and the burning event that charred that bone piece took place between 178.6 and 174.4 ka, and with available dating resolution are basically contemporary. No possible natural process could explain the rings: bears were hibernating in the cave system and can sometimes damage stalagmites while stumbling in the dark, but not only are the rings much larger than any bear nests, this wouldn't explain built walls.

The mystery of Bruniquel only deepens the more one thinks about it. It's not a place for staying like other caves or rockshelters: so deep into the hillside, continuous illumination would have been needed. That means not only a herculean effort in terms of fuel collection, but also choking smoke. Moreover, building fires on top of the ring walls makes little sense if this was a structure for living inside. Flowstone covers most of the floor between the structures, but there's no visible lithic or butchery detritus.

And constructing the rings was no casual feat. The total weight of arranged speleofacts is over two tonnes, and even assuming multiple individuals were involved, it must have taken at least six to seven straight hours to build.* Why would Neanderthals spend hours – maybe days – deep underground, breaking and hefting heavy rock, piling, balancing and burning them?

Laser-scanned overhead images best communicate the utter weirdness of this place. The harvested stalagmite stumps rise up like some sunken forest through the flowstone floor, and the rings unmistakably radiate purpose. But exactly what is baffling. The

---

* Estimating an average of one minute to find, alter and place each of the 400 pieces.

remote chamber is one of the most spacious parts of the cave, but is somewhat secret, located where walls pinch inwards and turn, before a large corridor continues for at least 100m (110yd).[*]

The burned areas are a further mystery: presumably some were to provide lighting, but intense heat damage on some of the speleofacts hints that it might also have helped in fracturing them. And were the bones fuel, food or something else? Most intriguingly, magnetic analysis, which can detect evidence of ancient heating, showed two things: other hearths may lie beneath the flowstone floor, and some of those burned zones have double cores. This latter feature most likely means hearths were re-lit, strongly suggesting Neanderthals made return visits.

Steeped in eldritch vibes, the significance of Bruniquel's ring chamber is enormous: the only monumental construction known to have been made by Neanderthals. Yet on further consideration, everything about it echoes the processes of fragmentation and accumulation central to so many other aspects of Neanderthal life. Exploring deep inside the earth's body, perhaps the festooned flowstone and thickets of knobbly, white protrusions looked like flesh, guts and bones; a stone carcass to be broken apart and put back anew.

Mind in the Body

Bruniquel laughs in the face of austere, survival-only explanations for Neanderthal behaviour. It surely was made by thinking, but also *feeling*, minds. Emotions in reality underlie virtually everything people do, no matter if logical explanations also exist. All human cultures also have a desire for transcendence. Whether expressed through painting cave walls, raising cathedrals, singing millennia-old sacred songs or climbing among airy mountain peaks, this urge is common to all peoples across time and space. Did Neanderthals experience similar impulses, perhaps motivating them to build a ring chamber in the darkness?

---

[*] The end of the cave system remains unexplored, but there must have been another entrance for the cave bears that hibernated after the main rockfall.

Musing about minds from 50 or 100 millennia ago is of course fraught with pitfalls, not least because even with living people, resolving the miracle of consciousness is like reaching for a shimmering mirage. Until we can trace how our own neurons and sensory systems mesh together to produce perceptions and emotions, doing so for Neanderthals remains out of reach. But that's not to say we can't make well-informed guesses.

As with our fellow great apes, their existence would be founded on emotion. Fear, pleasure, pain, excitement and desire would have flooded through every Neanderthal that ever lived; hearts hammering, guts clenching, loins tightening. But more tantalisingly, some apes seem to express more complex shades of feeling. In particular, chimpanzees have been observed responding with primal outbursts to natural phenomena like heavy rain or waterfalls. Ascribing any level of formal spirituality to Neanderthals would go far beyond the archaeological evidence, but they too encountered all of life's sensory marvels. Perhaps as photons from a salmon-belly sunset saturated their retinas, or the groaning song of a mile-high glacier filled their ears, Neanderthals' brains translated this to something like awe.

Feeling wonder inside oneself is one thing. Being able to share an awe-inspiring or transcendent experience is much more powerful. For a meta-physical life to emerge, language is key because it allows emotion and meaning to become crystallised. Whether Neanderthals had any kind of language is, of course, one of the most enduring questions about them. What does the most up-to-date brain science tell us? Compared to the entire *Homo* lineage, Neanderthals' brains are, like ours, enormous. Despite being less bulbous, the average capacity of their skulls was a bit greater. That means more neurons: the connective plumbing between different areas. But it's about more than bulk.

What matters more is organisation. Neanderthals' flatter foreheads had less room for the frontal cortex area, intimately connected to complex thought processes like memory and language. And they also had smaller cerebelli, another area involved in concentration, communication and language. In living people, reduced cerebelli does seem to indicate poorer skills, and in Neanderthals the connections to other language-related zones were also smaller. However, as with debates in Chapter 3 over possible cognitive trade-offs from their larger visual system, it's extremely hard to be sure whether in Neanderthals size truly equates to proficiency, or if their grey matter compensated in other ways. Notably, our own brains have slightly shrunk since early *H. sapiens*, with no apparent shrivelling in cognitive capacity.

When we add in bodies and archaeology, the odds start to stack up in favour of some spoken communication. While there's been a lot of back-and-forth debates over this, today it appears that Neanderthal vocal cords could make pretty much the same range of sounds as ours. There were perhaps some subtle differences in vowels including 'ah', but their breath control wasn't appreciably poorer, giving them the ability to utter lengthy sound combinations. Moreover, though their inner ear shape was slightly different, it was similarly finely tuned to sound frequencies generated by speech. If this anatomy in humans is regarded as specialised for language, then Neanderthals cannot have been so different. The same thing is true of brains: in your head right now, Broca's area is busy understanding the words on this page, and it was also well developed in Neanderthals, with neurons that would have lit up as well-practised hands knapped a Levallois core, or even when children watched elders butchering.

Complementary evidence for language comes from the fact Neanderthals seem to have had similar rates of handedness. Tooth micro-scratches and patterns of knapping on cores confirm they were dominated by right-handers, and this is also reflected in asymmetry in one side of their brains. But when we zoom in further to genetics, things get increasingly thorny. The FOXP2 gene is a case in point: humans have a mutation that changed just two amino acids from those in other animals, whether chimpanzees or platypi. FOXP2 is definitely involved with cognitive and physical language capacity in living people, but it isn't 'the' language gene; no such thing exists. Rather it affects multiple aspects of brain and central nervous system development. When it was confirmed that Neanderthals had the same FOXP2 gene as us, it was taken as strong evidence that they could 'talk'. But another, subtler alteration has been found that happened *after* we'd split from them. It's tiny – a single protein – and though the precise anatomical effect isn't yet known, experiments show it does change how FOXP2 itself works. Small chinks of light like this are fascinating, but we're far from mapping out any kind of genetic recipe where adding this, or taking away that, would make Neanderthals loquacious or laconic.

Taking everything on balance, it's very likely Neanderthals spoke in some form, but about what? Numerous animals call attention to things, and some primate vocalising even contains contextual information: the type of predator *and* where it's located. But more subtle communication, such as describing things that already happened or have not yet come to pass, requires understanding of order and time. Certainly there's ample archaeological evidence that Neanderthals were organised in terms of who went where and when, so some level of discussion for collaborative activity is probable.

Could they tell stories? The tales we spin weave together past, future and even magical creations. It's possible to argue that all of these concepts are manifest in composite tools: syntax-like objects made of ordered parts from numerous places, added at different times. In making and using these, Neanderthal imaginations moved far beyond the here-and-now, and with birch tar even included a 'supernatural' substance.

Perhaps the most crucial requirement for storytelling, whatever the subject, is the desire for connection. Standing above a glass-still pool, there's no doubt Neanderthals recognised their reflection – as do dolphins, elephants and apes. With this ability comes empathy and understanding of others' points of view, and all combine in shared systems of meaning. Language is effectively commonly understood sound symbols, and even captive apes can learn to express simple ideas – such as 'give ball' – using graphical symbols. But they never use this skill to casually chat, despite this defining everyday human communication. Neanderthals very likely used symbols too, at least including gestures, in addition to learning animal tracks: essentially graphic signs for each species. They certainly laughed, probably joked, and may have memorised chronicles, of a sort. And returning to the Bruniquel rings, we're confronted by a creation redolent of deeper meanings.

An odd coincidence: just around the river bend from Bruniquel is the Montastruc rockshelter. A couple of years after Falconer witnessed the La Madeleine mammoth engraving in 1864, Montastruc produced more stunning Upper Palaeolithic *objets d'art* including two carved reindeer, probably swimming. The last chapter explored how these creatures were central to some Neanderthals' lives, and they must also have seen them crossing rivers, yet in the 150-odd years since the Montastruc carving's discovery, no comparable artefacts have emerged from any of their sites. On the other hand, the past three decades have seen an explosion in archaeological evidence – beyond Bruniquel – for symbolic aspects to Neanderthal life.

As with every human culture, their quotidian experience would have been imbued with associations: hearing whinnies implied horse, while smelling smoke meant fire. But did more abstract, symbolic significances also exist, such as red = blood? Primate visual systems are primed for vivid colours, particularly reds, as well as lustre. Bright, shiny things catch archaeologists' attention too, and that snap of recognition can preserve a precious remnant from oblivion. Did Neanderthals also have a magpie lust for shine and glitter? When things with those qualities but no obvious practical function are found, it's hard not to instinctively assume an aesthetic motivation for their presence.

The simplest cases are manuports, meaning 'carried in by hand'. They're always rare compared to bones or lithics, but known across the Neanderthal world. Examples include a quartz crystal at Abri des Pechêurs, south-east France, or a fossil shell at Pech de l'Azé I. Shiny things snag the gaze, while fossils mimic living things in an unexpected substance, so we must assume Neanderthals' curiosity was piqued. They also picked up things with unusual tactile qualities, such as pumice stones found at some Italian sites. And these curiosities were sometimes also moved long distances: the Pech de l'Azé I fossil was carried at least 30km (20mi.), and given *everything* hefted by hand must have been thought important, this was no thoughtless choice.

But were such things symbolic? Bowerbirds instinctively collect natural gewgaws and create exhibitions of 'bling', but they're aimed at attracting females – a kind of peacock's tail – and there isn't the same impression of material curiosity. Yet we can't assume that our own standards of meaning were always shared by Neanderthals. The beauty of rock crystal lithics at Abri des Merveilles, south-west France, may suggest they were significant objects, but in fact Neanderthals sourced and knapped them exactly like other rocks.

What's needed to infer something deeper is special treatment, or repeated associations and patterns in behaviour. Cioarei-Boroşteni Cave in the southern Carpathians might have something along these lines. Excavations in the past 20 years uncovered a hard ball-like object, just large enough to fill your hand, yet remarkably dense. Scanning revealed it to be mineral geode, possibly opal. Where Neanderthals found it is unknown; the local river passes through volcanic regions where geodes might be found, but it's so heavy that it's not likely to have rolled far downstream.

Already an intriguing object, when surface carbonate crust was removed, tiny specks of colour emerged. High magnification showed red ochre* patches, overlain by an unidentified black material. The geode is an anomaly at Cioarei-Boroşteni, but use of pigment isn't. From an overlying level, red and black residues were found inside eight bowl-shaped sections of stalagmite and calcite crust. They're very small – mostly about 6cm (2.4in.) across – and it's not clear if Neanderthals shaped them or simply used already broken pieces, but they appear very much like containers. What the pigment was used for can only be imagined, but Cioarei-Boroşteni's significance lies in showing that Neanderthals, over considerable periods of time, were interested in applying colour to unusual things. That, fundamentally, is a definition of art.

Colour

Evidence for Neanderthal symbolism has seen a pigment boom in the past decade. The odd bit here and there was noted since the earlier twentieth century, but recent analytical advances have identified pigments at more than 70 sites, just in Europe. As well as red and yellow minerals, Neanderthals were collecting and using various black substances. But what for? Colour is central to visual displays for social communication across the animal world; but before we get carried away, practical applications are also possible. Minerals can be used for sunscreen, insect repellent, hair management or even

---

* Ochre is a generic term for natural red or yellow mineral pigments constituted by iron oxides, goethite and clays.

antiseptics; ochres in particular can be used in hide working or as hafting glue additives, while as mentioned in Chapter 9, black manganese may be useful in fire-lighting.

There's certainly plenty of evidence pigments *were* used: many nodules have wear traces, sometimes from being rubbed on soft things, or due to scraping in a way that produced richly coloured powder. Astonishingly, we know that already between 250 and 200 ka Neanderthals were making liquid red ochre. Minute analysis of red-stained sediments from the open-air locale of Maastricht-Belvédère, Netherlands, showed they were ochre splashes. The nearest source was between 40 and 80km (25 and 50mi.) away, but it's not known if this was carried as a block or already powdered. Presumably it was mixed on site, either with a container or by mouth.

Some older central African ochre quarries exist, presumably worked by early *H. sapiens*, but Maastricht-Belvédère is the oldest known *use* of pigment. For Neanderthals, as time goes by pigment gets much more common in the archaeological record. Most impressive, in levels around 60 ka at Pech de l'Azé I, there are some 500 pieces of black manganese, at least half bearing varied wear traces. And while there are far fewer pieces overall just next door in Pech de l'Azé IV, the fact that they extend through nine levels points to continuity in behaviour.

Combe Grenal is also remarkable for long-term pigment use, with around 70 blocks through 16 layers. But here the colours and uses shift, and they appear linked to different techno-complexes. Quina layers mostly have grey-black minerals, with wear ranging from scraping to grinding; one even seems to have been used as a retoucher. After this, mineral use becomes rarer, but unworn red pieces appear. Then during a phase of Discoid assemblages, more reds, browns and yellows turn up but they're different chemically, and must have come from other sources.

In some south-west French sites it's possible to make out that, once again, Neanderthals were focusing on quality. They must either have been systematically searching large areas for the richest manganese minerals, or selecting the best pieces from individual sources.

Intriguing new research at Scladina, Belgium, shows that in some situations, colour was the key factor. Sometime after 45 ka, Neanderthals there had brought more than 50 flat, dark-grey pieces of siltstone at least 40km (25mi.) over a high plateau from another river basin. No use traces were visible,[*] but the graphite-rich stone has no pyrotechnic value and is extremely soft, and when rubbed leaves clear black marks.

Neanderthals' choosiness and willingness to systematically collect pigment tells us that whatever their use, it was far from thoughtless. Most interestingly, some sites indicate connections between pigment and shells. Cueva de los Aviones sits beneath an eighteenth-century fort at the port of Cartagena, southern Spain. Remnant deposits by the old cave mouth contain hundreds of shells, probably collected for food, but beneath cemented sediment on two dog cockles, red ochre was visible. Small holes near the tips of both shells have been claimed as artificial,[†] but whether or not that's the case, they're not anomalies. A horse bone and three other shells – thorny oysters – also bear pigment, and remarkably they show Neanderthals were colour mixing. Analysis showed that the 'recipe' contained haematite, goethite, black carbon (probably charcoal or burned bone), limestone and sparkly pyrite.

The Cueva de los Aviones discovery stimulated speculative headlines about Neanderthal cosmetics and jewellery, but even if that's not true, it's a hugely important find. Despite their small size, the shells could have been containers, though the mixing must have been in something else. These Neanderthals were experimenting, combining substances to create different visual effects. Moreover, they had to have obtained the ingredients from different rock outcrops, the closest at least several kilometres away.

Recent dating of flowstone overlying the Cueva de los Aviones sequence gave an unexpectedly ancient result of 115 ka; far in excess of radiocarbon dates. If confirmed this would push complex pigment use well back in Neanderthal history.

Elsewhere, Neanderthals were applying pigment to shells that definitely weren't food waste. Microscopic analysis of a tiny fossil mollusc from the Discoid Level A9 at Fumane Cave identified pure red ochre inside surface micro-pits, but *only* on the outside. What's

---

[*] Possibly due to excavators washing them before their significance was realised.
[†] Naturally occurring holes on other shells look slightly different.

more, the pigment was sourced up to 20km (13mi.) away, while the nearest rocks with such fossils are located over 100km (60mi.) from Fumane.

Combined, the fossil shell and pigment are more than two unusual materials: they have a distinctive new meaning, a specialness. Tantalisingly, researchers could also see wear on the shell's lip where something soft but abrasive had repeatedly rubbed sideways, suggesting that it had once been strung or attached by thong or thread. This was an aesthetic artefact, the colour meant to be seen.

This single object sparks a cascade of scenarios, all imagined but archaeologically supported. Sometime around 46 ka, a Neanderthal noticed stone shells eroding from limestone, and picked one up. She carried this tiny piece from place to place; kept it so long its surface became polished. Her fingers pressed in special red colour, leaving a stain on her skin. Eventually, one day it dropped from its usual place – folded in or tied onto leather – and unknowingly or not, she left it behind in the mountains.

Had the shell been excavated from an early *H. sapiens* site, seeing this as symbolic behaviour would be uncontroversial. Whether it had practical as well as aesthetic use, it was unarguably precious.

Pigment was probably a lot more common among Neanderthals than the tiny surviving traces would lead us to believe. They may also have used it on canvases rather larger than shells. In 2018 new dates from three Iberian caves full of Upper Palaeolithic paintings were announced. Each sample covered – or was next to – red pigments, and they were so old that nobody but Neanderthals are known to have been around. At Ardales Cave, Malaga, various stalagmites and flowstone formations in different areas of the cave have obvious red daubings, sometimes only visible as a layer inside broken formations, gradually hidden over time. The new dates seemed to fall into two phases, the younger of which had minimum ages of 36 ka and so could be *H. sapiens*. But a handful of older dates had minimum ages before 45 ka, and one was an astonishing 65 ka.

By itself, Ardales would already be hugely important, but the other sites were truly unexpected. At La Pasiega, Cantabria, one sampled area of millimetre-thick crust covering a vertical red line pointed to it being painted well before 60 ka ago. The third locale was different again. Maltravieso in central Iberia was already well known for hand stencils made by spraying or daubing paint, a motif found in numerous other Upper Palaeolithic decorated caves. On the ceiling of an isolated area a faint example was revealed by photographic processing, and when adjacent calcite samples were dated, the oldest came out at more than 54,000 years old. While the Königsaue birch tar and the Le Rozel sands bear accidental prints, if genuine this would be the first intentional image of a Neanderthal hand; a spine-tingling thought.

These discoveries have set ablaze debates over whether Neanderthals really were the artists. Taken at face value, the chronology should place this beyond doubt: more time lies between some of the paintings and any sign of *H. sapiens* in the region than between you and the end of the last ice age. But to many, the dates are literally incredible, in the sense of being highly unlikely. They're extreme anomalies, even for other calcite samples in these locales. The same caves have panel after panel of art that, based on abundant, independent evidence, are believed to be Upper Palaeolithic. The line at La Pasiega is part of a larger grid-like design surrounded by other images, all of which have overlying crust dates younger than 12 ka, and in the entire cave no other dates are older than 22 ka. The situation is similar at Maltravieso.

Caves are extremely tricky in geochemical terms, and in the absence of an explanation for why closely adjacent crusts formed at dramatically different times, the suspicion is of contamination.[*]

Whether or not the dates in the Iberian sites are eventually verified, the implications for Neanderthal aesthetics that they represent may not in fact be so revolutionary. A painted line on a cave wall is not so different to slicing lines through animal skins, across bone or wood. And the hand negative, while undoubtedly striking, might not be a massive cognitive leap for hominins who probably already understood the *idea* of representation. Animal footprints are effectively symbols, and even simple tracking requires an 'idealised' form to be kept in mind. Handprints, then, are human tracks, and visible in daily life, whether butchery-bloodied or marked by soot. Moreover, a

---

[*] Natural contamination by uranium giving an erroneously old age can be ruled out with other methods, and ideally samples should be removed down to rock, then micro-sliced in laboratory conditions.

little-known site in France may provide independent evidence that Neanderthals elsewhere were marking and painting walls.

In 1846, two years before the Forbes Quarry skull was found, caves near the small town of Langeais, near Tours, were uncovered, probably by railway workers mining cliff-front deposits. One, known as La Roche-Cotard, was largely emptied out by 1913, but reinvestigations since 2008 discovered small patches of red pigment on the walls, along with finger marks dragged through soft, silty deposits. Geological study and archives from the early twentieth-century excavation suggest that the cave was filled to within 20 to 50cm (8 to 20in.) of the ceiling, then sealed off by 39 to 35 ka. Only Middle Palaeolithic lithics were found, along with fauna dating to between 50 and 44 ka – all of which implies that the pigment and finger flutings are the work of Neanderthals.

Certainly this site is far from ideal in terms of direct dating and context, but the use of pigment would also not be out of place with what we see elsewhere. Most interestingly, the largest spot is placed on an unusual, sinuous chert formation, emerging from the wall like entrails of the rock itself. And the co-presence of pigment and finger work connects to another realm of Neanderthal material engagement: engraved markings.

Markings

Neanderthals spent much of their time incising, scraping and creating markings on different substances, often as a by-product of other activities; think of butchery cut marks. But increasingly, it's also looking like sometimes the marking was itself the point, and occasionally they even engraved pigment itself.

In the same level at Les Bossats where large stones had been brought in, over 80 small, reddish-orange nodules were found. Using twenty-first-century analytical methods – including the world's only dedicated heritage particle accelerator – researchers showed that rather than local iron-rich formations,[*] Neanderthals had used minerals brought in from across the Loing River.

---

[*] Known as *crottes de fer;* literally 'iron turds'.

Available deposits range between 5 to 40km (3 to 25mi.) away, but only purer concretions had been selected. And most unusually, among varied wear traces including pounding, scraping and smoothing, some nodules had been incised with deep, parallel lines. They're different to scratches that would produce powder, and occur in groups of two to four. All the pigments were found within 10m (30ft) of each other, some even closely packed with lithics, bone fragments and burned materials into two small depressions, probably intentionally hollowed out. Almost identical objects when found at early *H. sapiens* sites are interpreted as graphic forms, and very likely symbolic.

These are rare things in Neanderthal sites; however, there may be one other among the manganese pieces at Pech de l'Azé I. But scored marks are known on many other materials. Scratches on cortex, the chalky outer skin of some lithic artefacts, can be accidentally caused. On the other hand, lines and pits on objects in some Italian sites must have been made before the core was knapped, which is hard to explain except possibly as a way to produce white powder. Indeed, among the red and black pigment at Combe Grenal, Neanderthals also brought in four pieces of chalk.

But a small engraving on cortex from Kiik-Koba, in the eastern Crimean Mountains, seems different. Microscopic analysis found that 13 roughly parallel lines were probably created with the same tool. Three are noticeably shorter and distinctly shaped, however, marking a change in desired form – another engraver or perhaps a change of tool. The cortex wouldn't have produced powder, and crucially the lines all begin and end *inside* the area of the artefact. Whatever the motivation, it's difficult to see this other than as an aesthetic project; rapidly executed, but with focused attention.

More commonly incised than minerals or stones are animal remains. While microscopic scrutiny has winnowed out various examples as butchery or natural markings, a number of others don't have such explanations. The most ancient is an elephant bone from Bilzingsleben, Germany, engraved with two sets of parallel lines at different angles. At around 350 ka, it's not much older than Schöningen and was likely made by early Neanderthals, but following this there are few other objects for the next 150,000 years. In contrast, three new finds have recently emerged, all dating between 90 and 45 ka, and all involve unusual species.

At Pešturina Cave, Serbia, 10 fan-shaped lines on a probable old bear neck bone have no feasible origin during butchering. Instead they're more like the Kiik-Koba cortex lines, terminating before the bone's edge, like a design within the space. The other two artefacts are small, but remarkable for their symbolic potential.

One was made on a broken, already old hyaena bone from Les Pradelles, and the other on a raven's wing from Zaskalnaya VI rockshelter, in the Crimean Mountains. Despite being widely separated geographically and in the method with which they were created, they share tiny sequences of evenly spaced incisions.

Five out of the seven notches on the Zaskalnaya bone are deeply sawn, but two that appear to have been added in-between are far shallower cuts, probably using the same tool but held differently. Without those two additions, the overall effect would have been perceived as uneven: they are about aesthetics.

The Pradelles hyaena bone is even more exceptional. On a surface just 5cm (2in.) long, a Neanderthal made nine parallel incisions, with extremely similar shapes. All cut in the same direction with a single tool and probably at the same time, the final mark looks to have been crammed into the narrowing width of the bone, as if its inclusion was more important than the overall appearance.

Then it gets weird: near the third line's base is a further set of eight minuscule nicks, in two pairs, each set intersecting at their origin points. Just 2 to 3mm (0.08 to 0.1in.) long, they're nonetheless regular, all made with the same tool – though not necessarily that which cut the larger series – and certainly not natural.

Both the Zaskalnaya and Les Pradelles markings go beyond most other Neanderthal engravings by showing regularity and structure. The raven bone hints at a desire to maintain a pattern, which might include paired markings. But the Les Pradelles bone is probably the first strong case for Neanderthal notation, counting things of equivalent value, with the small secondary marks somehow adding to or altering the meaning of the main series.

Neanderthal mathematicians are far from implied, but like many other animals, they must have had an innate ability to precisely recognise small quantities. This is more instant comprehension than counting, and would have been complemented by a general understanding of more versus less when dealing with larger quantities of things. It's believed that human skill with numbers evolved from these capacities, amply demonstrated in apes, and the process probably began early in the hominin lineage with small numbers. That's just what we see in the groupings of marks on the Les Pradelles and Zaskalnaya bones. Rather than an ordinal '1 to 100' understanding, it's possible Neanderthal numeracy was based in sets, like tally systems.

Fascinatingly, in children innate number recognition is present across the senses: hearing as well as looking can evaluate quantity. It's therefore possible that, given the diminutive size of the secondary incisions on the Les Pradelles bone, they were meant to be experienced through touch. Imagining Neanderthal fingertips rubbing across them raises the fact that all the engraved objects discussed so far were easily carried, perhaps even shared. But there's one exception.

Within Gorham's Cave, Gibraltar, 13 intersecting lines were deeply etched sometime well before 40 ka onto a raised section of the stone floor. They form a rough grid pattern, dubbed by the media as 'the hashtag', but would have taken far longer to produce than the average tweet. Experiments suggest that somewhere between 200 and 300 gougings were needed, and they were produced in a particular sequence. First, two deep horizontal lines were done, followed by five verticals, all cut in the same direction; next, one of the horizontals was deepened, and finally more verticals were added. Again, the impression is of sets and ordering.

The weight of evidence from more and more cases of pigment use and mark making is increasingly leading even sceptics to accept that Neanderthals had an aesthetic, symbolic element to their lives. Nobody will argue that the Neanderthal artistic oeuvre – even assuming the Maltravieso hand or Pasiega line dates are correct – is the same as that created by cultures across the globe today; but did they have 'art?' Our species likes to reserve such tendencies as a self-defining trait, but even captive chimpanzees, when materials and the idea of painting are supplied, enjoy colouring and marking surfaces.[*]

---

[*] Humans enjoy purchasing them too: pieces by the first chimpanzee painter, Congo, have sold for many thousands of pounds.

Ape works in fact show startling parallels with some of what Neanderthals did: they generally stay within the canvas, and show appreciation for symmetry or balance. Marks are made at roughly equal distances, and gaps filled or paintings lengthened to fill spaces. Chimps also revisit particular zones in a single session, overwriting existing marks with new ones, and some are keen on mixing colours. A personal style even emerges, with some chimps favouring different forms, including radiating, fanlike lines.

Most intriguingly, though intensely focused while painting, they often appear less interested in the resulting image. For them, the aesthetic – in its original sense of being perceived and enjoyed – lies in the creation, not the end product. Art as the *process* of bodily and sensory engagement with materials might be unfamiliar to classical Western sensibilities, but plenty of human cultures through time understand its transcendent power.

Feather and Claw

The Zaskalnaya raven that once soared over the Crimean Mountains connects us to another recently recognised realm of potential symbolic behaviour: what Neanderthals were doing with birds. As we saw in Chapter 8 there's plenty of evidence that they were eating them; the Zaskalnaya bone also bears marks from removing meat. However, there are hints that things sometimes went beyond mere survival. In particular, in a number of sites wings are more common than might be expected.* They're far from the meatiest parts, yet often have the most cut marks and come from unusual species: in the raven layer at Zaskalnaya, there's also a bone from a grey heron's wing.

Think back to the intriguing spatial separation of waste from processing wings in level A9 at Fumane Cave (the same layer as the ochred shell). While the presence of other body parts from chough and grouse point to use as food, raptors are *only* represented by butchered wings, whether lammergeier, greater spotted eagle, black vulture or even diminutive merlin. Clearly, through at least a century

---

* Some carnivores create natural wing-rich assemblages, but they can be separated by taphonomic analysis.

Neanderthals here were interested in these creatures and their wings, and this seems to have continued to some extent in later layers.

But bird feet may also have been a focus of attention. Again at Zaskalnaya, in the raven's layer the end toe bone from an eagle was found, while the older level A12 at Fumane produced an eagle claw. And even the grouse from Level A9 there seem to show an excess of foot bones, despite being hunted and brought back whole from the nearby pine forest.

Systematic studies have found a similar pattern of butchered raptor feet or claws – especially eagles – across a number of locales dating between 100 ka and 45 ka in France and Italy. In some places there's more than one: at the collapsed cave site of Les Fieux, a few kilometres south of the Dordogne valley, 20 bones from large raptors from multiple layers are almost all talons. Most tantalisingly, two of the biggest white-tailed eagle toe bones from the same layer were missing the talons; perhaps having been taken elsewhere.

Theories that talons were adornments existed, but became amplified after eight butchered white-tailed eagle talons from Krapina were proposed to be a necklace. Along with a toe bone, microscopic examination found smoothed-over cut marks and small, bright areas that resemble contact polish from rubbing against both soft and hard things. No other bones there have similar wear patterns, and while various birds are also part of the Krapina fauna, it's only eagles – three or maybe four birds – that were butchered.

However, proposals that the talons were originally strung together are difficult to support since even though they all come from the uppermost level,[*] it's a thick deposit and there's no proof they were associated with each other, much less as a single, impressive object. A recent finding of collagen fibres preserved beneath a thin silica film on one of the talons is intriguing, but it's not enough to claim as the binding for a claw collar.

Although these discoveries fired reconstruction artists' imaginations to produce images of Neanderthals sporting feathers and claws, can we be sure there isn't a practical purpose? Raptors and birds from the crow family were absolutely on the menu in some contexts, but the

---

[*] Above that containing the hominin fossils.

preponderance of talons does stand out: sometimes they're the *only* butchered bird bone. And careful analysis of traces from jointing, skinning and scraping wings and feet implies that this wasn't always to do with getting flesh or marrow, especially for small species like chough. It's not hard to imagine uses for entire wings or even feet and talons – from brushes to hunting camouflage to piercing tools – but none really convince given the effort required.

An often overlooked resource, however, are tendons. These stringy things can be used for all manner of purposes, and we know Neanderthals were systematically extracting them from hunted mammals like reindeer. In big raptors they're particularly large and strong, but experimental research showed that in many sites Neanderthals were removing claws from behind, and in the process cutting right through the tendons, while also apparently avoiding the shiny talon itself.

This duality is also seen with wings: sometimes it's probably about getting out tendons, but often the butchery marks point to Neanderthals actually seeking the primary flight feathers. Unlike fluffy down, these have no thermal properties, and neither would they be of any use as flight aids on spears of the size typically used.

Instead, it does look as if aesthetics or a symbolic interest in some cases was part of, or the primary, motivation. Feathers have been used for social purposes in societies across the world, and bejewelled grouse-foot brooches are still worn by some in the shooting community or for pinning Scottish kilts.* Is it so strange Neanderthals might have collected bird parts for beauty, or other associations?

Colour is one of the key things that makes feathers attractive, and it's notable that many species Neanderthals were interested in had distinctively dark plumage: blacks, dark browns, greys. Even talons and claws are typically black and glossy. Red is present too: the sooty feathers of male black grouse are set off by red 'topknots', while chough combine shiny black plumage with red or yellow bills, red feet and black claws. Significantly, in 2020 pigment was identified on the same talon at Krapina as the collagen fibre, and it's another mixed

---

* While these were Victorian gentlemen's hunting-related jewellery, they may originate as an older lucky charm tradition.

recipe: red and yellow minerals, charcoal and clay. This association strongly suggests that in some places, bird parts were part of the Neanderthal aesthetic palette.

But why these animals? Many of the species that seem to have received special treatment like birds of prey or the crow family were well known to Neanderthals, attending like familiars at their kill sites. Chough in particular may even have become habituated, since they live by caves, feed in pasture where large herbivores graze, and as can be seen at ski resorts, are partial to human rubbish. But birds in general may have had deeper resonances. They were always within sight or hearing, and far more than for most people today, the Neanderthals' existence would have been bathed in constant birdsong. Days beginning and ending with dawn and dusk choruses were filled by seasonal song, alarm calls or far-off cries of circling gulls or eagles. When night enveloped the land, owls echoed across valleys, nightjars comfortingly hummed and nightingales lent glory to the dark. Yet, like us, Neanderthals were left far behind by birds' effortless flight; perhaps they too dreamed of soaring into endless skies.

More than This

Aesthetics can be about altering substances and materials, creating sensual experiences or producing visual or other effects. Sometimes these might be applied to the self: whether or not the Krapina talons were really an Eemian necklace, the possibility that some objects like shells, talons or feathers were used for bodily adornment is real. We might also consider other animal parts like hair that leave little trace could also have been used: missing bones at Schöningen may indicate that as well as hides, Neanderthals took with them the swishing tails of horses.

Clothing was unquestionably worn by at least some groups, and while researchers might dream of a Middle Palaeolithic Ötzi – a frozen body complete with entire outfit – we must remember that Neanderthals from widely separated times and places might have been just as surprised by each other's attire as us. One thing was probably universal: their keen interest in material properties would have featured in their clothing choices, and this might have extended beyond function to appearance. Lissoirs today aren't just used for softening hides, but also burnishing: in addition to inducing a water-repellent effect – making soggy autumn hunts more comfortable – they add a shell-like lustre.

Tanning to make leather isn't necessary when hide working, though it helps with preserving and waterproofing. But if you want to add colour along a pink–orange–brown spectrum, it's perfect. Incredibly, an organic residue on a small stone flake at Neumark-Nord tells us Neanderthals did sometimes make leather. Chemical analysis identified high amounts of tannins, potent plant-derived substances that give colour to your cuppa and are the means by which bog bodies survive. Moreover, the tannins at Neumark-Nord came from oak, which along with chestnut is the best tree species for tanning in an interglacial world: once more, Neanderthals were selecting for quality.

This tiny brown scrap is a peephole to wet, stained hands moving around in a large container as the bark stewed. And since tanning even small, thinner hides like fallow deer takes a week or more,[*] this was a staying, not a walking time. While smoking will turn hides brown, tanned leather combines practical qualities and a range of brighter colours; apparently important to Neanderthals who took the time to produce pigment mixes. Oak isn't always available, and in cooler climates willow and birch bark or even berries work, while birch tar itself is also another option.[†]

But the evidence of bark tannin at Neumark-Nord has one last interesting connection to other Neanderthal crafts that might just have been about self-adornment. As well as the Fumane shell, another object bears traces of having been strung or threaded. A deep grove on the manganese retoucher at Combe Grenal cuts through earlier scrape marks, and has a smoothed inner surface that indicates something soft within it had repeatedly rubbed. Until recently, there was no evidence that Neanderthals made any kind of cord, but that changed in 2020 with the announcement of an astonishing find at Abri du Maras. Hidden on the underside of a flake was a natural

---

[*] Big, thick hides such as bison can take a year.
[†] Russian leather often uses birch tar and its famous aroma was used in Imperial Leather soap.

encrustation containing a 6mm (0.2in.)-long twist of plant fibre, either made from pine or juniper bark, or potentially the roots of these species. Most surprising of all is that technologically it's a classic three-ply thread: three individual strands each with fibres twisted in one direction, then all rolled together with the reverse twist. Furthermore, it's extremely fine: equivalent to threads in a hand-woven linen scarf.

Even assuming possible dehydration shrinkage, something so thin has limited purposes. Hafting bindings for small lithics is possible, but thread for attaching or stringing special objects is another. Unique finds like the Abri du Maras thread are almost unbelievable even to archaeologists, and rightfully subject to critical appraisal. But like the Neumark-Nord tanned scrap, the Fumane shell or any number of one-off flukes of preservation, they're what we have to work with. A balance must be struck between caution and not ignoring singular artefacts simply because they're rare or wondrous.

Whatever Neanderthals wore, whether pigment on skin, gleaming tanned leather, cosy furs or threaded red shells, it was always about more than function. Placing things on bodies is a powerful way to express status and identity, and observable in plenty of animals. Apes will sometimes 'drape' themselves with things, and in particular chimpanzees like to carry parts of their prey, leading to an observation of a wild individual sporting a knotted strip of monkey skin round its neck, complete with tail. The knot might have been accidental, but the wearing wasn't.

For Neanderthals, wearing clothing made from hides and furs must have mentally recalled the animals whose bodies they came from. Altering one's appearance, or making oneself stand out through objects or colour, also opens the door to more complicated things like communicating social affiliation to kith and kin.

Belonging to social categories, such as age and gender, might also have been entwined with the aesthetic and symbolic things Neanderthals made or wore. Earlier chapters have already touched on how difficult it is to define gender in the past, but there do seem to be hints that Neanderthals classified anatomically or genetically by sex lived in distinctive ways memorialised on their bones and teeth. The apparent connection between female bodies, intensive use of mouths for clamping and dragging as well as symmetrical arm development all point to hide working. This would echo many hunter–gatherer cultures with strong associations between women as hide-working artisans, making things just as crucial to survival as stone tools. Neanderthal ideas of gender probably drew on many things and didn't map precisely onto Western concepts of femininity, but it's an intriguing possibility that hide working and clothing itself might have been one point at which their material culture intersected with social identity.

Synthesising all the evidence for Neanderthal aesthetics and symbolic concepts results in an impressive corpus. But perhaps one of the most important results of the past three decades is that this burgeoning database makes it possible to discern conceptual commonalities between individual examples, as well as other aspects of their lives. Pigment is found with shells more than once; it's also incised with lines, as are bones and stones. And it's mixed with other substances to make something new, just as hafting adhesives were cooked or made up from recipes of pine resin and beeswax.

In some places, other unusual things were happening. At Les Pradelles not only was bone engraved with lines and notches, but this material was also used for large numbers of retouchers, and even directly knapped. And in the same layer as the raven bone at Zaskalnaya, Neanderthals were doing all that and also using red pigment, collecting wings and feet from large impressive birds, and even carrying tail bones of a dolphin all the way from the Black Sea.

One thing remains, which is to ask how all this compares to early *H. sapiens*. Neanderthal pigment recipes echo a 'paint kit' inside a shell at Blombos, South Africa, dating to 97 to 105 ka, and the inclusion of sparkly pyrite is similar to shiny sheets of mica found with ochre at Madjedbebe, Australia, dating to 52 to 65 ka. The Fumane ochred shell resembles finds from a number of early *H. sapiens* sites, but Blombos is remarkable for having an assemblage of shell beads very likely strung together.

Some Neanderthal engravings are clearly structured, but they're far from what was happening in another South African cave, Diepkloof. Earlier levels dating around 100 ka contain simple linear engravings on ostrich eggshell fragments that are indistinguishable from those

Neanderthals created on bone or pigment. But by 80 ka, incised fragments with complex grids and bounded rows of lines appear, and continue through multiple layers. In the roughly contemporary site of Blombos, there's also a famous piece of red ochre with a framed X-gridded pattern. So far we have nothing from Neanderthals so strongly ordered, nor any graphic tradition visible across multiple pieces, as at Diepkloof.

What Neanderthals do have in common with early *H. sapiens* prior to 45 ka is an absence of any unequivocal representational art, manifested by carvings or breath-taking creatures running across stone ceilings. The oldest known image of an animal was painted before 44 ka in Sulawesi, Indonesia; there are also handprints around the same age from Lubang Jeriji Saléh, Borneo, and around 41 ka a tiny woman sculpted in ivory was left at Vogelherd, Germany.

It's not clear whether these were independent artistic flowerings, or if, like their Southern African contemporaries, the populations dispersing into Eurasia before 80 ka brought with them a common artistic tradition. And perhaps the roots go back even farther. The very oldest graphic engraving is a clear zig-zag on the surface of a freshwater shell from Trinil, Java, made an astonishing 500,000 years ago. This raises the possibility that the Neanderthals' and our ancient, aesthetic heritage was a shared legacy from deep in the *Homo* lineage. We might have walked into a new continent and found it adorned by art already many millennia old.

The specific motivations behind Neanderthal aesthetics must remain unknowable. We might understand the primeval excitation of neurons by light, hue and texture, or how skin and mind both are exhilarated by the rushing scream of a flock of swifts up into the sky. We might even reach for what seem like obvious metaphors: liquid red ochre as blood from the earth. But truly trying to glimpse Neanderthal minds is like seeing shafts of sun filtering into a cave, hazed by the dust motes of millennia. We must also forget classical ideas about art, and appreciate that sometimes the significance and symbolism may have been in the *act* of transformation itself. Changing colour, marking surfaces, even taking feathers from a wing that flew could have had meanings that resonated most during the process of creation, rather than emerging afterwards.

And this returns us to the enigma of Bruniquel, a reminder that the tests we set for symbolic significance may well have had nothing to do with what was meaningful for Neanderthals. Monumental in scale and vision, it's the first great art project, and at the same time truly weird, as in the original meaning of *wyrd:* destiny-changing power. Hominins might have made nothing like it for another 160,000 years, and the 'why' behind those stacked circles of burned stalagmites is lost in darkness. But it marks a threshold in creative potential, and perhaps is even more unexpected than ochre paintings on cave walls. To our eyes today it's both compelling and beautiful.

CHAPTER TWELVE
Minds Inside

Lips shiver, parched tongue licks at trailing sweat droplets. Through eyelid cracks she sees dawn-glow on the rough wall, and beyond, her companions silhouetted. Riding the rolling, convulsing river of her body, their comforting touches on arm, wrist, said: 'We are watching, waiting.' Moving to meet a new aching crescendo, roughened old hands pull her to kneel. The world shrinks as she closes her eyes, and yet it inflates, a dark blood singularity where nothing exists except this moment, this enormous surging within. She feels/sees the baby squeezing down inside. The force ebbs but the others mutter and keep her up; and they're right, because muscle heave returns, unstoppable like a pounding bison mob, ragged herd-breaths filling the air – or are they hers? – as she unfolds into a new configuration.

Suddenly the birth scorches hot and – even as legs still thrash in her belly – she reaches down to find a tiny head: slick-pelted like an otter, fragile as a swan's egg. One last bone-cracking effort breaks free the cascade and a slippery form is

*scooped up onto her stomach by the hands all around. Her shaking arms clasp
downy flanks, softer than autumn leather worked through the snows until the
teeth ache. Cave-dark eyes gaze up at her, and she nuzzles the baby, breathes
deep of its scent: blood-loam rich and heady.*

What connects you and I today asking questions about pigment, feathers or engravings, and the makers of those objects, is that we are all beings of feeling, with hearts that swiftly beat from terror or joy. If Neanderthals found some things beautiful, is it possible to know what – or who – they loved? Even what terrified them? Once again this is a balancing act, teetering between the solidity of the archaeological record and the possibilities that spin off from it.

Thinking through fear is an easier starting point. Always present in Neanderthal minds must have been the predators with whom they co-existed. Even equipped with flame, weapons and the guile of 100,000 hominin generations before them, encountering cave lions or hyaenas would have elicited instinctual dread. On the other hand, cases of carnivore butchery indicate that Neanderthals were able to overcome this. There might even have been recognition of fellow top hunters, especially perhaps wolves, the most cunning of all pack creatures.

Other risks they faced were elemental. Lithic movements show that river crossings certainly took place, including huge spans like the Rhône. Even if shallow fords were chosen, danger lapped around bodies as arms held precious things out of the water; a mis-step may explain the Tourville-la-Rivière arm. And in a world where unexpected icy winds or soaking storms could be life-threatening, exposure might have been feared.

Even if Neanderthals had mastered flame within the safe confines of a hearth, the horror of wildfires was a different matter; perhaps a particular issue during arid phases of the Eemian. And conversely, the absence of fire might also have been disconcerting. When Neanderthals stepped into echoing deep caves like Bruniquel, illumination was vital, and its loss potentially fatal. Outside, northern winter nights would have been long and dark. Even warmed by embers and lit by the cold burning stars, dawn must have been greeted with relief.

What about happiness? Whether walking over springy steppe-tundra or through grassy forest clearings, Neanderthals very likely experienced the simple joy of sun-warmed skin. Other kinds of contentment are also guaranteed. Human appetites for pleasure are far from unique among our fellow apes, and we must assume that many Neanderthal sexual relations were consensual and fun, though others possibly not. Anatomically, pelvic dimensions point to vaginas very similar to ours, and as penises are tailored to fit, those too were probably more like living men's equipment than that of chimpanzees.

Luckily for all concerned, unlike chimps Neanderthal males lacked the gene for 'penis spines'. While in apes they're more like tiny hardened pebbles than spikes, their presence does affect copulation: marmosets have sex and orgasms that last twice as long when the spines are removed.

We should probably therefore picture Neanderthal sex as more leisurely and satisfying than chimp-style rapid thrusting bouts. Not forgetting clitorises – organs solely existing for pleasure – unluckily for Neanderthals, like us they probably lacked bonobo-like versions that make face-to-face orgasms easier. But masturbation in some form is pretty much guaranteed, whether during sexual encounters as is found among humans, or more generally for social bonding and diffusing tensions, as in bonobos where it takes place between pretty much anyone.

So much for sex; how about love? Some of life's most intense feelings come with 'first love'. Neanderthals certainly went through adolescent growth spurts, but did they also have hormone-fuelled emotional upheavals and crushes? Le Moustier 1 shows us that youths were already precociously strong – his upper arms are as thick as an adult man's today – making any teenage scraps a serious business.

The age at which girls began to menstruate isn't known, but it's possible it marked a shift in how others treated them. However, over a lifetime, the number of times a woman bled could well have been relatively few: depending on social dynamics, many women in traditional societies without reliable contraception tend to either be pregnant or breastfeeding. And while research with hunter-gatherer cultures on their experience of menstruation is rare, there's a trend for shorter periods too – sometimes just two or three days – compared with Western populations. One thing girls today probably share with Neanderthal teenagers is learning how to cope with discomfort from periods, and how to keep clean; today that information might come from female relations or peers, and the same thing was probably true then.

A fascinating question is whether Neanderthals understood what beginning to bleed really implied, and similarly, what intercourse might lead to, 10 months on. Unlike other animals, all human cultures understand that male–female sex is directly linked to babies. If Neanderthals had also grasped this, it would have had profound social implications.

Many theories have been proposed for how reproduction was organised. Male-dominated groups is one idea, supposedly backed up by the fact that at El Sidrón, all the males were from the same genetic population. In contrast, the adult women were from two different lineages, and researchers interpreted this as evidence they'd joined a group based around male dominance. But in fact, since the El Sidrón fossils were deposited in a jumbled mass having been washed from elsewhere in the cave system, we don't know if those individuals were even alive at the same time, much less formed a social group. Moreover, among hunter-gatherers, it's often girls who remain in the same group as their mothers.

Added to this is the basic fact that Neanderthal bodies don't have any extreme size difference between the sexes, as we see in gorillas, which means it's unlikely there were alpha males with harems. Instead, like much of humanity, Neanderthals probably had pair-bonded sexual relationships; at least, mostly. That means that, unlike many other primates, parenting was more likely a shared task, and adult partnerships were long-term affairs.

That food sharing was a fundamental part of Neanderthal life tells us they were used to distributing things as a way of maintaining social bonds. Maybe, then, desire and devotion could be one explanation for creating and carrying small aesthetically pleasing objects. Perhaps the Fumane shell had been found, carefully covered in red ochre, and given as part of an intimate relationship; a 'love token' from 50,000 years ago.

Precious Little Ones

A different but just as powerful bond also exists between infants and parents. Despite constant denigration of them as a species, growing, impelling forth and sustaining babies was something at which thousands upon thousands of Neanderthal mothers succeeded. The implication of a swelling belly was very likely understood, and if so, labour might have been anticipated with excitement and apprehension. What was birth like for Neanderthals? Today it can be a life-defining event, and while circumstances vary, hormones and extraordinary physical effort often produce extremes in emotional intensity. Human labours often begin nocturnally, and instincts to seek particular places or positions can take over.

Neanderthal mothers probably chose to give birth in sheltered locales, hidden from predators. Caves or rockshelters are the obvious option, and this might even be one of the motivations for seasonal visits to such sites. But unlike most mammals, human mothers often prefer to have some proximity to others, particularly the first time round.* It's even been theorised that a *need* for birth attenders makes *H. sapiens* unique. Our babies must twist on the way out, prolonging labour, making it harder to catch the baby, and increasing the risk of a potentially fatal obstructed labour.

Combined scans of the Tabun woman's pelvis with the Mezmaiskaya newborn's skull recreated their birth canals, showing no twist was needed, but their infants already had longer heads, so it was still a tight fit. Neanderthal girls certainly faced some risks from pregnancy and labour over a lifetime, and may well have witnessed death or injuries. However, maternal health has varied dramatically through history. The dire death rates in seventeenth-century Parisian hospitals, for instance, where you were more likely than not to die giving birth,

---

* Some traditional cultures including the !Kung of southern Africa celebrate solo birthing as a mark of courage, but in reality, especially for first-timers, others are present.

were down to high infection rates and crude medical interventions. In contrast, traditional societies with informal midwifery traditions, whether hunter-gatherers or not, can be safer.*

But even in other apes, birth attendance is about more than physical assistance, and intimate studies of bonobos reveal astonishingly human-like behaviour. Females will actively support the labouring mother, checking her progress by looking and touching. This isn't just curiosity or a desire to grab the baby. Female companions will stay much longer with the mother during labour than the day before or after, and they're visibly more excited *before* the baby arrives than after. Moreover, comforting social communications are aimed at the mother rather than to each other, and they're also obviously protective. Both flies and males are kept away (and the latter in contrast show no protective behaviour). Most amazingly, just before delivery some experienced birth bonobo attenders were actually observed to mime catching motions towards the mother, and helped hold the baby's head as it emerged as well as assisting the mother to change position.

Bonobos of course are famous for their female-dominated societies built around strong friendships. It's this, combined with previous experience of mothering, that seems to lead to their doula-like behaviour. In stark contrast, chimpanzees live in groups physically dominated by males, and the females generally lack same-sex adult friendships. They prefer to give birth alone, and will isolate themselves afterwards. That's because infanticide is a huge risk, whether by males or other females, whereas it's never been recorded in bonobos.

Most interestingly, one of the key reasons why bonobos are so different (despite females still moving between groups) is that, like Neanderthals, they don't fight over food. If resource scarcity underlies chimpanzee male aggression and therefore isolated births, then collaborative hunting and complex processes of food sharing in Neanderthals would imply the opposite. With prime foods taken back

---

* Although not without danger: in many places women today face a 1 in 16 lifetime risk of dying in labour, and for under-30s it can still be the leading cause of mortality.

from hunts to waiting mouths, aggression within groups was far less probable, and bonobo-style female friendships could develop. With more human-like tricky birthing, it's not so outlandish to imagine Neanderthals too had midwifery tendencies.

As a context of intense emotional interactions, and where transmission of knowledge and skills might affect survival, labour could be one setting in which evolution would favour increased social communication. Moreover, the potential was even greater when multiple generations lived together, giving plenty of opportunity to watch and learn child care skills from a young age. Chimpanzee youngsters mimic infant care by cradling special rocks or pieces of wood, and small human children will, given the chance, carry tiny siblings around. The presence of experienced mothers, and even grandmothers, can help buffer the overwhelming experience of becoming a new parent, especially with post-birth basics, such as what to do with the placenta.

A surprisingly large fleshy mass, primates often eat it, but this might be more about avoiding the attention of carnivores than nutrition. Humans tend to bury it, but whatever Neanderthals did probably varied according to where they were, and even social traditions. After the placenta is passed, women typically experience bleeding known as lochia that can be heavy and last for many days or even weeks. Although it's entirely possible for some new mothers to get up and carry on as normal, for many societies the period of time after birth includes rest and increased support from relations. Extra quantities of the materials normally used for soaking up menstrual blood would be needed for lochia, as well as extra food: all that milk making for energy-hungry newborns would've required Neanderthal mothers to eat at least 500 extra calories per day.

Once they came blinking into the world, Neanderthal babies would have been extraordinarily like ours. Their developmental milestones were almost identical, going from helpless curled-up creatures to rambunctious toddlers within a year. They would have looked just as cute too, a vital feature ensuring parental affection over their similarly extended childhoods. Much is made of whether they grew faster or not, but in comparison to other primates, the difference is negligible and youngsters would've been totally dependent for many years.

Remarkably, we're able to see that in some cases babies were still largely breastfed at around a year old, thanks to ingenious analysis. At the Grotte du Renne, Arcy-sur-Cure, particular proteins from growing or healing bone were found in an area where remains of a Neanderthal child about a year old had been found. What's more, isotopes showed the highest-ever nitrogen level from any Pleistocene hominin. They're too high to come from anything except eating masses of freshwater fish, munching on carnivores or much more probable, getting most food from breastfeeding.*

Even with affectionate, attentive carers, babies this young still faced risks. In many hunter-gatherer societies, sickness and infection are the biggest infant killers, and one of the tougher times comes with weaning. Along with first tastes of food other than sweet breastmilk, tiny mouths take in a cocktail of new pathogens and parasites. Typically full weaning usually happens somewhere between 2 and 4 years old, and as demand for milk lessens, the mother's fertility returns and a new sibling will soon be on the way.

The age of weaning in Neanderthals is another milestone used to assess whether they grew up faster, and dental isotopes are one way to measure this. Although the methods are still very new and some are debated, barium is one isotope that's been proposed as a breastmilk marker. Tracking levels in the tooth of a Neanderthal child who lived around 100 ka at Scladina, Belgium, records changing amounts. It seems to have been exclusively nursing for just over seven months before other foods were introduced, but then not long after the baby's first birthday, the barium abruptly ceases.

If barium is truly tracking breastmilk consumption, such suddenness would more likely indicate that the mother was seriously ill or died. It looks nothing like normal weaning seen in humans, primates or even earlier hominins. But interestingly, it also implies that there were no other nursing mothers able or willing to adopt this baby. More recent research using micro-sampling across a nearly 3-year-old's tooth from around 240 ka confirms that even in Neanderthals, the Scladina child's experience wasn't typical. A much more gradual

---

*Nitrogen in part tracks the place in the foodchain, and since babies are effectively eating their mother's bodies, this makes them look like hyper-carnivores.

weaning signal is present, with significant breastfeeding continuing until just after 2 years, when it gradually declines and stops a few months later.

Perhaps one of the most joyously celebrated milestones in a bipedal hominin's life is walking. Neanderthal babies would have been carried for speed and safety, and indirect evidence of this comes from the young child from Payre. The changes in isotopes indicate that the group moved while the baby was just a few months old, far too little to walk unaided. Moreover, it was during its first winter, when some kind of protective cover or carrier would have been needed. Independent walking is itself connected in some hunter-gatherer societies to weaning, since physical closeness in a wrap or in arms promotes continued breastfeeding. Carrying tends to stop once the child is able and willing to get about without too much risk – or when they're too heavy – between 3 and 4 years old. So if Neanderthal children were stopping breastfeeding around then or just before, this could also mark the point at which they were no longer routinely carried.

Just as we hold the hands of unsteady walkers, we help babies through weaning by offering special foods. That Neanderthal mothers and babies shared food is a given (along with the mess), but can we identify 'baby meals'? There may be a hint in the Engis child from Belgium, which has much higher nitrogen levels compared to adults from the same region. Aged around 5 or 6, it's very unlikely it was still consuming enough breast milk to explain this, and an unusual diet must be the answer. Freshwater fish is a possibility, but perhaps particular parts of reindeer or mammoth like brains, or even fermented foods, are more likely.

Other parts of their bodies show that as Neanderthal children grew up, they quite quickly began to be very active. What's not certain, though, is how – or even if – childhood was conceptualised as a life stage. Undoubtedly physical immaturity was understood and probably catered for, but did adults help youngsters learn how to be grown-ups? The presence of dental wear in very little children points to at least some imitation; the enamel on a 3-year-old's front tooth from Combe Grenal was already worn away, potentially from tooth clamping. But there's an intriguing clue to something more from wear on the El Sidrón 1 boy's teeth.

We already saw in Chapter 4 that dental micro-wear shows he'd learned to eat by using a lithic tool to slice off food held in his mouth, but compared to older individuals, his scratches are much narrower. Either this was due to lack of confidence and tentative motions, or perhaps more likely he was using smaller, thin-edged artefacts. It's possible he knapped them himself, but fine flaking isn't always easy, and someone else may have made what was the equivalent of a child's cutlery set. Given that peer-to-peer learning was probably just as important as for other hunter-gatherers, it's interesting that at El Sidrón the only other individual with comparable narrow dental scratches was likely a teenager.

If youngsters did have artefacts made for them, it's a good marker for the existence of some kind of childhood status; but does this mean Neanderthals had toys? Humans, like many other animals, learn virtually everything through play. This includes objects, and youngsters can find fun in pretty much anything. It's possible some items with aesthetic qualities – bright colours or shiny surfaces – could have entertained little ones, but the time taken to produce many of them, combined with the use of relatively rare materials or substances, argues against that.

Of course, many toys are often simplified or miniaturised versions of practical items. Specially sized artefacts just might be visible, however, in some of the Schöningen spears that are noticeably shorter than others. They're clearly not 'pretend' weapons, but perhaps sized for – or by – smaller hands. A lakeside ambush could've been a relatively safe setting for learners, where thrashing horses were disadvantaged enough to allow practice spearing by youngsters with eyes wide, heart thudding.

After the hunt would come butchery, another crucial skill. Most diminutive lithics can't be counted as playthings, since small size is often caused by resharpening. Genuinely micro objects like tiny Levallois points or bladelets were systematically produced, and probably not the work of children, because the techniques concerned were tricky. But that doesn't mean they weren't also useful for little fingers to practise with, just like the minuscule, carefully done cut marks on small birds at Cova Negra might fall somewhere on the spectrum of play and survival skills that children's games are really all about.

Look After Us

Childhood was an introduction to a life full of dangers as well as excitements. Injuries and illness, sometimes severe, are very common in hunter-gatherer societies, and Neanderthals were no different. Strikingly, however, numerous bones testify that some outlasted their maladies. Proving whether this happened thanks to the kindness of others is, however, complicated. The Devil's Tower child from Gibraltar was just 2 or 3 when his jaw was fractured, and very likely required help from adults to eat. But while assisting toddlers might be expected, the other cases in more mature individuals are harder to be sure about. The Le Moustier 1 teenager's jaw had also been badly broken at some point in older childhood and probably affected his ability to eat, but it's impossible to say if he just soldiered on or was cared for.

At the other end of the scale, obviously toothless Neanderthal elders like at La Chapelle-aux-Saints existed, but so too with chimpanzees who manage to cling onto life despite receiving no help or special soft foods. This is a reminder that assumptions about what *requires* care will depend on one's own culture and experience. What animals do varies significantly. Chimpanzees and particularly bonobos do give comfort to upset or injured individuals, but don't *consistently* help others or provide provisions. In contrast, highly social creatures like elephants and cetaceans act together to support their injured, sometimes physically, and strikingly, lions, wolves and even mongooses will occasionally bring food to incapacitated adults. What's noticeable here is that, far more than apes, these species are by nature *collaborative* hunters and foragers.

Single Neanderthals could likely not survive long outside their group. If injuries or sickness were a routine part of life, then individuals requiring assistance – along with infants, mothers or those born with different bodily conditions – would have been all around, not anomalies. Just as killing, butchering and eating animals involved shared action and rewards, then it makes evolutionary sense that needy individuals, at least sometimes, were helped.

A few really beat-up Neanderthals are hard to explain without something like this. The dreadful head injury suffered by the Saint-Césaire woman would probably have caused confusion and certainly massive blood loss, and she must have been helped at least temporarily. Similarly, Shanidar 3 might have been lucky and avoided a collapsed lung after their chest was stabbed, but they may have struggled with breathing and walking. They hung on for more than two weeks before death, and given the increased calorie needs of Neanderthal bodies, this seems a long time to survive if nobody was bringing them food.

But it's Shanidar 1 who is the most convincing case for long-term support. Probably partially blind, certainly deaf in at least one ear, with hugely reduced mobility *and* a chopped-off arm, he nevertheless endured to a ripe old age, even adding arthritis to his aches and pains.

Despite his mobility issues, Shanidar 1's leg bones record normal levels of activity, and he obviously adapted to a one-handed adult existence. It would be astonishing if he overcame all this entirely alone. He clearly didn't exist at the margins, and after recuperating, probably contributed to gathering and even small game hunting. But he was probably provisioned with meat from big game for the rest of his life, and also protected.

Maybe the true marker between how some animals care for each other and the lengths humans go to is in medical skills. Chimpanzees' habit of eating bio-active substances probably helps with parasites or mineral balance, but this might be about taste preference rather than truly understanding the bodily effects. Chimps will sometimes dab at their own wounds with leaves, and intriguingly, orangutans have been seen applying chewed-up leaves to their skin, of the same species used by local Indigenous cultures for pain relief. Even so, that's still a far cry from the way we draw on natural resources in myriad ways to produce tisanes, poultices, unguents and other treatments.

There's no clear-cut evidence of herbalism in Neanderthals, but their deep knowledge of plants means it's not entirely unfeasible. An important cue to plants' medicinal characteristics can be distinct, often bitter tastes, which Neanderthals could detect and tolerate. When probable yarrow and chamomile were identified as being consumed at El Sidrón, claims that these herbs were used as medicines attracted much media attention. But they're both also used as flavourings, so the motivations remain hazy. Really the proof of self-medication is indirect, from healing and lack of infection in horrific wounds like those of Shanidar 1, the Saint-Césaire woman or the giant skull wound and another potential amputation at Krapina.

By the time he died, Shanidar 1 would certainly have been wise as well as old. But was his age out of the ordinary? Very small populations of Neanderthals, combined with a high percentage of skeletons being young adult males, makes it seem on the surface like lifespans must have been pretty short. If true, that would have serious implications: prime-aged men would be dying off before they'd had much time to reproduce. But all this rests on assuming that the fossils accurately reflect the age range of the living. One indicator that this is not the case is a drastic under-representation of female bodies, and as we'll see in the next chapter, cadavers were sometimes ending up in the archaeological record after Neanderthals had been doing things to them. Given all this, the fact that there are relatively few identifiable elderly – over 60 years old – individuals can't really be taken as proof Neanderthals were dying well before their biological time.

Although there were likely never hordes of hoary old ones, just as in recent hunter-gatherer societies three living generations was probably normal for Neanderthals. And growing up with grandparents matters, for two key reasons. First, casual babysitting is extremely useful, as it frees adults to find more food or do other tasks. Second, in addition to reinforcing what children learn from parents, grandparents are more likely to be competent in complex skills, whether producing hafting recipes, tracking or even medical care.

Knowledge and experience from a lifetime accumulates and beds down into wisdom. Elders would have been fundamentally important to the wider group too, since those Neanderthals who made it to 40 would have been goldmines in times of crisis. They could remember old ways, where things used to be; what to do when herds failed or the weather behaved in ways out of anyone else's memory. This store of sagacity extended to generations before, through what they'd learned from their own parents and grandparents. Such insights could make the difference between surviving and perishing.

Neanderthal elders might also have been important to the group's social history. If they understood that sex would beget children, some notion of kin lineages would have existed too. Knowing who was related to who is obviously useful during everyday life, but if gatherings between groups – whether planned or not – did take place, it might have been even more vital. Among hunter-gatherers, aside from resources, one of the key motivations for travelling long distances is for socialising, including finding sexual partners.

Not all Neanderthals were inbred in genetic terms, and keeping track of who moved groups, and which were friendly or hostile, would be most powerful in the most senior Neanderthal memories.

If we accept that Neanderthals had some level of speech and could project their minds in time and space as they planned moves, portioned out carcasses or made composite tools, then the existence of storytellers doesn't sound so implausible. Tales emerge from lived experience, and in orally based cultures around the world it's largely via stories that collective wisdom is conserved. In hunter-gatherers, the commonalities in stories are striking, often merging ecological information, social norms and cultural origins. Details of the surrounding world are understood and rationalised through relational ideas. Sometimes ancient tales from widely separated cultures are remarkably similar. Celestial cosmologies from ancient Greece and the Aboriginal Australian Kothaka people both include three bright astronomical star groupings: Orion/Nyeeruna, the Hyades in Taurus/Kambugudha and the Pleiades or Seven Sisters/Yugarila Sisters. The stories from almost opposite sides of the world explaining these constellations are virtually identical: a male hunter pursuing a group of women, thwarted by another entity. The common theme of chase may have emerged in response to the way these stars rise and move across the sky. Seen this way, nature tells its own story, and people merely repeat it to one another.

We might imagine Neanderthal tales to have worked in a comparable way, and if cultural lore of all sorts was transmitted by mouth, it becomes a kind of time travel, potentially allowing a group to 'know' things that took place a century ago through the memories of elders. Sometimes stories might have been spectacularly ancient. Particularly strong evidence comes from some Aboriginal cultures, where knowledge of subtle changes in star brightness was maintained over centuries, as well as the 4-millennia-old experience of a meteorite impact. Most astonishing is the 10,000-year-old cultural memory in Australian coastal communities that the oceans rose at the end of the last ice age. Perhaps Neanderthals too 'remembered' how the world of their ancestors changed through millennial-scale climate shifts, and found stories in constellations no living person has seen.

The emotional existence of Neanderthals who lived, worked, ate and slept together was rich and based on cooperation. Their collective hunting was mirrored in the sharing of meat and the arrangements of their sites. Born into loving arms, they grew up into complicated social beings, driven as much as we are by passions both devotional and destructive. And like an amplification of sound, collaboration within Neanderthal groups may well have evolved in tandem with multiple generations and storytelling.

In a world un-hazed by light pollution, the night sky arcing above their heads and the regularly shifting moon may have become part of murmured songs around a fire. Certainly, place itself and how it's bound up with experiences and memories of relations with things, animals and others would be central to how they understood the world. When histories and lineages merge with land, what happens to the bodies of the dead becomes one of the most powerful ways to invest a place with social potency.

CHAPTER THIRTEEN

Many Ways to Die

Full fathom five, oblivion lies. White horses crest waves rolling beneath the boat's spotlights. Engine chokes off; destination reached. Court orders are checked a final time, and a nameless official emerges from below, carrying something. A heavy curved vessel glistens dully under electric illumination, while the surrounding sea waits like a black maw under a cloud-shrouded moon. Seconds tick to the appointed hour. A fat splash, and the urn – for that is what it is – drops swiftly out of sight. No pearls will there be in this skull, for all was obliterated by furnace and mill. Crew wait restlessly for a quarter of an hour, clock-watching for the moment of extinction: fashioned from salt, the death-jar has dissolved. Ashy bone-grist palls outwards; disperses, disintegrates, disappears. Annihilation achieved. Engines roar, eager to return to the living land, and soon no trace remains but a dirty oil-glint on the swell; then that too is gone.

All creatures are unified by death. But the manner in which they respond is far from identical. In the darkest hour between 25 and 26 October 2017, the remains of the reviled 'Moors Murderer', Ian Brady, were secretly disposed of somewhere off the Liverpool coast. A child killer, his 1966 trial was the first serial murder case after Britain abolished the death penalty. Brady inspired such public revulsion that when he finally died, funeral directors refused to deal with his remains. Finally, a judge secretly ordered a local cremation with no observance, music or flowers, and that the ashes be disposed of at sea, inside a dissolvable salt urn that would ensure that no possible marker of Brady's physical existence could be preserved. Even 50 years later, the state did its best to ensure his obliteration.

Most interesting, press reports afterwards saw no need to explain *why* any of these extraordinary steps were taken. Official statements mentioned wishing to avoid distress to relatives, but something more was going on. The authorities actually released a statement that his coffin had been excluded from public areas and burned in a back-up furnace, and that everything had been scrupulously cleansed; this was about moral pollution.

Death and its handling have always mattered. Along with art and symbolism, some of the fiercest debates about Neanderthals have concerned what they did with their deceased. Despite advances in the past three decades, and even a newly discovered skeleton, understanding what mortality meant to them is still controversial.

Some researchers maintain that to prove intentional mortuary (death-related) practices, particular features must be present in every case. Others argue that a wider context is needed. Balancing these views, using modern taphonomic approaches is crucial, but so is acknowledging our own preconceptions. The simple presence of hominin bones is obviously not reliable evidence Neanderthals placed them there. But equally, the lack of a casket-shaped hole doesn't mean they didn't lay the dead to rest.

Look across the globe and through human history, and it's obvious that mortuary practice has always ranged far beyond burials. Nonetheless, unpicking what archaeologists dig up is complicated because there are many ways for bodies – or bits of them – to end up in sites.

Fragments of Neanderthals jumbled in with natural open-air accumulations of animal bones exist, like the arm at Tourville-la-Rivière. It likely washed up on what were sandy banks of the Seine below chalk cliffs through flooding around 180 to 235 ka, though what circumstances led to this isn't clear. Random parts at butchery sites are interesting, given that these locales were probably only relatively short occupations. Bits of at least one Neanderthal adult and a teenager are among thousands of reindeer bones at Salzgitter-Lebenstedt, perhaps hunting casualties.

In some cases Neanderthals were the prey, or at least a carnivore's meal. On the finger bones of a child found at Ciemna, Poland, distinctive damage from the digestive juices of a large raptor are visible. Furred scavengers also took their share, shown by gnawed leg and foot bones in some hyaena dens. But in places where tenancy alternated through time between Neanderthals and predators, how body parts ended up being scavenged is hard to unpick.

Even for mostly complete bodies, never mind fragments, it's extremely difficult to prove that they were intentionally left at a particular place. Numerous reasons could lead to individual Neanderthals taking their last breaths in a cave: succumbing to sickness, injuries received elsewhere, even starvation or violence. The problem comes with inflexible tick lists of how the remains of the dead 'should' have been treated, which may overlook otherwise meaningful actions taken by the living.

Burials aren't a gold standard, they're simply the most obvious. Moreover, they're also a spectrum, going from a specially dug pit, to a natural hollow or niche, to simply covering with sediments. The myriad other ways bodies can be dealt with must also be looked for, even if they're not so easy to identify: being exposed, cut up, burned, curated, displayed, recycled or even eaten. First, however, it's worth understanding where these urges come from in the first place, with the origins of grief.

Death is emotionally overwhelming for our closest living relatives. Intense bonds are the bedrock of chimpanzee and bonobo societies, and their severing is traumatic. Exactly what happens varies, but a passing is never ignored, as has been claimed for Neanderthals. Each death totally occupies individuals – and often the entire group – for many hours, with extreme emotional expression and interaction with the body being the norm. The atmosphere is volatile and explosive, flickering from aggression to submission, reassurance to dominance. Stress is released through mating or vocalising, and most intriguingly sometimes the sounds are those associated with fear and threat, especially similar to 'stranger danger' calls.

After a frenzied chorus in the immediate aftermath, the collective noise pulses from calm back to frantic crescendos. Yet at other times quiet reigns. Particular individuals undertake vigils of attendance, sitting and staring silently at the body, even overnight. Sometimes the watchers' hands reach out, sometimes not, but in general the urge to touch corpses is unmistakable, whether by poking, patting, dragging, carrying, cradling or grooming. Interactions with bodies sometimes seem aimed at eliciting a response, whether looking in the eyes, nudging or even trying to play.

Yet circumstances matter too: sudden deaths inspire particularly intense emotional reactions, probably as a result of being unable to rapidly process change. And social ties also affect what happens. Close relations and friends of the dead often show the most extreme responses, and stay close to bodies.

Chimpanzees and bonobos patently have a hyper-awareness that death signifies a change of state, and at some point, things shift. Wounds that would normally be licked are investigated, but not tended to. Bodies metamorphose to materials, becoming things to be used for social displays. Once this happens, high-status individuals begin trying to control access to corpses, even if they'd shown little interest while the deceased was alive.

Infant deaths evoke extended responses. Babies who perish by natural causes may be carried by their mothers for weeks, more than 100 days in one case. Furthermore, mothers persist with interactions as if life lingered. Fur is groomed, faces touched, flies swatted away. But as time goes by, these instincts decay, growing less gentle and protective.

Bring Us the Bodies

The key lesson is that mortality is experienced by those left behind, and Neanderthals undoubtedly had emotional bonds as deep as apes.

Death for them must equally have heralded a tempest of overwhelming feelings and bodily interactions. The deceased did not suddenly become meaningless trash, but gained new social potency, perhaps even beyond that while alive. Corpses were like dark neutron stars, irresistibly drawing towards them passion and attention of many sorts. The archaeological challenge is seeing how this was expressed, from among the many thousands of bones and teeth from hundreds of individual Neanderthals.

In some ways bodies are like sites-within-sites, decaying in predictable ways as flesh, fat and bones come apart at different rates. Kebara 2 is a largely complete Neanderthal's upper body from Israel; look closely and you can see how small finger and wrist bones fell into the empty stomach cavity as it decomposed. Thanatology – forensic taphonomy – is vital in assessing how 'natural' any particular Neanderthal skeleton is.

But demonstrating burial must go further. Detailed criteria have been proposed: pits should be artificial and filled by one sediment deposit, very different to the surrounding level. Skeletons should lie complete at the very base of the pit, preferably in an extended position, and any associated objects should be unusual. Fulfilling these exacting characteristics certainly gives confidence for deliberate burial, yet they're so stringent that even some historical human cases might be rejected.

In fact, while not taking an unduly credulous position, any site with even a partly complete Neanderthal skeleton is a klaxon for something singular going on. That's because in general it's *vanishingly* rare to find entire animal bodies within cave assemblages. Now and then within carnivore dens a still-connected limb comes up, but whole skeletons are extraordinarily uncommon, and virtually unknown for big creatures. The only exceptions are bears that died while hibernating, or animals lost deep in cave systems or down pitfalls.

In a typical Neanderthal cave or rockshelter, however, in order to survive normal exposure and scavenger ravaging, bodies need to somehow be protected, or made unappealing. The Abric Romaní wildcat is a case in point. Its unusual intactness probably happened only because Neanderthals butchered it just before leaving, shortly after which the next flowstone phase began.

Similarly, special natural circumstances must be invoked for whole hominin corpses to be preserved. Theories range from rapid influxes of sediment to sliding into holes to becoming frozen, and therefore undetectable to carnivores. Yet in most cases, there's no actual evidence these things occurred, and moreover we'd still need to ask why it apparently only happened to Neanderthals.

Even weirder are situations where bodies appear embedded within archaeological layers. How could they have stayed intact over centuries while deposits full of highly fragmented animal bones and lithics built up around them? Either group members were aware of and avoided them, or they were intentionally covered up.

Diving deep into the evidence for claimed Neanderthal burials shows just how complicated things are. Many were excavated decades ago, and are increasingly being held up to critical reanalysis. Perhaps the most famous cold case – controversial for more than a century – began in spring 1908. This was, in Louis Capitan's words, the 'year of the Mousterians': in March, the emerging remains of Le Moustier 1 was from the outset viewed as a burial, then a few days before his final disinterment in August, another virtually entire skeleton was found about 50km (160mi.) east along the Dordogne valley.

We've already met the Old Man who lay inside one of eight caves near La Chapelle-aux-Saints. The most complete skeleton yet found, its excavators claimed it was a grave. The bones were rapidly exhumed and sent to Boule's lab in Paris, then all the dirt was shovelled back into the cave. There it stayed for 100 years, until a new project began to re-examine just how the Old Man came to be there.

Even if the contemporary drawings were made out of scale and after the event, they support the evidence from the bones that this was a remarkably intact body. The skeleton came from the base of a 'pit', shown in one illustration resting on small stones. It was no loose bony jumble, and missing parts are almost all mirrored on the other side, so the whole skeleton was apparently originally there. Only two other groups of connected animal parts were found from the cave: some reindeer bones in a possible midden towards one wall, and lower leg parts of an aurochs or bison, supposedly somewhere above the Old Man's skull.

Compared to the entire faunal assemblage, even including hibernating cave bears, the Old Man was less smashed up, damaged and weathered. What's more, it wasn't just in different condition to animals, but also to *other* Neanderthals. Missed by the original excavators (and a local resident who continued scratching around and sending his finds to Boule), recent research identified remains from at least two other Neanderthals: an adult and a young child.

The Old Man's body seems on taphonomic grounds to have had an unusual history, but to sceptics it's explained as an elderly Neanderthal who crawled into a natural concavity, rested his head against the side and died. His frozen or wind-mummified corpse avoided scavengers, or alternatively they couldn't access the body because the cave was intensely occupied by other, living Neanderthals.

But neither explanation adds up. The climate wasn't full-on arid and glacial, while if other Neanderthals were living alongside the decayed body, it's hard to imagine it wouldn't have become disturbed without being rapidly covered by something. Sadly, no photos of the skeleton *in situ* survive, but others show the head outside the cave, still partly encased in sediment. These reveal that the skull had shifted to the right of the neck, disconnecting the lower jaw and sending the chin up to the level of the nose, probably during decomposition. But importantly, the neck vertebrae were unaffected, implying that the remains were already partly supported by sediment before they rotted, otherwise the head would have tumbled off.

This leaves the mysterious pit. The original excavators didn't note any soil differences, but they were digging fast into a candle or gas-illuminated night, sometimes breaking parts of the skeleton. Notably, the pit had supposedly been marked by a surface depression, suggesting that the sediment inside was structurally different and didn't build up slowly along with the surrounding layer.

When meticulously re-excavated, twenty-first century researchers confirmed the pit had existed, was deeply sunken* and curiously one edge was cut by a fissure containing vertical fragments of reindeer

---

*A photograph from the new excavation mirrors one from Boule's 1909 visit: in place of a wicker picnic basket is a heavy-duty case for the laser 3D recording system.

bone. Warping and cracking in the cave floor pointed to freeze–thaw cycles, which can force materials downwards through cryoturbation: massive sediment disturbance involving frost heaving and sediment 'liquefying'. If this feature formed *after* an earlier archaeological layer was already in place, as indicated by the fissure's contents, it's less likely that the pit was a natural erosion feature.

This geological evidence doesn't definitively prove Neanderthals were gravediggers. However, in other places they clearly had the brains to make and fill pits, while the other main possibility – a bear hibernation hollow – doesn't fit either. Obsessing over its origin is something of a distraction in any case, since using natural hollows doesn't disqualify a skeleton from being an intentional deposit, as proven by the early *H. sapiens* cave of Cussac, south-west France.

In all, there's no natural explanation that can account for *every* feature of the Old Man at La Chapelle-aux-Saints. In 1908 it was a revelation, and thanks to modern reanalysis, today remains one of the best-supported cases for Neanderthal burial.

Recent studies might support some alleged burials as intentionally deposited bodies, but with others interpretations have only grown murkier. The site of Regourdou, south-west France, found by chance in 1954 during demolition of agricultural buildings, ended up as perhaps the most grandiose of claimed Neanderthal burials. After excavating independently for several years, one night the landowner found skeletal remains and called in the professionals. Despite lacking a clear original context for the bones, accounts of a curled-up skeleton inside a stone structure with offerings of bear bones nonetheless became repeated in the scientific literature.

Such a story was crying out for critical reanalysis, and researchers did just that, finding almost 70 new pieces of hominin bone among the fauna. This confirmed that the body, a young adult, was one of the most complete skeletons found. But other things at Regourdou didn't stand up to modern scrutiny. The bears had died while hibernating, and there's no evidence the Neanderthal was curled up or in some kind of rocky tomb.

On the other hand, Regourdou definitely has some strangeness. Aside from disturbance to the body by bears (and later by burrowing rabbits), anatomical connections between small hand bones indicate that the corpse had originally been whole. Yet most of the skull was missing – including the upper teeth – even though natural processes such as erosion that might have worn or dissolved it away aren't obvious. There are no cut marks either, which would definitely be expected if Neanderthals had removed part of the head from a fresh cadaver, although bears had gnawed one probably already dry bone.

This all implies that the skull came off after the corpse had fully rotted, but by what means and where it ended up are unknown. With considerable understatement, the researchers concluded that Regourdou is 'problematic'. Many unanswered questions remain: what were Neanderthals even doing deep in a cave system? There are butchery traces on some of the bear bones, so hunting hibernating animals is a possibility. The low number of lithics does resemble a kill site rather than a living locale, unlike the context of the Old Man at La Chapelle-aux-Saints. But if so, this was a rarely visited place: less than 1 per cent of bear bones have cut marks.

One answer might come from the fact that over time, Regourdou itself changed. What began as a deep bear cave became a natural pit trap after some of the ceiling collapsed. In the same layer as the Neanderthal, remains of other creatures like deer, boar and horse were found, victims of falls. Perhaps in the heat of the chase one unlucky day, a hunter suddenly felt the ground disappear from beneath them and went tumbling into the dark along with their quarry. Even more interesting, the research found a new hominin bone, hinting that a second individual had perhaps met a similar fate.

A final twist: lower down the same hill as Regourdou is another cave: Lascaux. Around 17 ka *H. sapiens* went into its darkness to paint vast horses and bulls across the ceiling, as the bodies of animals and a Neanderthal already 80,000 years old lay in a chamber somewhere far above them.

Perhaps one of the most intriguing Neanderthal CSI cases was only found in 1993. Somewhat older than Regourdou at between 170 and 130 ka, an incredibly complete body lay right at the end of a narrow tunnel in Lamalunga Cave near Altamura, southern Italy. Cemented in place in the 'Apse of the Man' and almost inaccessible behind stalagmites, researchers needed to be creative. In addition to recording the fossils with a GoPro camera and laser scanning, they extended a delicate sampling apparatus through a gap to grab a tiny piece of shoulder blade for analysis.

How did this Neanderthal end up there? The scattering of bones shows that the body simply fell apart as it decayed, but the original posture isn't clear, nor whether the individual even died there. Lamalunga wasn't ever lived in: there's no archaeology at all. Getting lost doesn't seem likely, as the body is only 50m (55yd) from the old cave entrance. If that had already been blocked off, the only way in would have been falling down chimneys open to the surface, but unlike Regourdou, Lamalunga Cave doesn't seem to have been a natural pit trap for animals. Something unusual happened, but all we know for sure is that the corpse decomposed slowly in the dark, bones shifting and clattering to the floor, before growing an eerie new skin of knobbled flowstone.

It's impressive enough that entire adult Neanderthal skeletons survived tens of millennia, but even more so for fragile babies. Children's bodies decompose faster, so without protection their bones are particularly prone to destruction. As Chapter 3 showed, we nevertheless have remains from Neanderthal youngsters across a range of ages, including remarkably complete infant skeletons. Unexpectedly, newborns' bones can resist decay slightly better since they're more mineralised. But even compared to very rare examples of carnivore infants occasionally found at dens, the numbers of young babies from Neanderthal sites is striking.

The Le Moustier 2 newborn mentioned earlier isn't unique. One of the shortest Neanderthal lives ended around 70,000 years ago, in the Caucasus Mountains. A week or two old at most, it's astonishingly well preserved. Just a whisker above the rock floor of Mezmaiskaya Cave, the tiny skeleton was found lying on its right side, knees flexed, legs drawn up, with its left arm bent slightly towards the chest. It almost appears to be napping. So intact that its minuscule tooth buds were still present, most bones were in the correct position and largely undisturbed, aside from some lower leg parts that probably became loosened when the surrounding hardened sediment eroded.

Explaining how this tiny baby was deposited and remained almost undamaged means either another series of special circumstances, or the involvement of other Neanderthals. Such a young baby would be highly unlikely to be forgotten, or simply dumped after death. It's true that if orphaned young chimps don't get adopted by other adults, they typically become ill, depressed and die. But even if this was an abandoned body, other questions remain.

Newborns typically don't learn to roll for many weeks, so the side-lying posture is surprising, and though rodents nosed and nibbled the exposed dry leg bones, there's no damage at all from large scavengers. Moreover, despite no evidence for a pit, compared to the rest of the cave's lowest layer, the sediments immediately around the skeleton were distinctive. Instead of any lithics or fauna, there were small charcoal fragments lacking elsewhere.

Whether this baby died at Mezmaiskaya or somewhere else, the balance of evidence implies that it was intentionally placed and protected. We can almost certainly envision a mother – likely still bleeding, breasts swollen with milk – giving up the small, soft body into not much more than a scraped-out hollow, before covering it over, perhaps with the remains of a dead hearth, and finally leaving.

In contrast to the Mezmaiskaya baby, which was excavated relatively recently, little can be said about how the Le Moustier 2 infant ended up in the ground. Peyrony's claim of a surprisingly large pit is hard to evaluate since he apparently made no drawings and took no photographs.* He recorded that the filling sediment looked like a mix of the three layers it cut through, which would support a dug feature, and the un-weathered bones certainly point to rapid covering. No information on the body's position has survived, but it was just as complete as Mezmaiskaya. Once again, either very unusual natural circumstances were involved, or the Le Moustier 2 infant was deposited by other Neanderthals.

We should expect Neanderthals coped with the emotional wreckage of losing youngsters in many different ways, some of which may only be perceptible archaeologically through subtle indicators. Across

---

* It was Peyrony's workmen who uncovered the remains, as he was busy managing multiple sites including La Ferrassie.

ancient human cultures, while some deceased babies were buried, others were placed in middens, walls, wells, under floors or even inside jars. The 'correct' approach varied according to each society's understanding of infancy. The Roc de Marsal child – whose tooth contributed to our understanding of growth rates – might represent something like this for Neanderthals, where a body was deposited but not in a way that could be called a burial.

Reanalysis of this 1961 find over the past decade was aimed at examining the largely complete skeleton's odd resting place: a hollow feature in the limestone. While the hollow is entirely natural, how the body got in there isn't obvious. It lay face down, slightly angled downwards, leading to suggestions that the corpse had slipped in somehow. But the legs both face right and are tightly flexed, while the left arms trails down into the hollow. It's hard to see how that posture fits with either being washed in, or moved by flowing debris, and furthermore the climate at the time wasn't extreme enough to mummify or freeze the body.

Certainly the Roc de Marsal child was yet another example of a fully connected corpse when it entered the hole. Aside from dangling finger bones becoming muddled by earthworms, and the lower legs being damaged due to exposure, it's remarkably complete, without carnivore damage. This cannot be claimed as a burial, but neither is it satisfactorily explained by natural processes. The cave contains no evidence for massive, rapid sedimentation, so the dark material around the body – identical to that in other hollows – must have built up slowly.

Perhaps a lost or abandoned child crawled in and died, albeit in an odd position. On the other hand, being aged between 2.5 and 4 years old argues for a strong parental bond. If the child died and the group was present, the urge to interact somehow with its body may have been intense. This could easily have included moving and placing it in a small contained space.

There's a final interesting fact. Along with typically fragmented lithics, animal bones and teeth, the sediment around the skeleton also contained three unusual objects. Hyaenas, birds and complete bones are all very rare throughout the rest of the site, but inside the hollow along with the Neanderthal body were intact reindeer and partridge limb bones, plus an entire hyaena jaw. They may not be overtly special like shells or pigment, yet their association to the skeleton only adds to the list of questions here.

Places of Death?

If individual skeletons are difficult to decipher, the phenomenon of multi-body sites is even more so. We already saw that remains from more than one Neanderthal were at Regourdou and La Chapelle-aux-Saints, and reanalysis elsewhere shows the same thing. At Le Moustier in 1910, Hauser found an isolated skull fragment and possibly a tooth (the locations of which are now a mystery), and even Feldhofer harboured more bodies than was once believed. In 2000, researchers relocated refitting pieces of the original Neanderthal from the surviving cliff-foot rubble heaps, and also found parts from two other individuals.

Scraps are one thing, but what should we make of places with more than one (mostly) complete skeleton? They've sometimes been called cemeteries, but that implies a persistent tradition over many generations. If Neanderthal skeletons come from separate layers with clearly dissimilar archaeology, a continuous practice is hard to countenance. But on the other hand, particular sites are undeniably over-endowed with the dead.

This is visible even among the Neanderthals' immediate forebears: parts from nearly 30 individuals came out of the natural pit at Sima de los Huesos, Atapuerca. They probably arrived in the chamber *en masse* from somewhere higher in the cave system, but the sheer numbers, both adults and adolescents, is extraordinary. Furthermore, the only accompanying artefact is a pink quartzite biface dubbed 'Excalibur' by the excavators. At the dawn of the Neanderthal world, it seems that concepts of places for the dead were already emerging.

Probably the richest of all is Krapina, with some 900 pieces of bone from across the skeleton; however, they're so smashed up it's tricky to calculate how many Neanderthals they represent. Just based on the teeth, it's at least 23 – the largest number known for any site – but using other methods it may be nearly 80.

Krapina is exceptional, but two other places also have parts from at least 20 individuals. At La Quina, they're spread across a number of adjacent locales and layers, but L'Hortus, also in France, is especially intriguing. Located in the middle of a near-vertical 100m (330ft) cliff, this cave's interior is narrow and chute-like; not a comfortable home. Nonetheless, this didn't put Neanderthals off from using it for very short stays, although in later layers something seems to change and bones – including many children and a baby – begin accumulating in one section of the site. This went on for centuries, maybe millennia, but why is an enigma. There don't seem to be particularly exceptional preservation conditions, so was it something more?

There are other places that were never extensively lived in but still contain many bodies. Not far from Cueva de los Aviones in Spain is the Sima de las Palomas, a deep shaft that is filled with deposits between MIS 6 and MIS 3. Nearly destroyed by nineteenth-century miners, it lay abandoned until a local naturalist abseiled down in 1991 and spotted hominin bones stuck to the wall. These were remnants from what had been an enormous, cemented rocky pile. Over the next 25 years archaeologists undertook painstaking excavation using a specially built scaffold, and established that this wasn't a hole down which hapless Neanderthals fell, but something more complicated. Aside from very few lithics and some burned animal bone, parts from at least 10 Neanderthals had accumulated, probably over centuries and perhaps even longer.

Most striking are three quite intact individuals, all coming from a period apparently falling between 45 and 55 ka. Lowest was an adult, then a child, and uppermost was a very short adult woman, and importantly the skeletons were unarguably still articulated. There are also distinctive postures: one of the woman's legs was extended, with the other crossed beneath, while her arms were bent and hands up near her face. The hands of one of the other individuals were in a similar position.

Certainly this was never a living site, but it seems a stage beyond Regourdou and wasn't an easy place to accidentally fall into. The site is located on a massive marble mountain that rises up from the flat coast, with the shaft itself like a wound in its flank, and something repeatedly attracted Neanderthals here.

A rather different place, but with nearly as many Neanderthal bodies, is the La Ferrassie rockshelter. Unlike Sima de las Palomas, this was certainly a living site, which makes the accumulation of at least eight individuals especially unlikely by chance. The probable male LF1 was largely complete and the first to emerge in 1909 after road building in front of caves uncovered the rockshelter. The next summer, LF2 – smaller and probably female – appeared just 50cm (20in.) to the west. Following this and up to the 1920s, another five sets of remains came out from beneath the main rockshelter, from which up to 1,000m³ (35,000ft³) of sediment were removed.

There are also parts of many children: the youngest of all was LF5, a two-month premature infant, while LF4b was a newborn, LF6 roughly pre-school age and LF3 around 10 years old. Six decades later a different team uncovered a toddler, LF8, at the rear of the rockshelter.

In recent years excavations have been renewed at La Ferrassie, aiming to understand how at least some of these Neanderthals came to be there. They relocated LF1 and LF2's precise positions, partly thanks to matching up sediment still encasing one of the latter's feet. Both skeletons were close in depth, and date around 47.3 to 44.3 ka. The other individuals also appear to be at roughly the same stratigraphic level, and LF3 and LF4 were separated by even less space than LF1 and LF2. This means that La Ferrassie has some of the best evidence for multiple skeletons reasonably close together in time, and when the faunal collections were checked in 2019, new teeth from at least another two adults were identified.

Some of the flashier claims about La Ferrassie over the years haven't held up, such as special objects in graves, or claims for circular carvings on a massive limestone slab covering LF6. These are natural features, and similarly what were called 'pits' in other parts of the site are likely depressions caused by freezing processes.

But could any of the skeletons have been intentionally deposited? Most were roughly oriented east–west, and LF1 was lying not quite flat on his back, with right arm raised and the left extended down straight. Both legs were bent and angled to the right (also seen in LF5 and LF3). The head faced left, lower jaw gaping open and slightly separated from the skull. This odd sprawling position was potentially explained by the new research. Sediments in the western zone of the rockshelter containing LF1 and LF2 had come from a platform above, outside the caves. Both bodies may have slipped downslope extremely gradually.

La Ferrassie might not be the cemetery it's been claimed to be, but something out of the ordinary must explain why so many bodies accumulated, even over many centuries. They particularly stand out compared to the handful of later *H. sapiens* scraps here, and the number of youngsters is especially striking. Most important, like La Chapelle-aux-Saints and elsewhere, the hominin remains were in quite different condition to the animals: more complete, with little weathering and no gnawing. La Ferrassie hints at some sort of long-term tendency for dealing with death that involved depositing bodies here, but another site has even more impressive evidence.

You may have heard of the massive Shanidar rockshelter in Iraqi Kurdistan, for claims that one skeleton was accompanied by funeral bouquets. That theory is now regarded as unlikely – the pollen was probably naturally accumulated – but Shanidar is still remarkable because it produced parts from at least 11 Neanderthals, many obviously originally whole skeletons. Ten came from excavations between 1953 and 1960, while fieldwork over the past few years has uncovered another individual.

Understanding Shanidar is tricky, partly because most of the fossils weren't dug up to modern standards, but also because even today the site has a reputation for dangerous rock collapses. Such accidents might explain some of the skeletons, including Shanidar 1 (S1), S2 and S3. Notably, the peculiar position of S5 could have been caused by a rock breaking the individual's spine, forcing the head back on top of the rock. Confirming this scenario is now difficult even though more parts of the skeleton were excavated in 2015/2016. But if a rockfall was responsible, it must have been sudden: even though S5 was probably aged over 40, this Neanderthal didn't have the same physical challenges that might have slowed down S1.

Very singular circumstances, however, are found for five of the other Shanidar Neanderthals. Even if the flower story wasn't proven, the associated body, S4, had an interesting posture: curled up almost in a foetal position on its left side in a rocky niche, the knees drawn right up and the left hand apparently close to the face. Sceptics would point to the lack of any pit features and propose that this individual simply died in that spot. However, the local climate wasn't cold or dry enough to naturally preserve the corpse, implying it was in some way protected from erosion or disturbance. But there's no obvious natural source of rapid sedimentation.

This is where things get weird. The S4 skeleton was highly fragile, and the excavators decided to remove it as a single sediment block in 1960 for later excavation at the Baghdad museum. As they prepared, it was realised that fragmentary parts from two other adults and a young baby (S6, S8 and S9) were right below those of S4, so close that some of the remains had become mixed.* One of these other adults also appeared to have been curled up. But there was more to come.

When excavations resumed nearly 60 years later, the area around the block was relocated, plus more fossils that had been left sticking out of the trench wall. Meticulous work revealed them to be the entire upper body of a mature adult, lying almost directly below where the other Neanderthals had been. Moreover, its posture, although flatter on the back, is strikingly similar to S4. The crushed but complete skull lay on its left side, with the right arm across the beautifully curved ribcage, hand clenched. Beneath the chin, the left arm was tightly bent up, with the wrist flexed back on itself, almost as if they were sleeping.

Given the relative position of the block and the uppermost location of S4, it's quite probable that the new upper body actually belongs to either S6 or S8. Such a tight cluster of at least four individuals is already unique in Neanderthal sites: the only place where so many skeletons are closely associated in their original positions. And it gets better still. Examining the context of the new bones, researchers found they were surrounded by rapidly deposited dark-brown

---

* Unfortunately the spatial relationships between the fossils inside the block weren't possible to fully reconstruct because they'd endured what must have been a bumpy ride to the museum on the roof of a taxi.

sediments, probably the rotted remains of the body but also with some intriguing plant remains. And in what's the best evidence yet of a burial pit, everything was contained within a depression with a distinct, curved base. Originally this might have been a water channel, but micro-morphology suggests it was deepened by artificial scooping.

Most remarkable is an artefact that may have been intentionally placed with the body. Among the new bones there were only two lithics, one of which was found within the chest cavity. It's oriented vertically, but probably slid that way as flesh below decayed, and is just a few centimetres from the left hand, as if it might once have been held in the fingers.

Taken together, although Shanidar isn't exactly a Neanderthal necropolis, there's absolutely more going on than the remains of those who perished by rockfalls. Even in those circumstances, there's great likelihood that survivors might have physically interacted with bodies: sudden deaths are especially traumatic and might have provoked attempts to move or position corpses. But the cluster of three adults and a baby really suggests that these bodies were either put into the same small space over a very short space of time, or that Neanderthals returned to the same spot repeatedly. In relation to the latter theory, it's very interesting that two unusually large rocks stacked on top of each other are located quite close to the new skull. They weren't part of a rockfall, but would have extended upwards through the sediments, remaining visible as the other bodies also accumulated.

Whole Neanderthal bodies resting in the earth have always drawn our attention, but what about the bony tatters? Are they simply the lost, abandoned or other unfortunates? Difficult-to-interpret 'orphan' skulls and other body parts are widespread in caves, but also in the open air.

Looking just at Germany shows quite how rich and varied some landscapes with Neanderthals can be. River gravels produced the Steinheim skull, another from Warendorf, plus parts of an adult and a very young child at Sarstedt. A large piece of skull comes from up in the Wannen-Ochtendung extinct volcanic crater, and Germany is also well known for travertine spring and tufa sites with Neanderthal remains. Along with the hunted rhinos at Taubach, adult's and children's teeth were found there as early as 1871, and similarly, teeth and skull fragments of at least three individuals came from Bilzingsleben.*

Some open-air locales had unusual preservation processes. Alongside perfect imprints of oak leaves, fragments of least six Neanderthals were excavated from the travertine quarry at Weimar-Ehringsdorf. Though most were skulls, more complete remains of a child included a jaw, parts of the chest and an arm encased in the travertine, not far off from parts of an adult. In such places that must have attracted herbivores and carnivores alike, it seems unlikely a small corpse would have avoided becoming dispersed, unless it was perhaps submerged in a pool.

Elsewhere, explanations for other body parts remain elusive. At 'Ein Qashish, just north of Mount Carmel in Israel, Neanderthals were active in the landscape over long periods of time between MIS 4 and MIS 3. The odd hominin fragment is present in early levels, but a later layer contained much of a Neanderthal's lower body: an almost connected left thigh and shin, part of the right lower leg and a back bone. All came from within the same couple of metres and probably represent a single young male individual.

The surrounding landscape – wetland fringes of a floodplain with seasonal pools – would have been attractive, and the 'Ein Qashish individual may have been out hunting. One potential cause of death could be an old ligament injury that probably gave him a limp, making him a target for furred hunters.

Yet there's no carnivore gnawing, and his bones still have their grease-rich ends. What's more, the layer in question is well preserved, including refitting lithic scatters, and it seems unlikely that erosion would have removed the upper half of the skeleton. But open-air locales weren't simply hunting 'drive-thrus'. Places such as Les Bossats tell us Neanderthals did many things outside caves, including processing pigments. At 'Ein Qashish in the same layer as the bones there are ochre fragments, possible limestone anvils/grindstones and

---

* Another probable Neanderthal skull was found in 1816, which if recognised would have been the first known hominin fossil.

a marine shell brought from at least 10 to 15km (6 to 10mi.) away. None of them are directly associated with the bones, but they're extremely unusual for this region. Had such a collection come from an early *H. sapiens* site, one might well see claims for symbolic activities.

Neanderthals must have felt deep urges to interact with bodies, but how this panned out might have depended on where the deaths happened. Out under the sky, the 'right' way to deal with death may have been different to inside caves or rockshelters. Although specific evidence for body placement is sparse, places like 'Ein Qashish may reflect some kind of active Neanderthal involvement.

The Closest Cut

We can't be sure that Neanderthal remains at open-air locales were intentionally manipulated, but recent research has found that other things than burials were happening with bodies in caves and rock-shelters. More and more cases of unquestionably butchered Neanderthal bones are being identified, even including some of the original Feldhofer remains. In general, Neanderthals were covering the full range of what they did with animals: skinning, dismembering, jointing, defleshing. Sometimes also comprehensively snapping and smashing bones. Neanderthals have actually been known to do this since 1899, because the first identified case was among the hundreds of bones at Krapina. All parts of the skeleton across multiple individuals there had been processed, with skinning and defleshing – even of skulls – plus some smashing. The discovery of this early in the history of Neanderthal research, and its interpretation as cannibalism, laid the foundations for an aggressive reputation.

In contrast, nobody noticed for decades that Le Moustier 1 had been butchered. Original publications, Klaatsch's diaries and surviving photographs* of the site show the odd position of the skeleton. The skull was face down, angled back with the lower jaw slightly detached.

---

* Hauser recorded that 22 photos were taken of every stage of excavation, but only a few have been found.

In contrast to the diary entry and a sketch by Klaatsch suggesting some of the upper body parts were in anatomically correct positions, the photos show long arm bones sticking out from the back of the neck; this wasn't an articulated body.

Modern analysis revealed that as well as being disordered, the Le Moustier teenager was butchered. His skull was skinned and defleshed, tongue removed, jaw cut off, sliced and possibly battered, and the meat from one femur was also removed. Interestingly, however, his body hadn't been left scattered, and the skull and lower jaw were actually next to each other. Moreover, excavation photos and documents record that his face and forehead were lying directly against an unusually large flat stone compared to other rocks in the surrounding sediment.*

Though separated by many tens of millennia, Krapina and Le Moustier are later Neanderthal sites. It's during this period, after about 130 ka, that body processing seems to be, if not common, then no longer rare. In many contexts there's little difference in how hominin and faunal remains were treated, with similar focus on marrow-rich body parts. A newly identified site, Sirogne, south-east of the Périgord, is a good example, with Neanderthal teeth missing their roots, most likely from jaws being smashed for marrow.

Perhaps surprisingly, however, direct proof Neanderthals were actually *eating* the processed bodies is scarce. Tooth marks aren't common on animal bones, so those on hominin remains are especially noteworthy. At least one leg shaft from Krapina has shallow, paired grooves looking exactly as if someone had gnawed it like a corncob.

In general, however, tooth marks are extremely scarce. This is particularly noticeable when compared with cannibalism among *H. sapiens* from various periods in time. Around 15 ka, bones from at least six people – possibly just a few generations of one group – were butchered at Gough's Cave, south-west Britain. They were processed similarly to – and jumbled among – the hunted animals, but a whopping 65 per cent have processing traces and nearly half also bear tooth marks.

---

* Possibly a massive flint flake or natural limestone plaque; unfortunately it is now lost.

Burning of hominin remains is also very rare among Neanderthals, and it can be ambiguous. At Zafarraya, Iberia, although three burned pieces were found in association with a hearth, just as with arguments over cooking animal bones this might have been accidental. Krapina also has burning, however, and given the abundant evidence of body processing and eating, it's a fair suggestion that some of the dead here were cooked.

Let's assume at least some cannibalism was happening. The question then is, why? As cases mount up, twenty-first-century analysis is pointing to more subtle interpretations than simply calorie chasing. At Krapina, although whole bodies were obviously present, the most nutritionally rich bones don't appear to have been selected for processing. Something similar was happening at El Sidrón: previous chapters have covered the biology and behaviour of the 13 or more individuals here, but they were also probably cannibalised. The bones had been very intensively processed, with traces of dismembering, slicing and hammering.

Yet despite some bodies being taken apart, it doesn't look exactly like typical animal butchery and isn't systematic or targeted to the richest parts. The bones had no carnivore damage or weathering, and some parts from the chest, arms, hands and feet were still connected. Moreover, the representation of elements is strange: facial bones are mostly missing, but there's a hyoid (an especially fragile bone that supports the voice box), and toes are oddly numerous. However, there is one pattern across the El Sidrón bodies: it's the youngsters who have the most cut marks, which is hard to explain if this was solely about nutrition.

Let's come out from among the bodies and think about cannibalism itself. For any species living in small groups at low population densities, *regularly* eating each other is a highway to extinction. And it's not even good economics, since compared to animals of roughly equal size, hominin bodies are surprisingly poor in terms of nutrients.

Could it be starvation? Hunter-gatherer lifestyles can be challenging, and in some recent populations famine was known and feared, though rare. Were times so much more desperate for Neanderthals? Some cannibalism cases have been linked to glacial conditions. For example, dismembered and filleted remains of two adults, two teenagers and two different children come from Level 25 at Combe Grenal, probably dating around 70 to 65 ka and full of cold-associated reindeer.

But things aren't that simple. As we saw in Chapter 10, Les Pradelles is most certainly a Quina reindeer-hunting camp, but new evidence shows it wasn't dramatically colder than Wales or Scotland today. And it's important because among the butchered animals, there are also remains of at least nine processed Neanderthals. They include adults and children, and were treated in a nearly identical way to the reindeer, with defleshing and bone ends smashed off, presumably for marrow.

This assemblage has been claimed as another case of nutritional cannibalism, but this doesn't stack up. Not only was it not extremely cold, but the immense amounts of butchered reindeer surely imply that, at least seasonally, food wasn't an issue.

At the opposite end of the climate spectrum, interglacial body-processing sites have also been argued to result from starving Neanderthals unused to hunting forest creatures. Krapina is dated to the Eemian but probably after the climate peak, and may even include the LEAP period of extreme environmental disruption after 121 ka. But the bones come from two different layers, so it's highly probable that the butchery practised there covered a longer period in time.

Intriguingly however, there is another cannibalism site that may be contemporary with Krapina. Level 15 at Moula-Guercy, a cave in south-east France, has very similar dates and at least some of the fauna such as porcupine might be connected to aridity. At least half the remains from six Neanderthals – an older man, an adult woman, two teenagers and two children – were butchered. The Moula-Guercy processing is quite intensive, with skinned skulls, removed tongues, dismembered joints and limbs, defleshed legs and systematically smashed bones.

But there are arguments against climate-driven interglacial famines. Sites like Neumark-Nord make it clear Neanderthals had adapted to forest hunting, taking out fallow deer with precision spear strikes, and even leaving good meat and marrow behind. Huge elephants were being targeted too; even if the Lehringen carcass and spear site is scavenging rather than a kill locale, these Eemian Neanderthals clearly had access to a lot of food. Smaller prey were certainly also available, including tortoise or beaver. There also doesn't seem to be higher rates of malnutrition in any of the body-processing sites than elsewhere, nor compared to some recent hunter-gatherer populations. The only clear pattern with regard to environments is the total lack of any body processing from outside Europe.

Rather than being connected to particular climates, might cannibalism simply imply that Neanderthals ruthlessly chomped down on weaker individuals? Children and elders would be most at risk, but they don't outnumber butchered teenagers and adults. Plus one man at Moula-Guercy is among the largest known Neanderthals, and surely would have been risky to attack.

Antagonism towards strangers is another explanation. Goyet in particular has been proposed as an example of aggression-motivated cannibalism. The butchered parts are lower legs, skulls and thighs, which does match economic selection for meat/marrow. Abundant cut marks on more than a third of the remains record dismembering, defleshing and even less common gutting gashes on the pelvis and ribs. In addition, there were huge amounts of bone cracking – the only complete bone is a fingertip – and probable crushing of the long bone ends, all of which points to consumption. However, the theory that these were 'foreigners' who were attacked and eaten relies on interpreting the isotopes as reflecting non-local individuals, rather than Neanderthals who simply had a very large territory, fitting lithic transfer data.

Other places confirm that those being eaten were from the local area. Anatomical quirks on the bodies from Krapina and El Sidrón suggest the dead were from closely related, probably local populations. And at Moula-Guercy, isotopes indicate that at least one of the butchered individuals probably grew up within just a day or two's walk, in the same region that many of the lithics were coming from.

If groups were competing for the same land then conflict could erupt, but the generally wide age range of processed bodies would imply either mass slaughter or ambush killings over a long period of time. Furthermore, butchered bodies don't have higher rates of violent deaths, and we'd need to assume Neanderthals tended to be aggressively territorial. As discussed earlier, the collaboration and food sharing at the heart of their societies argues against this.

In fact, cannibalism and butchery may well have had primal motivations, but not necessarily rooted in ravenousness or belligerence. Chimpanzees offer fascinating parallels. Though they hunt much more than used to be believed, and social disagreements can be very violent, killing is rare.

Murders nearly always involve other groups, but only if the odds are very stacked, and victims are almost always adult males or infants. Killing *within* groups is extremely unusual, but infanticide is known. Babies tend to be targeted during times of extreme emotion, either following conflict with strangers or triggered in males by some kind of post-hunting bloodlust. Cannibalism is sometimes part of these social dynamics, meaning that the consumption of a body for chimpanzees is inherently about something beyond nutrition.

Sometimes corpses are enthusiastically consumed and shared just like hunted prey, or they may be treated aggressively; adult bodies are even struck with branches or rocks. But in other cases the interaction is calmer, more exploratory, and eaters only take small amounts. Pieces of killed adults are occasionally eaten, and females within the same group seem especially prone.* However, babies are the most often eaten, and they're also more completely consumed. Infanticides after hunts are particularly likely to involve communal cannibalism, sometimes including their mothers.

Bonobos once again provide an intriguing counterpoint. There are no recorded infanticides, yet several cases of mother–infant cannibalism exist, which also featured meat sharing. In one situation, following a baby's natural death the group spent an entire morning eating much of the body, before the mother carried away the remnants on her back.

This demonstrates two things crucially important in understanding Neanderthal cannibalism and body processing. First, there is no need to invoke aggression as a default. Second, after consumption the dismantled scraps aren't transformed into waste, but still treated as representative of, or connected to, the deceased.

For both bonobos and chimpanzees, the bodies of the dead evoke many emotions. Even if the process often begins with trauma and

---

* Including focusing on the victim's genitals.

confusion, typically corpses shift to a liminal status; not alive, but equally not a lump of meat. They're more intensively manipulated than hunted animals, and carried for longer. In some – if not all – cases, the eaters must know what and *who* they're consuming. Cannibalism is very probably a powerful means by which individuals and groups process the impact not only of killings carried out on emotional impulses, but other deaths too. In other words, it's about grieving.

Such contexts are just as likely – if not more so – in Neanderthals. Moreover, we might consider that both chimpanzees and bonobos use tools in their interactions with the dead. Corpses are poked at with sticks, as if to wake them, but even more astonishingly, in one case the dead individual's teeth were picked. After the death of a male named Thomas in a Zambian sanctuary in 2017, his adoptive mother Noel refused to leave the body. During her attendance she began intently cleaning Thomas's teeth using a grass blade, watched closely by her teenage daughter.

Teeth picking with grass or wooden splinters for chimpanzees is inherently an intimate, caring act. The very first individual was reported to do this about five decades ago. Named Belle, she was part of a sociable group of orphaned youngsters, was especially keen on being the picker, and quite remarkably was observed removing a loose milk tooth for her best friend.

Shift these scenarios to Neanderthals, and add into the mix their far greater cognitive sophistication, and lives that revolved around using lithics. Suddenly it's not difficult to envision how skills in carefully taking apart hunted carcasses might be transposed into a grieving process that involved butchery and cannibalism as acts of intimacy, not violation.

By examining Neanderthal cannibalism and body processing in detail, there may be evidence that points towards this kind of interpretation. Most strikingly, the butchery itself isn't always identical to what happened with animals. It's sometimes more intense, even if the method is similar. This trend actually goes back into the Lower Palaeolithic, visible at Gran Dolina, where bodies were also butchered at twice the rate of fauna. There was greater attention to heads, removing tongues and brains, and even fingers and toes were skinned.

Similar patterns are there in some Neanderthal sites. The Krapina remains seem more intensively smashed than the fauna and at Moula-Guercy half the Neanderthal remains have cut marks, compared to under a quarter of deer and even less for other species. Plus the bone smashing is extra-focused, potentially using anvils, and only hand and foot bones remained unbroken. Another difference is that only animal bones bear burning traces.

Idiosyncrasies are found elsewhere. At Les Pradelles the reindeer and hominin butchery rates are the same – about 30 per cent – but while the animal parts were clearly selected and brought to the site based on their nutritional value, this isn't the case for the hominin remains, with few limbs and many skull pieces. Something quite different was going on with the Neanderthal bones here, and it's more than where they came from. Despite not being obviously separated from the animal bones in the site, the condition of the hominin remains is much more varied, showing more carnivore damage, including teeth that had passed through hyaena stomachs. This points not only to a different accumulation process compared to the fauna, but also in what happened to them at the site, potentially including being exposed.

Eternities

While each context of body processing and cannibalism is unique, it's possible to explore conceptual connections between them. Neanderthals could probably have butchered anything blindfolded, and like chimps with toothpicks they drew on familiar skills in their desire to connect with the dead. Perhaps this, together with consuming the body, was a comfort amid the confusion and fear summoned by death. And there may be hints that, just as we do, some Neanderthals tried to keep a part of the deceased with them.

As researchers have increasingly scrutinised faunal bone collections for worked objects including retouchers, examples made from Neanderthal remains have been discovered. At both Krapina and Les Pradelles, fragments of thigh bone were used, while at Goyet four retouchers were made on shards from the thigh and lower leg. Moreover, there the Neanderthals appear to have specifically chosen hominin bones, in spite of their lesser suitability than other species and skeletal elements.

It seems that, as with most animal examples, the retouchers at Goyet and elsewhere were used while the bone was still fresh. Neanderthals weren't simply picking up random fragments, and the choice was potentially happening during or not long after the body processing itself. Moreover, at Goyet the retouchers were particularly well-used, with multi-phased damage. They were kept in the hand for some time, even if it's not clear how long.

So far all examples come from body-processing sites, and there's no evidence that these – or indeed any – retouchers were carried between places. But it's a tantalising possibility, and a reminder that we don't know how the majority of isolated Neanderthal bones and teeth arrived at particular sites.

If death was bound up with emotions, is it possible there was a personal element to butchering, eating and using bodies? It's very likely that at least sometimes the deceased and the butcher had a relationship in life. Faces are well known to be the key focus for human communication and individual identity, and so it's intriguing that heads seem to have had extra attention in some contexts. At Moula-Guercy, unlike the deer, *all* the hominin skull fragments had butchery traces, and were particularly fragmented. Similarly, the Le Moustier 1 skull was comprehensively taken apart, but butchery traces are only present on one other body part (his right thigh).

This brings us to one of the most singular artefacts from any Neanderthal site. In 1906 La Quina was the first place where prehistorians worked out what bone retouchers were, but it's also well known for producing – across several layers and areas – remains of at least 22 Neanderthals. Among them are several pieces of a fractured skull, probably from the same young adult. One cut-marked fragment also bears distinctive battering damage from use as a retoucher.

Neanderthals, with their exceptional knowledge and appreciation of anatomy across species, would absolutely have known what they were handling. Even more than the Goyet retouchers, the choice to use this object was not accidental or casual. Its shape and thickness are well outside other retouchers at La Quina or elsewhere, and furthermore it's the only known skull retoucher from any species, anywhere. This object was selected *despite* its unsuitability, and while there are other unusual retouchers in the same layer (a reindeer jaw and horse tooth), the skull represents the only Neanderthal bone from this level.

Sceptics might still be looking for something that goes beyond practicality to prove that reducing the dead into constituent parts was a practice that resonated in social and symbolic ways. Stunningly, it exists, and also involves a skull. The most complete cranium from Krapina bears a series of 35 mostly parallel tiny cut marks running from slightly above the brow ridge over the forehead towards the rear of the skull. Just 5mm (0.2in.) long, they don't fit any butchery pattern, are totally unique at that site and have no parallels in any other hominin skulls, whatever the species.

Yet they do recall something. They represent the longest series of sequential markings made by Neanderthals, even more than on the hyaena bone at Les Pradelles or the raven at Zaskalnaya. Their placement on a hominin bone, and moreover a skull – the most symbolically resonant part of the body – is extraordinary. What it most closely mirrors is behaviour of *H. sapiens* people who lived more than 100,000 years later at Gough's Cave. There, in addition to the body processing and cannibalism, bone modification was also going on. Skulls were carved, possibly to act as vessels, but most remarkably on a smashed long bone fragment someone etched a delicate repeating design formed of small cuts.

Is it possible to move from individual objects or skeletons, and say something about what death meant to Neanderthals? The complexity of their interactions with bodies – whether depositing whole or partial skeletons, butchering them or using them as tools – dovetails with growing skill and diversity in other aspects of behaviour including hunting, material technologies and aesthetics. The greater frequency after 150 ka is also probably not just down to better preservation, but an amplification in social practices. Additionally, it's noticeable that places with Neanderthal remains often contain parts from multiple individuals.

There may also be trends in what happened to whom. Neanderthal communities were certainly constituted by commonly understood categories including age and gender, probably reproductive status and sociability and skill levels. These characteristics would have influenced how individuals treated each other in life, and perhaps in death too.

One of the most remarkable patterns is the apparent paucity of female skeletons. This isn't due to difficulty in identifying sex in fossils; where genetics are available, the anatomical classification has been confirmed. Age also shows some patterning, with the very young and elders more often found as individual skeletons than processed and recycled bones. Krapina for example is full of butchered adults, but there are no infants.

On the other hand, children seem associated with the potential multiple-body deposits. The Shanidar 9 baby was in the same small space as one adult male and two likely adult females, while Garrod believed that next to the Tabun woman's upper left arm there was an infant,* though no remains emerged during excavation of the sediment block in London. And speaking of the Near East, while there are plenty of skeletons, so far there are no known cases of body processing.

This chapter might leave you with the impression that Neanderthal fossils which had odd things happen to them are all over the place, but it's more complex than that. Some sites have a few pieces, others many skeletons, and there's also no clear correlation in regard to body processing between or within sites: the Le Moustier teenager was butchered, but the baby was not.

Then there is the fact that plenty of locales with rich archaeology have no Neanderthal remains at all. Hugely conspicuous is Abric Romaní: occupied for tens of millennia and excavated to high standards, yet none of the hundreds of thousands of bones are hominin. Other Iberian sites that Neanderthals seem to have used in similar ways *do* have hominin remains: a child's tooth and skull fragment were found in 2016 at Teixoneres, and Cova Negra contains bones from at least seven individuals: two adults, one older child, and four little ones.

---

* In a letter written to colleague and friend Gertrude Caton-Thompson a month after the discovery, she wrote: 'We found vestiges of a very young infant close to the left humerus.'

Just as intriguing are sites where things shifted over time, and in some cases like L'Hortus the presence of bodies might have influenced what else Neanderthals did there. Choices between depositing or fragmenting bodies might also have been linked to the fact that Neanderthals needed to be highly mobile and may not have expected to return to a particular locale for many months. Far from denoting different mortuary traditions, some of the variability we see may reflect contextual decisions linked to the way that loss could be managed depending on movement and season.

How to Be Dead

In essence, it's now very hard to maintain that Neanderthal bones all accumulated by random processes, or in the case of butchery, that it was just about filling hungry stomachs. Once the breadth of mortuary practices is understood, the borders between what they and early *H. sapiens* did with the dead begin to look fuzzy. Neanderthals were also potentially engaging with bodies before we did: the Tabun woman, who was laid out nearly flat on her back, may date as early as 140 to 170 ka.

But just as with the aesthetic traditions, there remain divergences. No intact Neanderthal skeletons come from open-air sites; though they're also rare among *H. sapiens* until after 30 ka, from then on spectacular burials exist. Double or multiple burials are also more common in *H. sapiens*, including two 27,000-year-old newborn ochre-covered twins from Krems-Wachtberg, Austria.

The identity of the dead may also be different. Among *H. sapiens* contexts, while adult males are just as overrepresented and there are few women, notably elders aren't more common, undermining claims they lived longer. Strikingly, there are also far fewer very young children and babies.

An even clearer difference concerns body posture, which seems more formalised in *H. sapiens*. There are more early examples with strongly bent limbs than in Neanderthals, but over time a sarcophagus-like pose becomes more common: flat, with straight arms and legs. In contrast, Neanderthal bodies tend to either be partly or wholly lying on one side; sometimes with foetus-like flexed legs but in other situations they're bent or extended asymmetrically.

Spectacular grave goods are definitely an *H. sapiens* phenomenon. Just a few minutes downriver from Le Moustier, Peyrony discovered that the La Madeleine rockshelter contained a young child's burial dating around 11 to 9 ka. Like the Moustier 2 baby, he recorded that it was found in a pit, but there the likeness ends. The child was laid flat, with an 'aura' of red pigment, and around the head, shoulders, knees, wrists and ankles were *thousands* of tiny animal teeth and shell beads. They would have taken months to cut and grind to size, and wear patterns show they had rubbed against each other over a long period. They came from Atlantic and Mediterranean beaches in addition to fossil shell sites, pointing either to vast territories, or extensive exchange networks. Most significantly, they are miniatures of those in adult burials, strongly suggesting that children received objects special to their stage of life.

The remarkable La Madeleine burial summons up a vision of a child that once laughed and ran as its embroidered clothing shimmered and tinkled. No Neanderthal sites have anything like it, but what's the best evidence for special things placed with their bodies? Claims for 'grave goods' are often quite subjective. Goat horns near a boy's remains at Teshik-Tash or articulated horse and panther feet at Sima de las Palomas are unusual but not clearly associated with the bodies. The odd stones under Le Moustier 1's face certainly were, but with such an old excavation it's impossible to say more. The atypical chert flake just a few centimetres from the curled fingers of the new Shanidar remains is tantalising, but the most convincing case is another Near Eastern Neanderthal.

Amud 7, a baby no older than 10 months, was excavated from a cave near the Dead Sea in the 1990s. Exactly like the Mezmaiskaya infant, it lay on bedrock on its right side, and despite some crushing by sediment, even fingers and toes were in the correct position. What marks Amud 7 out is that nestled right against its hip bone was a large red deer jaw. That species is common in the cave, but *complete* bones are rare. There is no intervening sediment, indicating that the heavy, perhaps still-fleshed jaw was placed directly on the corpse before it decayed.

Yet this isn't exactly burial 'bling'. Neanderthals seem to have had an aesthetic interest in coloured minerals, shells and perhaps bird parts, but there's no sign of anything like that with the dead. On the other hand, not all early *H. sapiens* people got richly endowed graves. The mortuary practices of the first Upper Palaeolithic cultures in Europe between around 45 and 30 ka were actually more akin to Neanderthal bodily interactions: they kept bits and pieces of skeletons, including pierced teeth. Truly jaw-dropping burials like a double child grave at Sunghir, Russia, don't really appear until more than 10 millennia *after* the first *H. sapiens*.

And the 'golden age' burials tend to blinker us to things these people did that mirror more ancient, Neanderthal-like traditions. At Sunghir, ochre-covered bones were being brought from other sites and placed alongside skeletons. Even cannibalism was happening in the Upper Palaeolithic. Brillenhöhle Cave, south-west Germany, is a few thousand years older than Gough's Cave, and contains partial remains of four intensively butchered adults and a baby. This is assumed to be a funerary ritual, not murder.

Outside the Eurasia bubble, however, there's an interesting reversal. Early *H. sapiens* populations in Africa produced masses of evidence for complex behaviour, but there are hardly any skeletons. Only two sites have relatively complete bodies, one of which dates to around 70 ka, with pre-existing quarrying pits containing the bones, though there are no associated artefacts. The other site is Border Cave, South Africa, which contains a possible infant burial of a similar date. However, it was excavated in 1940, so precise associations between bones, a claimed pit and a single shell – originally strung and covered in pigment – aren't clear. If the Neanderthals had a mortuary record of this paucity it would most certainly be held against them.

Death is worth an entire chapter because it's so deeply woven into how we define and differentiate ourselves from other animals. Neanderthals neither ignored corpses nor treated them like rubbish. They were not unmoved in the face of death, and the need to process – if not rationalise – emotional trauma most likely came through interaction with the corpse itself.

Neanderthals as fellow beings struggling with mortality paints the rest of their existence in very different shades. How they dealt with their trauma was diverse, and included depositing bodies, as well as taking them apart and bringing their raw components back into life via consumption, using them as tools, or marking them in special ways.

By focusing on burials as the best measure of mortuary meaning, we devalue the uniquely Neanderthal ways of doing things. Similarly, starvation or violence as primary explanations for cannibalism stem from recent Western taboos. In fact, bodily consumption as grief management is little discussed but does exist. In 2017 a tabloid story reported a British woman who regularly eats the ashes of her mother, and this isn't the only case. If it sounds outlandish, consider that curating bodily relics – from locks of hair to bony ossuaries – has long been embedded in Western society, and in the Christian Eucharist ritual the bread and wine are literally believed to become the body of Jesus inside the mouths of the faithful. Catholics say this is not about death, but life; maybe the same was true for Neanderthals.

The most important lesson is to take Neanderthals at face value, rather than squinting through a lens tinted by our own expectations. Those tiny cut marks across the crown of the Krapina skull stitch together many elements of their existence: bone as food, material and canvas, as well as the lithics that made them. To our eyes they're not aesthetically striking, but surely were significant to their maker/s. Fragmented and marked bodies mirror the broader pattern of Neanderthals taking apart, moving around and redepositing many materials. In so doing they stretched and distilled actions, memory and identity through time and space.

Just as hearths were cores for movement in sites, the presence of the dead may have impacted processes of place making at a landscape scale. Locales associated with the dead can have unique social potency, hinted at in the way chimpanzees, bonobos and even elephants appear to either revisit or avoid locations connected to deaths and corpses. If Neanderthals were differentiating places and entire landscapes through the things they chose to do, then involving dead bodies in this would only be an extension of existing behaviour. We can even imagine that the enormous variety of environments they lived through bled into how they responded to mortality. What did it mean to die in a world of beech forests, rather than one where massive herds of reindeer churned the tundra?

There's only one conclusion to draw from all this. If mortuary traditions extend beyond our own species, and even back to our last common ancestor with Neanderthals, then so too does a key definition of humanity. No formalised spiritual framework was needed; Neanderthal 'funerals' probably ranged from ardent and anarchic to methodical and precise. Just as the extinguishing of life drags out a primal keening in us, they too were motivated not only by fear, but also love. And it's these emotions that underlay the end of our entwined story: annihilation and assimilation.

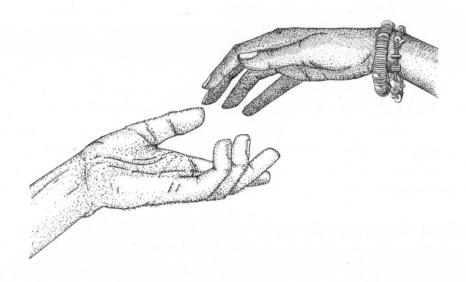

CHAPTER FOURTEEN

Time Travellers in the Blood

They walk away from the face of the sun, following paths the land reveals. Things they know appear in fresh guises: trees clothed in new leaves, beasts with unfamiliar furs. Even the rocks change beneath their feet. And they perceive Others. They are there in soft-scuffed dirt of trodden path, in burned-stone smell lingering round fractured rocks, in far-off smoke sinews rising to low cloud.

Trails meet, as they always will. Tense, testing dances take place beneath dripping autumn canopies, next to rushing rivers, in front of dark cave mouths. Sometimes fear erupts and blood leaps. Other times hands reach, fingers trace hair, skin and lips. Special long-held things — best stone, fatty cuts — are passed back and forth. With whispers in the firelight, other things are given as thighs slide together. Bellies swell and tiny faces push out under the stars, boundless limpid eyes opening to look upon the world as if simply returning to it. Breathing woodsmoke-spiced air, tiny fists uncurl as golden milk flows. The old ones remain in their bones, as new life pulses in flesh. The people make futures that stretch far ahead, spiralling through the years, centuries, millennia.

For most of the past 160 or so years, researchers have been toiling up a mountain of bones and stones in pursuit of more knowledge about the Neanderthals. That changed rather suddenly in the last two decades as ancient DNA shifted from pipe dream to reality. Genetics can illuminate many shadows archaeology cannot, so the chance to study it in Neanderthals felt like reaching the summit of a rocky ridge and suddenly seeing an immense, unexpected vista. Anchored in time and space, every sample offers a peephole onto unique information about ancestry and connectedness for individuals and the populations they came from. Zoom in, and DNA reveals biology beyond bones, and can even uncover entirely new sorts of hominins.

The surge of technological advances, new finds and theoretical shake-ups has been precipitous, and even for experts it's easy to feel overwhelmed. But amid the white coats, bone dust and test tubes, this is a story of intimacy: DNA's deep-past panorama looks onto a world of ancient communities that moved, interacted and interbred.

The Neanderthals had their own history, forming a complicated tapestry of lineages with genetic legacies scattered across thousands of kilometres. Appropriately enough, it was the original Feldhofer individual who was first to be sampled in 1997. At that point only mtDNA could be reliably extracted, and the result bolstered the evolutionary theory dominant at the time, which proposed that Neanderthals had arisen and remained genetically isolated in Europe.

Subsequent studies seemed to support this, and because mtDNA is inherited only from the maternal line, researchers could calculate the point at which the genetics of different individuals converged. This gave a rough date for the Neanderthal 'Mitochondrial Eve': a kind of great-great-great … grandmother.* Surprisingly, the result was less than 130 ka, and since Neanderthal fossils go back hundreds of thousands of years before this, something clearly wasn't right.

As more bones were analysed, it became obvious that each sample could dramatically change the overall picture. Initially the mtDNA showed that Neanderthal populations were small and homogenous; individuals dating around 50 to 40 ka from Spain, Germany and

---

* Mitochondrial Eve does not mean the *first* female Neanderthal, but the last common female ancestor for all Neanderthals.

Croatia were genetically very similar. But with more data, glimpses of regional diversity emerged. Sometimes geographic proximity matched relatedness: several of the Goyet individuals had DNA much more similar to each other than to any other Neanderthals. On the other hand, when the genetics of the second individual at Feldhofer were studied, they plotted closer to the Vindija lineage in Croatia than the original Neanderthal from the same German cave.

Descendants from deep population branches were still surviving scattered across western Eurasia by 50 to 40 ka. The Teshik-Tash child in Uzbekistan, for example, was shown in 2007 to be connected to European lineages, while even farther east at Okladnikov Cave in the Altai region of Siberia, another child's mtDNA held a bigger surprise. Dating around 45 to 40 ka, this was the furthest east any Neanderthal had yet been found, and unveiled a far more expansive Eurasian realm extending from the Mediterranean to Siberia.

But at some point there had been one or more massive upheavals. Some of the Spanish and French Neanderthals turned out to have mtDNA more like the Okladnikov child's than a lineage centred on El Sidrón, Feldhofer and Vindija. And the reverse is true: the Mezmaiskaya 1 baby in Russia, thousands of kilometres from Europe, is closer to Italian Neanderthals than to the Okladnikov child.

But mtDNA was always only going to tell half the story. Fuller and more complex accounts of their heritage needed nuclear DNA, and when technological advances brought this in reach, a Neanderthal genetics gold rush kicked off. Unusual freezer-like conditions inside Denisova Cave, in the Altai region of Siberia, opened up a 'Wild East' frontier, as DNA there was in exceptional condition. A sample extracted from D5, a toe bone, provided the first 'high-coverage' Neanderthal nuclear genome: our introduction to the recipe for another kind of human.

Dubbed 'the Altai Neanderthal', the toe had belonged to a woman who died around 90 ka. She came from a truly venerable lineage that had separated from others some 40,000 or 50,000 years before. And totally unanticipated, the mtDNA of Okladnikov, geographically nearest to her, was not closest genetically. Instead, it was the newborn baby from Mezmaiskaya, in the Caucasus thousands of kilometres west, who matched best.

The Denisova DNA bounty hasn't stopped. Since 2016, six more Neanderthals have been sampled genetically, using DNA gleaned from bones and even the cave earth itself.* Some have mtDNA grouping with the Altai bloodline, but others don't, including one individual, D11, who lived around 90 Ka.

These results have revealed a deep structure within the Eurasian Neanderthal population as a whole. Two main branches split, then remained isolated in Europe and Asia for millennia. Moreover, the Altai woman's descendants, sort of like long-lost cousins of all the other Neanderthals, had apparently disappeared and later been replaced by a twig stemming from the European branch. Just as with the mtDNA in Europe, it seems that at a regional scale, multiple nuclear DNA lineages existed that were either contemporary but not mixing much, or quite rapidly replaced each other.

All this implies that there were continental-scale movements of lineages, certainly towards the east, but also perhaps in the other direction. This was likely to be an incremental process rather than anything like modern notions of migration, but the fact that it happened at all points to enormous, long-term upheavals. For any region, we can't assume continuity between early Neanderthals pre-MIS 5, and those afterwards.

Recent genetic analysis of the Forbes Quarry skull has enhanced the family portrait. As well as confirming her female sex, it showed her nuclear DNA was equally close to high-coverage genomes from individuals from Chagyrskaya in Russia and the Vindija genome in Croatia. This makes her part of the population ancestral to them both.

At the same time, her DNA was still different to the Altai branch, indicating that the split from the eastern cousins was probably as deep as had been suggested by D5's dating, around 170 to 130 ka. This period comes roughly at the time that the MIS 6 glaciation ended, with rapid warming towards the Eemian climate peak. At least in some regions the archaeological record appears to show that Neanderthals were shifting technologically and culturally during

---

* Some of the sediment DNA is described as 'diffuse', and might possibly come from build-ups of faeces.

MIS 5, which is also when we see some mtDNA sub-populations emerge. And after the deep freeze of MIS 4, Neanderthals in Europe were certainly expanding their range, leading to the recolonisation of 'Western Doggerland', otherwise known as Britain. Perhaps some of these movements were echoed in a diaspora towards the east.

The genetics revolution has also been the cause of truly astounding discoveries. Denisova is today world-famous not because of the Altai Neanderthal lineage, but because of a minuscule bone: a girl's fingertip. Known as D3, her mtDNA didn't match *any* hominin group, and she turned out to be the accidental ambassador for an entire 'ghost' population nobody knew existed. More and more DNA from these hominins – referred to as Denisovans – has since been sifted out from bones, teeth and cave dirt. They extend at Denisova from around 50 to 150 ka, but as a population diverged from Neanderthals before 600 ka. In evolutionary terms they're closer to each other than either were to us, though not by much. Moreover, their DNA is more diverse than the Neanderthals', so either there were many more of them, or their overall population didn't suffer so many internal extinctions.

What were Denisovans like? For nearly a decade researchers had only the barest hints of their appearance. DNA indicates some had brown eyes, hair and skin, and their teeth weren't identical to Neanderthals'. But other physical remains are so limited – the D3 fingertip plus three teeth – that not much else could be said. In 2019 researchers tried to 'reverse engineer' Denisovans by looking at unique aspects of their genes involved in body growth. Although we won't know for sure until (if) we find skeletons, their heads may have been even wider than Neanderthals', and fingers also longer.

Beyond anatomy, however, things get very tricky. There is archaeology at Denisova, but the layers there have obviously been deformed by natural freezing processes, and hyaena digging could also be an issue. Moreover, genetic estimates for the date of some of the fossils don't match the ages for other artefacts in their layers, hinting that some of the hominin remains may have slipped out of their original context. Working out who made what may therefore not be possible.

Everything points to Denisovans as an Asian species. Remarkably, proteins from a jawbone at Xiahe, high on the Tibetan plateau – and 2,200km (1,370mi.) south-east of the Altai – are either Denisovan or from a close 'sister' population. But we also know they and the Neanderthals lived in the same cave, albeit at different times. Did they ever meet? The answer is a resounding yes. Tantalising hints in D3's DNA suggested her ancestors had at some point interbred with Neanderthals, but the real shocker was yet to come.

Re-enter D11, the tiny limb fragment from a young teenager who had lived around 90 ka. It had originally been found in 2012 and only recognised as hominin via protein sampling four years later. D11's mtDNA placed her as a Neanderthal, but this came only from her mother. Her nuclear DNA instead showed her father had been Denisovan.

'Denny', as she was nicknamed, is the only first-generation hominin hybrid ever found. So improbable that the researchers did not initially believe it, the implications are staggering. It was assumed that interbreeding was rare, and direct evidence would only ever lurk in the genetic gloom, many generations back from the bones we study. Actually finding the child of a union between different kinds of hominin implies it can't have been that uncommon.

In fact, Denny's DNA contained vestiges of yet more interbreeding. At least one of her father's ancestors had encountered Neanderthals too, albeit thousands of years and many, many generations before.

In a last surprise, this ancient Neanderthal ancestor wasn't from the same genetic population as Denny's mother. She was part of the eastern twig of the European lineage, also found at Okladnikov. In contrast, the Neanderthal ancestry in Denny's father connected much further west, to the El Sidrón-Feldhofer-Vindija grouping.

Everything at this remarkable site makes it abundantly clear that, far from being static, these hominin populations saw huge flux over time. The most recent research even suggests there might be mixed ancestry in *every* hominin there. What makes Denisova so exceptional? No Neanderthal fossils or DNA have ever been found from farther east, and no Denisovans from farther west. Perhaps this cave was literally at the edge of their two worlds.

Once We Knew You

There is another kind of hominin whose potential genetic connections to Neanderthals have been the subject of speculation and fantasy for more than a century: us. In 2010, hot on the heels of the Denisovans' debut, came a second revelation: contrary to the mtDNA, the first Neanderthal genome showed they *had* directly contributed to our own ancestry.

In the absence of interbreeding their DNA should have been equally dissimilar to that of everyone alive, but instead people lacking sub-Saharan African heritage had significantly more matches to Neanderthals. The only reliable explanation was if some *H. sapiens* had met and had children with Neanderthals *after* dispersing from the African continent.

This news caused a seismic shift in human origins, reverberating through many foundational assumptions about both species. Initially it was assumed this interbreeding must have been chronologically recent, probably in Europe around 40,000 years ago. A decade on, things are far more convoluted, and a brief tour of *H. sapiens'* early history is helpful.

Though hominins were present in Eurasia well before 1 million years ago, the oldest *H. sapiens* fossils are certainly African. However, old notions about a particular 'cradle of humanity' have now been superseded. The most recent fossil and genetic evidence suggests we evolved from an anatomically diverse meta-population, connected across many regions of the continent.

During the crucial period between 800 and 600 ka, when the ancestors of Denisovans and Neanderthals split off from what would become 'us', the fossil record is frustratingly scarce. But after this it seems the anatomical features shared by everyone today evolved over a long time, in different African regions. Brains grew rapidly after 500 ka, but skulls and bodies developed more slowly in a mosaic fashion. The people of Jebel Irhoud, Morocco, living around 300 ka already had big brains and flat, modern-looking faces, but more archaic upper and rear skulls. The oldest *H. sapiens* skulls, pretty much like extant humans, date to around 200 to 150 ka in East Africa, around the same time that 'classic' Neanderthal anatomy was also coalescing.

One of the biggest recent changes has been that more and more early *H. sapiens*-looking bones are now known outside Africa. While skeletons from Skhül and Qafzeh in the Near East were found during the 1930s and later dated to 90 to 120 ka, they seemed to be anomalies. Today the opposite is true. In 2018 an upper jaw fragment from the Misliya rockshelter in Mount Carmel was dated to an incredible 177 to 194 ka. Though fragmentary, there's enough to be sure it isn't Neanderthal.

The following year, even earlier dates around 210 ka were announced for a partial skull at Apidima, Greece, which was claimed to be *H. sapiens*. This site is tricky, however. A cliff shaft filled with jumbled sediments that may have come from a huge adjacent slope deposit means that exactly where the skull came from isn't clear, and what's more, other researchers point to Neanderthal-like features.

Certainly such a massive age and its location along the Mediterranean coast would imply an unexpectedly early dispersal, although the environmental context is quite comparable to North Africa. However, it's now clear that early *H. sapiens* were already many thousands of kilometres into East Asia probably before 100 ka, adapting to completely different ecologies. To reach China somewhere around 120 to 80 ka, Sumatra by 73 to 63 ka and cross to Australia at least 65 ka, they must have walked over mountains, across deserts and through jungles – and probably also rode the waves in watercraft.

Much of this wasn't known in 2010. Back then it still looked as if early *H. sapiens* lived in the Near East at places like Qafzeh without moving further for tens of millennia, and were then replaced by Neanderthals after 90 ka. It wasn't until the first Neanderthal genome came out that everything changed. Now 10 years on, things are *far* more complicated, and more interesting.

Current data finds between 1.8 and 2.6 per cent Neanderthal DNA in everyone except those of sub-Saharan heritage;* but it's not equally distributed. Western Europeans tend to have the least – 2 per cent or

---

* They also have some Neanderthal DNA, but it seems to have arrived from later interactions with Eurasian *H. sapiens* migrants.

under – while Indigenous Americans, Asians and Oceanians, including Aboriginal Australians and Papuans, have up to a fifth more. We now also believe that there were *multiple* interbreeding episodes, and in some cases they left their mark on Neanderthals too.

Neanderthal nuclear DNA contains glimmers of very ancient encounters, and recent research implies that interbreeding was effectively the norm from *way* back. The Neanderthals' common ancestors with Denisovans – Neandersovans – received DNA from a 'super-archaic' Eurasian hominin that had probably been around for 1.5 million years. After the Denisovans went their separate way, other shadowy signals for early mixing appear, this time with *H. sapiens*. It's there in the Altai and European Neanderthal lineages, meaning that it happened before their own deep split around 140 to 130 ka.

Another pointer may come from the first really old Neanderthal fossil that produced mtDNA. At Höhlenstein-Stadel, south-west Germany, indirect dates for a man's thigh shaft date around 100 to 120 ka, and his mtDNA looks *nothing* like that from later Neanderthals. This could be explained if he was from a lineage that became genetically isolated as far back as 270 ka. If accurate, this single individual would totally change the view of Neanderthals having very limited mtDNA diversity. However, there's another theory. Perhaps this mtDNA looks so very different because it wasn't originally Neanderthal at all, and instead was inherited after early encounters with *H. sapiens*; moreover, there are now also hints that something similar was going on with Neanderthal Y chromosomes too, potentially even earlier. This sounds bizarre, but similar processes are known in animals: polar bears' mtDNA seems to have been totally replaced by that of brown bears during interbreeding around 130 Ka.

These speculations need more early samples to be certain of what happened, but later interbreeding is more easily identified. The contact phase that apparently left the largest genetic mark in us took place between 75 and 55 ka. Remarkably, it's actually mirrored in DNA from an early *H. sapiens* bone found by the Irtysh River, halfway across Siberia in the Ust'-Ishim region. The partial leg shaft once carried the weight of a man between 46.8 and 43.2 ka, who had traces of Neanderthal ancestry from interbreeding going back 7,000 to 13,000 years before he died. Later research resolved this into two different interbreeding episodes: one between 54 and 50 ka, and the younger at least 5 millennia later.

At first glance the older phase could fit calculations based on the Neanderthal genome, but there's a catch. No sequenced Neanderthal genomes so far exactly match the DNA present in living people. It definitely didn't come from the Altai lineage, but it's no more similar to either Vindija or Mezmaiskaya 1 on the European branch. This might mean that the interbreeding with the source population that had the most impact on us happened in a region for which we've yet to get any Neanderthal DNA.

It also implies that their branch had split off by 80 ka, matching genome-based calculations that place the interbreeding between 90 and 45 ka. We can refine this archaeologically to *before* 55 to 60 ka, since today's Aboriginal people carry Neanderthal genes and were already in Australia by then. Put everything together, and it looks as if both interbreeding phases in the Ust'-Ishim man's DNA are too young to be those seen in living Eurasians.

Multiple periods of late interbreeding are also supported by other data. We can see that early *H. sapiens* populations in Eurasia had already separated into different lineages by 55 ka. The higher amounts of Neanderthal DNA in some people today probably comes from extra hybridising episodes within some of those lineages, which then passed into Asia and beyond.

We also now know that interactions potentially happened closer to Europe. Just after the 'Ust-Ishim results, DNA from another early *H. sapiens* fossil was published. This man died hundreds of kilometres westwards, at the Peştera cu Oase, Romania, between 42 and 37 ka. His genetic ancestry was nearly as mind-blowing as Denny's, because about 11 per cent of it was Neanderthal. This means he had a Neanderthal ancestor within just four to six generations.

That's the same gap between you and the prehistory pioneers peering at the Feldhofer skull in the 1860s. And just like 'Ust-Ishim, the Oase man's heritage also appears to contain multiple interbreeding phases, with another around 2 millennia before his death.

Taken together, there are at least three and potentially six periods since 200 ka when Neanderthals made babies with us.* That all this has been discovered in under a decade, from so few fossils, strongly implies that contact and hybridising happened a lot more often than we'll probably ever know.

One odd pattern sticks out, however. No late Neanderthals, even those from Vindija who are geographically close and only slightly older than the Oase man, show *any* genetic input from *H. sapiens*.

But this is a reminder that the locations of fossils today aren't reliable markers for where things happened several generations back. Perhaps when one of Oase's great-grandparents encountered Neanderthals, they were living much further east or south, and indeed, fossils from the Near East and Central Asia are yet to give up DNA. It's also possible there were reproductive complexities, making our DNA more likely to be rejected in Neanderthals hybrids, or disappear faster in their population.

Zoom back out from the genes to the bodies, and we come to the question of exactly how, and why, Neanderthals were having sex with other hominins. To account for the number of interbreeding episodes and the percentages of surviving DNA in us, there could have been hundreds of individual sexual encounters and resulting hybrids; perhaps more. Victorian scholars undoubtedly secretly wondered about inter-species relations, their imaginings tinged by cultural mores and preoccupations.† But understanding how Neanderthals and early *H. sapiens* felt 50,000 years ago is far harder to assess.

Various animals are known to extend sexual interest beyond their own kind, from leg-humping dogs to dolphins getting too friendly with swimmers. Among humans bestiality isn't common, averaging 1.5 to 4 per cent of people, but it's widespread. Easy access is the main

---

* Based on dating estimates from genomes, the oldest Ust'-Ishim phase is separate; the younger Ust'-Ishim phase chronologically overlaps with the oldest in Oase, but the more recent Oase phase is too young and so must be a third case.

† By the 1870s it was believed that male apes were sexually attracted to human women, and in a footnote to *The Descent of Man* Darwin described an experiment from a century earlier involving an 'ourang' and a female sex worker, aimed at determining if hybrid offspring were possible.

factor, explaining why rates can be double in some agricultural communities. Motivations, however, are extremely varied by culture and personal situation. In some hunter-gatherer societies, sexuality is embedded within cosmologies where hunting animals is part of the cycle of life and death. Typically, however, there isn't direct sexual contact with prey.

None of this really fits with Neanderthals and us. They walked upright, carried tools, probably wore garments and had some kind of speech. It's highly improbable that there wasn't a mutual recognition on both sides that the beings before them were people, albeit of a new sort.

There is no unambiguous evidence of *how* sexual encounters happened, only their consequences. Bearing in mind that the different phases of interbreeding happened over a vast range in time and place, there must have been many different dynamics. There are hints in the DNA that couplings might have involved more Neanderthal men with *H. sapiens* women than the reverse, but other explanations for the data are possible.

In theorising the social contexts behind all this, there's been a tendency to assume rape as a primary mechanism; an unpleasant residue from the days when prehistorians and the public regarded Neanderthals as more beast than potential beloved. Chimpanzee males will engage in coercive sex, but not with unknown females (whom they prefer to kill). It's theoretically possible *some* of our Neanderthal inheritance may derive from non-consensual circumstances, but xenophobia rather than xenophilia needn't be the default assumption.

Pleistocene hook-ups may just as plausibly have been more similar to how bonobos deal with unfamiliar faces. Bonobos are fundamentally friendlier: unlike chimps, they contagiously yawn even when watching strangers, as we do. They're more open to positive interactions with other groups, and moreover territorial border patrols and killings of non-group members are unknown. We should perhaps ask why the idea of enthusiastic partners driven by desire and even emotional attachment is regarded as more of a fairy tale than other explanations.

Perhaps a more pertinent observation is the fact that, however they were conceived, hybrid children were raised to survive. Presumably more often than not, infants stayed with their mothers, and they were fed, cleaned, kept warm; loved. These babies of mixed heritage grew up, understood the cultures they were born into, and went on to have children of their own.

Legacies

The legacies of those babies who became parents and then grandparents many times over is that a fifth – perhaps as much as half – of the great twisting genetic recipe making Neanderthals 'Neanderthal' endures today. While at most just 2 to 3 per cent of any living person's genome is Neanderthal, it's still a significant amount. Can we trace the biological, or even psychological, effects of assimilating their essence?

The actual number of genes we're talking about is tiny, and natural selection certainly removed much of what came over in each hybridisation phase. Nevertheless, genes from Neanderthals (and Denisovans) make up a substantial portion of the small 'active' part of our genome. Some of it almost certainly helped us.

This is very much cutting-edge science, and so current knowledge of what this means for our bodies, health or even minds is still ragtag. Studies matching an individual's genetic Neanderthal pedigree with their medical records have suggested links with digestive system problems, urinary infections, diabetes and over-clotting of the blood.

Inventing evolutionary explanations for these is tempting, but researchers are still at the very start of understanding how particular genes function in us, never mind how archaic versions might have worked. It's also important not to forget that, just like our own genome, many Neanderthal genes were copied randomly and were potentially neutral in their effects.

There may be some cases, however, where the genes we ended up inheriting make sense in regard to the unfamiliar Eurasian world *H. sapiens* entered. Without question, dispersing populations would have encountered new pathogens: not only diseases, but also bacteria. Living people with dual ancestry from Neanderthals and Denisovans seem to have 'preferred' the Neanderthal version of some genes involved in the skin's defences against infections. Similarly, a gene protecting us against bacteria that cause stomach ulcers came across from both Neanderthals and Denisovans, but people carrying two Neanderthal versions have extra resistance.

Eurasia presented other challenges for *H. sapiens* without hundreds of millennia adapting to its lower levels of UV and seasonal winter darkness. East Asians and Europeans share Neanderthal versions of keratin-related genes that make hair, nails and skin. It's possible they were more useful than forms we'd developed in tropical environments. But on the other hand, Neanderthals had diverse hair and skin pigment, so things must have been complex. Body clock genes are another area where we kept Neanderthal versions, and this is likely to do with the fact that circadian rhythms are strongly linked to day length and light levels. Perhaps Neanderthals passed on something that helped *H. sapiens* learn to cope with the especially long, dark winters.

Adjusting to colder climates would have been a major issue, and even if bodies were buffered by clothing, Neanderthal genetics might have helped us too. Some of their surviving legacy in our genome is connected to metabolism, and therefore to thermal efficiency. One gene affects how fats move into cells, giving carriers today a higher risk of Type 2 diabetes. But in hunter-gatherers this may have helped with energy management and coping with starvation situations. Something similar might also explain genes promoting fatness, and another connected with addiction. Once, these could have been advantageous in encouraging consumption of feel-good, fat-rich foods.

Large swathes of our genome have no Neanderthal contributions, which may mean that what we already had was worth keeping. Was this because the Neanderthal versions were also bad for them? In general their DNA doesn't appear dodgier, but some riskier variants have been identified.

One case relates to pollution. Hearths and even micro-charcoal in Neanderthal dental calculus tell us they sometimes lived in smoky situations. A mutation in all living people makes us between 100 to 1,000 times *less* susceptible to smoke and charred food toxicity. Since inhaling smoke from open fires or poorly vented stoves is the main cause of death globally for children under 5 years old, this is no small issue.

Another example of possible inferiority in Neanderthal genes is fertility. The parts of our genome relating to X and Y chromosomes have a clear lack of Neanderthal contributions. And at least one

Neanderthal male, El Sidrón 1, carried three forms of genes that today are linked to miscarriages of male foetuses. This led to speculation that hybrids were more likely to be female, and even that mixed children could have had genetic disadvantages.

But as geneticists have spent decades learning, DNA doesn't behave in simple ways. Genes are often more like herbs or spices in a recipe, their flavours varying depending on the other ingredients and the method of cooking. As research advances on how the genes of living people work, we'll be able to tell more nuanced stories about Neanderthal legacies in our bodies.

The same is true of minds. Identifying DNA markers for cognitive differences in Neanderthals has long been a key aim of ancient genetics. Could there really have been a 'light-bulb moment' when some novel genetic mutation or combination greatly increased *H. sapiens'* tendencies towards more formalised artistic traditions, or flashy burials? Again, the reality is inconveniently uncertain. Some Neanderthal genes we inherited are involved in basic brain functions like energy management, but socially expressed differences are the key issue. Living people with particular Neanderthal genes may have higher rates of mood disorders or depression, yet the effect is tiny in statistical terms, and we don't know if those genes functioned identically in the past.

Particularly interesting are Neanderthal versions of genes that affect brain structure. Some seem involved in expanding the back of the skull, building greater amounts of brain matter and more intense surface folding. If the Neanderthal versions still persist in people today, then they either didn't affect the survival of hybrids and their offspring, or they were actually advantageous.

Other 'Neanderthalised' areas of our brains are even more connected to advanced thought processes, including learning sequences of finger movements, plus conceptualising and calculating relative amounts and numbers. Suddenly those engraved sequences of lines and notches on various bones start to look rather more significant.

Most unexpectedly, Neanderthals also returned to us far more ancient forms of genes that we'd lost long before. It seems that some of the genetic inheritance from our common ancestor with Neanderthals dropped out of early *H. sapiens* populations over time. Parts were then knitted back into our genome through interbreeding encounters before 100 ka. But it wasn't all welcome: the ancestral version of the FOXP2 gene didn't stick, indicating that the version we'd evolved in the meantime was important.

The flip side of interbreeding is that some early *H. sapiens* genes should have moved into Neanderthals too. However, at the moment we don't have information on this, because no late Neanderthal genomes show any *H. sapiens* input. This fact underlines how each new genome and lab study is crucial, and work is ongoing to expand the samples.

Larger numbers of genetic samples have dramatically altered the view that Neanderthals were defined by a tiny meta-population. As mentioned earlier, some initial analyses suggested far lower genetic diversity than living *H. sapiens*.* Theories implicating inbreeding – consistent reproduction between closely related individuals – in their disappearance emerged, and even seemed to be supported by unambiguous cases. At Denisova, the Altai Neanderthal woman's parents must have been one of the following: double first cousins (sharing both sets of grandparents), an aunt with a nephew, a grandparent with grandchild, or even half-siblings. By many cultural definitions, that's more like incest than inbreeding. Further analysis of her DNA also found relatively close, if less extreme, relationships between her ancestors over many generations. A similarly small genetic population is indicated at El Sidrón, and a 2019 study presented a long list of unusual skeletal quirks shared by the individuals there; a feature also seen at La Quina, another site with many skeletons.

Why does inbreeding matter? Occasional very close pairings don't dramatically increase health risks, but over the long term it can concentrate damaging mutations, and increase problems like poor immunity. Most *H. sapiens* historic and living cultures have taboos against coupling too closely to parents, and plenty of animal species appear to follow similar rules.

But the picture has changed with more data. When the high-coverage genome from Vindija was sequenced, it didn't have significant markers for inbreeding in previous generations, and neither were this

---

* Sub-Saharan Africans have much richer DNA than Eurasians, who apparently suffered a genetic bottleneck – drastic shrinking of the population – at some point within the last 80 ka.

individual's parents close relations. This means that rather than being the norm for Neanderthals, where inbreeding and even incest was happening, it was probably about lack of choice rather than preference. The Vindija genome also revealed that not all late Neanderthal populations were shrinking, and population estimates for early Neanderthals would double if the Höhlenstein-Stadel mtDNA is not an import from extremely ancient interbreeding with *H. sapiens*.

The very latest studies have revealed further complexity. In 2020 a high-coverage genome from Chagyrskaya, Siberia, didn't show inbreeding between parents, but came from a reproductive population just as small as the relatively nearby Altai woman's, averaging about 60 individuals for many generations. In stark contrast, the earliest *H. sapiens* genome from Ust'-Ishim has more diverse DNA than any Neanderthal sampled so far. This implies that the interconnectedness of *H. sapiens'* social networks may well have been different right from the start.

The revolution in our knowledge of Neanderthals unleashed by ancient DNA in just 10 years is astounding. Artefacts had long suggested deep splits within their populations, but genetics opened up a world where Neanderthals from different lineages moved across continents. It wasn't just *H. sapiens* who were explorers.

The most radical result came from realising that their essence endures at the cellular level, coursing through our veins, tousled by wind in our hair. Their legacies affect not only what, but *who* we are. Yet so far we've sampled fewer than 40 Neanderthals – and have only 3 high-coverage genomes – from among the *thousands* of skeletal parts in museums, representing hundreds of individuals. The next decade will see the door onto their complex history and biology that's currently ajar pushed further open. Some questions such as the frequency of interbreeding will receive more refined answers, but others, such as who raised hybrid babies, require integration with the archaeology. What's clearer than ever before, though, is that the 'end' of the Neanderthals was a process involving bodily and probably cultural assimilation.

CHAPTER FIFTEEN

Denouements

Sun-blink, tail-twitch. Weight shifts from hoof to hoof. Enfolded in the herd's calm sweat smell, each eye stares, looking across the narrow valley to the white mountains rising in the east. Small pulse-tides surge and ebb at shadows or noises. Heads back down, bison tongues lick the dew, pull in great curls of grass and herbs to be slowly ground down. Curls of smoke play around the edges of the meadow, pricked apart by pine needles on their way down the hill, spread thin by the breeze until the matrix of soot molecules is barely there.

But it's enough: nostrils flare, pupils dilate, bodies stiffen and a staccato of snorts explodes. Tails curl upwards, waving their agitation as figures emerge from the trees. The herd stands its ground, secure with distance. But they have not seen these tall ones before, with their new smells and colours. The people slowly spread out along the brush-edged meadow, while the bison watch, seem uncertain. This is not the way of things. A moment of stillness stretches — then tensed arms raise, flick, and a flock of reed-thin weapons flit out like birds, carrying death on the wing. Tiny stone tips bury themselves forearm-deep into

furred stomachs, necks, then hooves stumble, flanks crash down. The unhurt
bison scatter, hearts battering ribs, even as those same parts of their kin are being
sliced into on meadow grass slick with blood. This new people, this new hunt,
this new way to fear will soon move on, duskwards.

'The Last Neanderthal' has long been imagined: a lonesome soul, their death blinking out the species at a single point in space and time. Though today we know they won partial immortality at the cellular level, their disappearance from the fossil and archaeological record is real. What we still don't understand is how those facts interconnect. Finding answers is extremely challenging: hominin bones are rare, and despite many advances in dating, the highest resolution for individual radiocarbon measurements is around 500 to 2,000 years, far beyond the generational timescales we're interested in.

Researchers have zeroed in on the crucial period when the final Neanderthal fossils and Middle Palaeolithic layers are found. Recent re-dating of anomalously young bones from a number of sites have all resulted in older ages. For example, since the 1990s some of the Vindija remains had ages around 33 to 28 ka, but reconsideration of the taphonomy and use of collagen amino acid analysis to ensure purer samples pushed them back a good 10 millennia. Similarly, at Spy up in Belgium, dates between 38 and 34.6 ka have been refined to over 40 ka. All this makes the extraordinarily young dates from Gorham's Cave in Gibraltar − as recent as 28 to 24 ka − look much less likely, especially as they were done on charcoal, a tricky material, well before modern purification techniques.* Consolidating data from numerous sites points to 40 ka, if not slightly before, as the point beyond which no reliable evidence for Neanderthals exists.

That's the 'when'; what about the 'where'? Europe historically was assumed to be the Neanderthals' heartland and likely place of their last stand. But their true range is far greater: Denisova cave is more than

---

* For the period 40 to 50 ka, just 1 per cent contamination will reduce the real age by over 8,000 years.

twice as close to Mongolia's capital Ulaanbaatar than to Le Moustier in France. While it doesn't have evidence for late Neanderthals, other sites from the region do suggest that they survived nearly as late as in Europe.

Denisova may be the easternmost known site for Neanderthals, but perhaps there wasn't really a border. The steppe and taiga that stretched between Belgium and Beringia – the vast area of land linking northern Asia to Alaska – was an environment they were supremely familiar with, and between 60 and 45 ka their populations in Europe were definitely expanding, including recolonising Britain.

Perhaps there was also a push towards the dawn horizon, and Neanderthal feet once stood on the shores of the Pacific. Repeated intermingling with Denisovans shows that the presence of other hominins in East Asia wasn't necessarily a barrier to movement. Some researchers also see Neanderthal-like traits in some Chinese hominin remains, though they look in other ways like early *H. sapiens*. Moreover, aside from generic Levallois technology, most intriguingly, someone between 47 and 42 ka at Jinsitai Cave in China was making artefacts extremely similar to Sibiryachikha assemblages from Chagyrskaya and other Neanderthal sites some 2,500km (1,550mi.) west in the Altai. It's not entirely unreasonable to imagine that the last breaths to fill Neanderthal lungs were inhaled not at the southern tip of Europe, but somewhere in the vastness of Central or East Asia.

The last clearly Neanderthal remains are one thing, but are there any cases of hybrid fossils? During the 1980s and 1990s before genetic evidence of interbreeding, researchers debated whether some Neanderthal bones younger than 50 ka seemed rather less heavily built. There were even claims of *H. sapiens*-like features: hints of jutting chins or more rounded skulls. One site considered in this way was Vindija, yet the genome turned out unequivocally to be fully Neanderthal.

As Chapter 14 discussed, the range of time and space where interbreeding could have happened has also dramatically extended, making it less likely that very late European Neanderthals would be the only candidates to show hybrid features. The Near East region makes sense as a contact zone, being geographically between Europe and Africa, but establishing that Neanderthals were there at the same time as early *H. sapiens* is tricky. Between 200 and 90 ka they may have alternated, but a partial *H. sapiens* skull from Manot Cave in Israel dating before 55 ka implies that late Neanderthals at Amud and elsewhere were roughly contemporary with that population.

In fact, even though the Manot skull somewhat resembles Upper Palaeolithic *H. sapiens* from Europe, it also has an occipital bun: one of those bumps above the neck occasionally seen in recent and ancient humans, but found in virtually *all* Neanderthals. Until genetic extraction from Near Eastern fossils becomes possible – warm climates make it harder – little more can be said.

Right now, the Oase jaw remains the only fossil representative *anywhere* for late interbreeding. However, since it took place up to six generations before his birth, how it manifested physically will be diluted.

Bones and genomes have been at the forefront of recent research into the last Neanderthals, but was DNA the only thing they exchanged with us? Layers containing their distinctive techno-complexes also disappear between 45 and 40 ka. What comes afterwards has caused probably the most vexed of all debates. Across Europe and Western Asia a smattering of strange assemblages overlie the last recognisably Neanderthal layers. They appear to combine Middle Palaeolithic flake-based technologies with a much more Upper Palaeolithic-style concentration on blades and bladelets. Additionally, there's many more shaped bone, antler and ivory objects.

As Chapter 6 showed, Neanderthals obviously knew how to make blades and bladelets, but these were never their main focus, and similarly, *formed* bone artefacts are very rare. What's more, the intermediate cultures also contain indisputably symbolic objects including pierced stones and animal teeth as well as curious carved rings.

Precise chronologies vary geographically: the oldest dates come from the eastern fringes of Europe at around 45 ka, but in its western reaches they're somewhat younger, and persist until about 41 to 40 ka. But in stratigraphic terms, there doesn't seem to have been an overlap. Within any site, Middle Palaeolithic assemblages are always below intermediate ones, which are then followed by classically Upper Palaeolithic layers. The intermediate cultures seem to sit like ephemeral interregnums between Neanderthal and *H. sapiens* dynasties. Their distinctiveness means that prehistorians tended to christen them by a variety of names, often after the type site. There's the Szeletian in Hungary, Bohunician in the Czech Republic, Uluzzian in Italy, Bachokirian in Bulgaria and the cobbled-together Lincombian–Ranisian–Jerzmanowician identified in Britain, Belgium and Eastern Europe.

The million-dollar question is who made them. The Proto-Aurignacian, which follows intermediate cultures in Europe, has produced *H. sapiens* mtDNA from a tooth at Fumane. But skeletal remains are vanishingly rare before this, and frustratingly, many key sites were either excavated over 40 years ago, or have obvious signs of disturbance or mixing between layers. With greater understanding of taphonomy, the potential for freeze–thaw shifting of sediments within has become obvious, and so untangling what these cultures really mean requires archaeological contexts of exceptional integrity, and a battery of high-resolution analytical methods.

One of the first intermediate cultures to be recognised was the Châtelperronian, from France and northern Iberia. Mid-nineteenth-century rail works between a coal mine and foundry uncovered fossils and artefacts in the Grotte des Fées, near Châtelperron, central France. Over the century that followed, similar assemblages elsewhere were classified together, but it was assumed that Neanderthals were too intellectually inferior to produce the blades or bone artefacts they contained.

Then came an astonishing find. Roughly midway between Poitiers and Bordeaux, mushroom farmers had been hollowing out the Roche-à-Pierrot cliffs with tunnels. Building works uncovered something far more precious than fungi: archaeological deposits beneath a collapsed rockshelter. Professional excavations commenced and, completely unexpectedly, in 1979 Neanderthal bones emerged from what looked like a Châtelperronian layer.

Known as Saint-Césaire 1,[*] that skeleton wasn't alone. Further north in France at the Grotte du Renne, Arcy-sur-Cure, bones and teeth spread through a series of Châtelperronian layers were also

---

[*] It's Saint-Césaire '1' because there are also unpublished teeth from a second individual, and new infants under study.

Neanderthal. These revelations presented a paradox for leading theories that saw the Châtelperronian as something that was made by *H. sapiens*, who had replaced Neanderthals because they were culturally more advanced. Two competing explanations emerged. Perhaps the Châtelperronian was actually an independent Neanderthal invention, converging on Upper Palaeolithic-like features by chance. Or alternatively, it was made by Neanderthals but resulting from some kind of cultural hybridisation. Possibilities ranged from full contact, to Neanderthals spying on Upper Palaeolithic groups or picking through their trash and then figuring out how to copy them.

Today things have grown more complicated. Nearly 100 Châtelperronian sites are now known, from the Paris basin down to northern Iberia, dating somewhere between 44 and 41 ka. In France it rapidly follows the youngest Middle Palaeolithic layers, but south of the Pyrenees there seems to be a gap of around 2,500 years before it appears. It was definitely over fast everywhere, lasting perhaps 1 millennium in any particular region; the same separation in time between you and the earliest printed money.

Most crucially, excavations of new sites without taphonomic problems have revealed a rather different cultural picture. Middle Palaeolithic flakes and tools are *only* present in Châtelperronian assemblages from old excavations or places where there are signs of disturbance. This means that the apparent 'transitional' character in the technology is far less supported.

Detailed studies of these 'clean' Châtelperronian layers show it was a true laminar world. Blades were retouched on one side opposite a sharp edge to make Châtelperron points, and the makers were highly selective: blades not up to scratch in size terms were rejected. A key locale is the Quinçay rockshelter, about 100km (60mi.) north-west from Roche-à-Pierrot. Of the more than 450 cores, fewer than 1 per cent had flake scars. Refitting confirmed the dominance of blade production, which were intended specifically to be retouched into points, of which there were more than 300.

Open-air Châtelperronian sites show the same thing. Canaules II, near Bergerac, has a clear separation from underlying Middle Palaeolithic archaeology. It was a mass-production workshop, containing thousands of near-pristine artefacts from a very thin layer.

Almost a third were refitted, and once again specialised blade blanks for points were undeniably the goal.

Even more significant, Châtelperronian laminar technology doesn't match the way Neanderthals made blades or bladelets, and more closely resembles Proto-Aurignacian approaches. Sometimes waste flakes from preparation or maintenance of blade cores *were* casually used, occasionally even retouched. But in total contrast to Neanderthals, Châtelperronians had no *systematic* interest in flake production.

Some researchers have seen similarities between tools called 'backed knives' found in some biface and Discoid assemblages, and proposed them as evidence for a direct technological 'ancestry' to Châtelperron points. But others have noted that backed knives are technologically completely different, with their blade-like parallel scars formed by chance when knapping across the surface of flake cores. What's more, in a number of sites there is a final Middle Palaeolithic Levallois phase in-between layers with backed knives and the Châtelperronian. This implies a significant separation in time, making a direct connection even less plausible.

Today Saint-Césaire and Arcy-sur-Cure remain the *only* Châtelperronian – or indeed, any intermediate culture – sites with Neanderthal associations. Despite new DNA identifications, both locales are very problematic. The Grotte du Renne was excavated over 30 years ago with good practice for the time, but lacked precise location recording and sediment studies. This means that skeletal fragments from at least six Neanderthals only have the layer and grid square recorded. Most were towards the bottom of the Châtelperronian layer but others came from higher up, which was taken to mean Neanderthals were present through its entire duration.

However, as well as stone blocks and technologically Middle Palaeolithic artefacts occurring well up into the Châtelperronian, there are also Châtelperron knives and bone awls (carved piercing tools) in the underlying Middle Palaeolithic layer. This is strongly suggestive of disturbance or movements between the two deposits. So far, refitting of lithics has been limited, but it identified that fragments were also moving several tens of centimetres between the different Châtelperronian layers. Furthermore, radiocarbon dates produced anomalously ancient results – older than 48 ka – from well into the Châtelperronian.

Put together, the Grotte du Renne contains worrying evidence that things were being displaced both within and across the crucial layers. The most recent research used ZooMS analysis to identify more Neanderthal remains, including the breastfeeding baby girl already mentioned, and they date to around 42 ka. These new bones may relate to those already known from an infant, including skull, jaw and upper body parts, which might imply relatively little disturbance. But given the other evidence for objects moving, it's not entirely out of the question that the Neanderthal bones were jumbled upwards from an original Middle Palaeolithic context.

While it's been proposed that the Châtelperronians themselves could have caused the trouble by digging, geo-thermal processes from freezing sediment can also move things over 1.5m (5ft) vertically, and there's plenty of evidence that the Châtelperronian occurred during an exceptionally cold period. What's really needed to be secure in interpreting Grotte du Renne is a complete refitting analysis.

The Saint-Césaire Neanderthal, in contrast, seemed a more solid case. When first found, it was removed as a block of sediment 1m (1.1yd) across to be excavated in the lab. Full details on the position and condition of the skeleton have, however, never been published, although direct dating produced results around 42 to 40.6 ka; potentially an underestimate due to low collagen.

But Saint-Césaire has also recently been subject to critical reanalysis, raising more red flags over whether the Neanderthal here was genuinely in an intact Châtelperronian layer. The highly crushed bones themselves indicate complex taphonomy and erosion, with the entire uppermost side of the face missing, despite the teeth remaining, but meticulous research on the artefacts published in 2018 also suggests things aren't as simple as they once seemed.

While only about 15 per cent of the 40,000 lithics excavated in the 1970s were recorded in 3D, it was possible to digitally reconstruct the stratigraphic boundaries and reassign other artefacts to their correct layer. The results showed that almost all the lithics from the Châtelperronian layer weren't related to blade production at all, but Levallois and Discoid.* Additionally, all the Middle Palaeolithic-type

---

* That is, 90 per cent of 4,400 lithics that had technologically distinctive features.

retouched tools were made on flakes, not blades. Even more striking, while much of the layer was jumbled and pointed to mixing, all the lithics from within the skeleton's sediment block were technologically Middle Palaeolithic.

A gigantic refitting programme found that just 4 per cent of the lithic fragments could be reunited, compared to Canaules II, which is nine times higher. This already suggested that the layers weren't intact, which was confirmed by spatial refit data that revealed objects had moved several metres along the cliff and down the slope. Adding in the fact that everything in the supposed Châtelperronian layer was far more battered, it looks as if some kind of massive sediment flow had come off the cliffs and mixed things up.

The researchers proposed a new explanation for Saint-Césaire: there *had* been a Châtelperronian layer, but it was thin and right on top of a rich Middle Palaeolithic layer. Geological disturbance later thoroughly mixed the two. The skeleton is, however, still something of a mystery. The body must have been deposited *before* the mixing, since the lithics and rocks around it are just as damaged as everything else, and this would also explain the erosion of the skull's left side. But although the 1970s field journal shows it was found at the base of what they believed to be the Châtelperronian layer, it's impossible now to be sure if it was sticking up from the Middle Palaeolithic layer, or had been really deposited during the Châtelperronian.

It seems that neither Grotte du Renne nor Saint-Césaire are entirely secure contexts connecting Neanderthals to the Châtelperronian. This means that, at present, we don't know who made it. And it also means that in France and northern Spain, the culture of the last identifiable Neanderthals was very much in the mould of what they'd been doing for tens of millennia: Discoid and Levallois assemblages. And while in a number of places they *were* showing interest in pigment, fossil shells, markings and some shaped bone tools like lissoirs, the Châtelperronian at Grotte du Renne and elsewhere unarguably contains artefacts that go beyond this. Finely manufactured bone tools include tubes made from bird limbs, pierced and grooved deer, fox and wolf teeth beads, and enigmatic carved, polished and engraved mammoth ivory rings. Still, vague hints at possible cultural contacts exist, but with ideas moving in the *opposite* direction: perhaps it was

Châtelperronians who picked up an interest in large raptors from Neanderthals, shown by the butchered eagle toe at Cova Foradada in northern Spain. And perhaps they learned from Neanderthals how to make lissoirs, then decorated them with their own V-shaped engravings.

The Châtelperronian hogged attention for a long time, but debates over possible Neanderthal authorship of other intermediate cultures have also evolved over the decades. One is the Uluzzian, largely found in Italy, understanding of which has advanced thanks to two terrifying natural disasters between 46.5 and 39.7 ka. The first originated on the tiny island of Pantelleria, a rocky volcanic massif last active in 1891 that tumbles like a pebble off Sicily's sole. A huge eruption happened between 46.5 and 44.5 ka, blowing out a massive caldera and pluming ash up into the sky. Prevailing winds sifted these deposits over large parts of Italy, and they're visible in archaeological sites as the 'Green Tuff'. North-east from Pantelleria are the famous Phlegrean Fields, near Naples, Italy, which also hosted an enormous volcanic eruption somewhere between 40 and 39.7 ka. The ashes were even thicker and more widespread than the Green Tuff, fluttering down onto people in southern Italy, through the Mediterranean and as far as parts of Russia. Known as the Campanian Ignimbrite, or CI, this layer is also distinctive enough to identify microscopically and chemically.

The Green Tuff and CI have incredible value to archaeologists because they're very brief temporal markers bracketing the Uluzzian. Initially viewed as an indigenous Neanderthal development, renewed research into the Uluzzian is suggesting things aren't so simple. There are far fewer sites than the Châtelperronian – under 30 – but they're found all over Italy, bar the north-west. They also extend eastwards into the Balkans and Greece. Based on the Green Tuff, the CI and radiocarbon dating, it seems that the oldest Uluzzian only began after 44.5 ka.

The best-known site is Cavallo Cave[*] in Apulia, the hottest and driest Italian region today. Odd, crescent-shaped lithics found here and at other sites from the 1960s onwards came to define the Uluzzian around the same time as the Châtelperronian was being discussed.

---

[*] Also known as Uluzzo 'A'.

But several decades of detailed technological studies show them to be two completely different phenomena.

Uluzzians didn't have systematic, sequential technologies like Levallois or Discoid. There are a few centripetally worked cores, but largely they combined casual knapping with an unusual technique. Known as 'bipolar', this involved balancing one end of a core on a stone anvil, then whacking straight downwards. Doing this gives hardly any control over the shape of products, which also tend to have splintered ends. But it's fine for less-than-ideal stone, like the slabs and small imported pebbles at Cavallo. Bipolar flakes are just as ready-to-use as Discoid flakes or blades, and if what you're after is very small, flat segments, it's ideal.

And this *is* what the Uluzzians wanted. Their stand-out artefacts are crescent-shaped tools called lunates, made on flat flake or blade segments by retouching inwards towards the thickest point. This left a curved 'back' opposite the long, sharp edge. Uluzzians also had organic technology, though it's not very common. In addition to bone retouchers they also made cylindrical objects, finely pointed at one or both ends, often quite small and some tiny: a couple are less than 5mm (0.2in.) wide. Where identifiable, all are horse or deer and some were repeatedly resharpened. They probably weren't weapon tips, but were likely awls, for making holes in medium-tough materials like leather and softer stuff such as furs; some of the smallest might even be fishing tackle.

There's also some evidence for aesthetic and symbolic artefacts. Tiny shells, some apparently pierced, come from Cavallo. Other species with tubular shells were snapped and sawn to make tiny segments, probably used decoratively. So far, however, there's no carved bone or antler objects, beads or decorated or painted objects.

The Uluzzian is interesting because it shows the complexity in picking out technological differences and similarities to the preceding Middle Palaeolithic. Bipolar knapping is sometimes found in those assemblages, but it's never the headline act. Most strikingly, there are also Middle Palaeolithic layers at Cavallo that contain bipolar artefacts but also Levallois, showing that Neanderthals were perfectly able to use the poor-quality stone for trickier knapping methods. And in contrast to Neanderthals, who selected between rock types for different activities when given the choice, Uluzzians were so focused on bipolar knapping, segments and lunates that they did it across all different rock types.

What was this obsessiveness about? Understanding the function of the segments has been key, and has led to a remarkable conclusion. A few were hafted and used as tools to cut and scrape vegetable and animal materials, but impact damage strongly suggests most were weapons.

Some acted as tips, others perhaps barbs along a shaft. Their tiny size – averaging less than 3cm (1.2in.) long and extremely narrow – points away from spears and towards darts, or even arrows. Blocks of red and yellow pigment have been found at two Uluzzian sites, and since most lunates at Cavallo had red residues especially on their curved backs, it seems ochre was involved in hafting, for a very particular kind of hunting.

But what the Uluzzian does share with the Châtelperronian is recent controversy over who made it. In 2011, analysis of two teeth found in the 1960s at Cavallo identified them as *H. sapiens*, but based on anatomy not DNA.

Unfortunately due to their condition they can't be directly dated, and there have also been criticisms of the security of their original context. One was supposed to come from a hearth at the bottom of the Uluzzian, where it partly cut down through the Green Tuff into the Middle Palaeolithic layer. The other was apparently from 15 to 20cm (6 to 8in.) higher, but since they were excavated 60 years ago and never fully published, their precise locations aren't certain.

Moreover, the original excavators recognised extensive disturbance in the site from ancient digging and recent looting, and erosion that in places extended through the Uluzzian layers. It's difficult to be sure the teeth were unaffected by this, and though most researchers accept that the teeth are *H. sapiens*, without direct dates and ideally DNA, not all regard them as reliable evidence for who the Uluzzians were.

Another parallel with the Châtelperronian comes from tantalising hints at Middle Palaeolithic cultural connections that aren't about stone knapping. At the cave of La Fabbrica, west-central Italy, glue residues have been analysed on Uluzzian artefacts and shown to contain a mix of pine/conifer resin and animal fat. Neanderthals aren't known to have used that combination, but in 2019 new analysis of residues on the Cavallo segments found a triple-ingredient recipe. It combined ochre, plant resin and beeswax, and as we saw in Chapter 7, the latter two ingredients *were* used by Italian Neanderthals.

It's impossible to be sure whether this is a remarkable convergence, or evidence of some cultural contact, but certainly the rest of the Uluzzian doesn't really share Middle Palaeolithic features in either lithics or organics. Neanderthals did occasionally shape bone tools, but there are no parallels approaching the very thin, small pointed forms in the Uluzzian.

So the Uluzzian doesn't seem to have directly evolved out of the Middle Palaeolithic, yet neither is it laminar like the Châtelperronian, and most Upper Palaeolithic cultures. What it does have in common with the latter is a focus on artefacts *designed* to be retouched (whether Châtelperron points were also hafted weapon tips is an interesting possibility).

In Italy, the last Neanderthals were gone no later than 43 to 42 ka, and whoever made it, the Uluzzian itself was over after 1 or 2 millennia at the most. But elsewhere in south-eastern European regions, there are signs that things weren't always so neat. In some places, the CI ash horizon is found *beneath* apparently Middle Palaeolithic layers, suggesting Neanderthals were potentially still around a couple of centuries after 39 ka. And at Buran-Kaya, a little farther along the Crimean Mountains from Zaskalnaya, another intermediate culture – the Streletskayan/Eastern Szeletian – is found below a Middle Palaeolithic level that dates to 41.1 to 43.9 ka. Later sites have *H. sapiens* remains associated with the Streletskayan, so here too it looks as if Neanderthals did cling on after others had entered their lands.

The past couple of decades have proved that *secure* evidence for genuinely hybrid cultures associated with Neanderthals is very thin indeed. It's not that they weren't savvy enough to make Uluzzian lunates or Châtelperron points, but the real difference is conceptual. Those objects were systematically made to exacting standards and methodically retouched because they were part of an integrated system of hunting using composite, mechanically assisted weapons: light spears, dart tips or even arrows. This is quite unlike what we see in Neanderthals, where even if some hafted weapons existed, they were thrusting or javelin-like throwing spears.

Potential convergences in other materials are also not especially strong. The fact that Italian Neanderthals and Uluzzians both used resin and beeswax adhesives is very intriguing, but it's isolated. In contrast, the Uluzzian and Châtelperronian – covering a couple of millennia at most and many times fewer sites – contain more shaped bone tools than the *entire* Middle Palaeolithic. Even more striking is the frequency and variety of aesthetic and symbolic objects in intermediate cultures compared to the Middle Palaeolithic. They're still much rarer than in later Upper Palaeolithic cultures, but nothing like the pierced teeth, bones and stones, decorated tools or carved objects are known to have been made by Neanderthals.

There's a last, even more mysterious culture that's worth discussing, from south-east France. What makes it so perplexing is that it's some 10,000 years older than the Châtelperronian, and *potentially* of Neanderthal authorship. Chapter 9 already introduced the key site, Mandrin Cave, where high-resolution soot chronologies accumulated on the walls. What the cave also contains is the richest and best-studied example of the Néronian. It's not just unusual because of its technology, but also because it's sandwiched between typical Neanderthal-associated assemblages. Preceding it is a Quina level, and afterwards come another *five* Middle Palaeolithic layers, dating to about 47 ka.

At the moment, there are no fossils associated with this strange, old culture that give any answers as to who made it. A non-distinctive skull fragment comes from the type site of Néron Cave, 70km (40mi.) upriver from Mandrin,* but it's not got enough collagen for radiocarbon dating, and is therefore also unlikely to be suitable for DNA analysis. This means that it's the archaeology that is centre stage, and it's really something. The Mandrin Néronian – under 20cm (8in.) thick and around $50m^2$ ($60yd^2$) – has produced 60,000 objects, plus probably millions of tiny pieces of knapping debris. Technologically it looks completely unlike anything else during this time period in Western Europe, combining blades, bladelets and Levallois-like points. Crucially they're made in sequence on the same cores, proving they were an integrated technological system.

---

* Néron Cave was discovered in the 1870s by the artist and early archaeologist Viscount Lepic, a friend of Degas; it's actually located right next to Moula-Guercy, although the latter's archaeological deposits are much older.

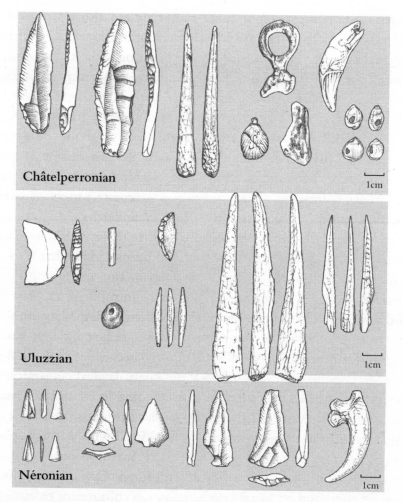

Figure 8 *Objects from the Néronian in south-east France, and two 'intermediate' cultures after the Middle Palaeolithic: Châtelperronian and Uluzzian.*

And the richness of the assemblage is extraordinary: there are some 1,300 points, which astonishingly is more than all European Middle Palaeolithic sites combined. While they vary in shape, they were systematically made apparently in three sizes, some left unaltered, others retouched steeply.

A third are less than 3cm (1.2in) long and therefore microlithic, but others are so diminutive – 8 to 15mm (0.3 to 0.6in.) long and 2mm (0.08in.) thick – researchers called them nano-points. Use-wear analysis confirms even the tiniest were damaged by high-speed impacts, but because weapon shafts must be smaller than the stone tip, they're too small to be used with spears. Moreover, experiments suggest they're too light to achieve useful range without a mechanical boost: we're looking here at something like darts thrown with an atlatl, or for the nano-points, arrowheads.

All these features make the Néronian a lightning bolt out of the blue. In all known Neanderthal assemblages, even if competently made, blades were never overwhelmingly dominant. At Mandrin Cave, about 75 per cent of all artefacts relate to laminar production and points. Similarly, Neanderthals did make very tiny flakes, including using Levallois methods, and bladelets in many contexts, but it's typically a response to the stone resources available. Across several hundred thousand years, there's only one object that *might* be related to propulsive technology: the Salzgitter bone point. It's small, obviously shaped and thinned at the base, and must surely have been hafted, though how isn't clear. But in Europe after Mandrin Cave, there are no comparable small lithic points designed for propulsion weapons for more than 10 millennia.

Comprehensive dating shows that the Néronian layer was deposited probably 50 to 52 ka, and the soot chronology points to not more than several decades, even just a few years, between it and the preceding Quina layer. Even if there was any technological similarity, there's not enough time for one to develop into the other. Beyond a handful of other Néronian layers in the same region, nothing remotely like it exists for thousands of years and hundreds of kilometres.

But what it does resemble are some so-called Initial Upper Palaeolithic (IUP) cultures in the Near East and the borders of Europe. They date around 45 to 50 ka, older than the European intermediate cultures, and the Bohunician of the Czech Republic is especially relevant. It includes Levallois points made on blade cores, and at some sites the lithics are miniaturised.

At least some IUP cultures were made by early *H. sapiens*. The Bachokirian in Bulgaria has bones dated to around 46 ka, although technologically it's not as close to the Néronian. It's also clearly several thousand years too young. In theory there could have been older 'cryptic' dispersals of early *H. sapiens* into Western Europe; alternatively, an ancient hybrid population is another possibility. The DNA from

'Ust-Ishim and other early *H. sapiens* bones do indicate that interbreeding was happening before 55 ka, so perhaps populations connected with this moved into Europe from somewhere in Asia. But their numbers must have been tiny, as they left no other trace until the Rhône valley.

But after all this, there is a glimmer of a cultural connection to Neanderthals. The Mandrin Cave Néronian produced one of the largest butchered golden eagle talons from anywhere in Europe. Focusing on raptor claws is not an Upper Palaeolithic trend at all, so the fact that it's there in the Middle Palaeolithic, Néronian and one Châtelperronian site is intriguing, given that in lithic terms, they're very clearly different to each other.

Scenarios about the Néronian itself will remain speculative until DNA is extracted, potentially from the sediments. But even if it turns out that Neanderthals weren't responsible, it's still extremely interesting because of what it implies about their population dynamics. The Mandrin Cave soot chronology shows an extremely rapid shift from a preceding Quina-based tradition to the Néronian: no more than a human lifetime, or even faster. Then the Néronian itself looks very brief: a thin layer corroborated by only 18 or so occupations within the soot archive.

After it ends, the cave was abandoned for many generations, perhaps millennia. Yet the Néronian wasn't the end of the line. When fires burned once more in Mandrin Cave, they were sat around by people – presumably Neanderthals – making basically Middle Palaeolithic artefacts again. In the first post-Néronian phase, the amount of blades and points dramatically drops off, while flake production increases, although they're often still unusually small. The second phase covers four layers, and flakes clearly become larger and essentially look like any other Middle Palaeolithic context.

The post-Néronian is also notable in two ways. The lithic territory, based on stone sourcing, dramatically shifts, shrinking in size and no longer crossing to the western side of the river Rhône. What's more, the soot chronology points to over 90 occupations and therefore a period of stability.

One way of understanding this overall sequence is that, whoever they were, the Néronians displaced the local Neanderthals so thoroughly that the region was emptied out for many generations. But it didn't last, and it didn't extinguish them. At Mandrin Cave that only came thousands of years later, when the final Middle Palaeolithic layer is directly followed in less than a century by the Proto-Aurignacian.

Endling Dreams

What happened at Mandrin Cave tells us that the end of the Middle Palaeolithic was far from a simple process in cultural terms. And it would be another 10 millennia until the last known intimate contact between Neanderthals and *H. sapiens*, as shown in the DNA of the Oase man. Frustratingly, we don't know exactly where his ancestors met Neanderthals – in two centuries, they could have moved hundreds or thousands of kilometres – and there are no associated artefacts with his jaw. What should we even call somebody like Oase? How many Neanderthal ancestors, and how close in time, defines hybridity? Was his mixed ancestor the only one in their group, or part of a wider pattern? Did they become part of stories passed down until his own generation?

Until more intact sites containing fossils with extractable DNA are found, these thoughts must remain mostly musings. But at least one thing is clear. There wasn't wholesale merging of populations, or of cultures. No Neanderthals across their whole range during the crucial period between 80 and 40 ka have any genetic hints of hybridising, and not all early *H. sapiens* individuals show it either: neither the mtDNA from Bacho Kiro, nor a Proto-Aurignacian tooth at Fumane almost the same age as Oase.

But the genetic patterns in living people tell us assimilation to some degree did happen. Though Neanderthals remained physically distinct even in their last visible skeletal remains, the scale and repetition of interbreeding, plus the range of retained genes in us, means they were – and are – human. Biologically speaking, individuals who can mate and create viable offspring are the same species. Chimpanzees and bonobos, who are both physically and socially quite different, have only been separated since around 850 ka; roughly the same time that our own ancestors separated from the lineage that would lead to Neanderthals and Denisovans.

Modern zoology's concept of allotaxa may be more appropriate for what Neanderthals were to us: closely related species that vary in bodies and behaviours, yet can also reproduce. Yaks and cattle are an example, and it was certainly happening in Pleistocene fauna too: different types of mammoths sometimes hybridised, while living brown bears were recently found to preserve a small percentage of cave bear DNA. And in recent cases recorded between polar and grizzly bears, biologists also observed 'back-breeding' going on between hybrids and both parent species.*

The fundamental conclusion to draw about the end of the Neanderthals is to expect the unexpected. Despite enormous advances in chronologies, technological analysis and species identification over the past few decades, in many ways there are more questions than ever. Some of the more intriguing uncertainties that remain include what caused the splitting, dispersing and perhaps even replacements visible in the Neanderthal meta-population from MIS 5. Climatic impact is a possibility, with the rapid temperature rise towards the hothouse'n'hippos Eemian peak, followed by a world in flux, experiencing massive temperature jumps of 11 to 16°C. The population changes are mirrored in the archaeology too, with a proliferation of techno-complexes and regional traditions between 125 and 45 ka.

Another theme that needs far more unpicking is the processes that led to encounters between Neanderthals and other species, particularly early *H. sapiens*. We are prone to paint ourselves as victors, but outside Africa we nearly went extinct at least once, and suffered a major population crash around 70 ka, just before the majority of interbreeding with Neanderthals. Moreover, despite dispersing populations obviously spreading all the way into Australia by 65 ka – adapting to arid deserts and wet mountain forests, even an ocean crossing to Indonesia – there's no clear sign of *H. sapiens* in Central or Western Europe until more than 20,000 years later.

Perhaps that land was already taken, and Neanderthals were successful enough, at least for a while, to prevent others coming in.

---

* All traced back to one female polar bear that had cubs with two male grizzlies, and each went on to mate with their hybrid offspring.

Yet there is the Néronian 'joker in the pack', reminding us that what we can make out archaeologically is far from the whole story.

A further irony is that long-standing claims that early *H. sapiens* possessed some intrinsic superiority are far from tenable. The Oase man's people went extinct in Europe, and are instead closer to today's Indigenous Americans and East Asians. Even more striking is the 'Ust-Ishim man. He lived either just before or just after the deepest genetic division into what would become ancient eastern and western Eurasian *H. sapiens* lineages. Yet he's unrelated to almost any living people.* Furthermore, during the 25,000 years after Oase, it appears that successive Upper Palaeolithic populations totally replaced each other, and were then replaced in their turn by later prehistoric cultures. Parisians, Londoners or Berliners today with ostensibly European heritage have very little connection even to Mesolithic people just 10,000 years ago. The vast majority of their DNA comes from a massive influx of Western Asian peoples during the Neolithic.[†]

This means that many of the first *H. sapiens* populations are more extinct than the Neanderthals; not a great sign of evolutionary dominance. Further paradigm shifts doubtless lie ahead with more ancient DNA samples. Our present evidence of interbreeding may be somewhat like the early history of exoplanet discoveries, where such objects were assumed to be rare, yet several decades on it appears there are more planets in our galaxy than stars. Today we know Eurasia was *always* a melting pot, home to hundreds, perhaps thousands of hybrid children. Neanderthal sites may be secretly hosting more evidence of them, or their close descendants, scattered among unidentified bone fragments or cave sediments.

---

* One study indicates potential descendants among Indigenous Eastern Siberians and East Asians.
[†] Neatly skewering those who try to claim white supremacist connections with the Upper Palaeolithic.

The fate of the Neanderthals has monopolised enormous amounts of attention, yet it may be the least interesting thing about them. Perspectives benefit from being flipped: their last 100,000 years was a time of great challenges, but rather than a prologue to a Neanderthal swansong, it was the scene of engagements with new opportunities. By 20,000 years ago, we were alone on the surface of this planet. Nonetheless, the Neanderthals still lived, after a fashion. Even as our encounters fell out of all memory, our blood and our babies still contain the fruits of interactions with the universe's other experiments in being human. Bones and stones long waited underground for us to rediscover our shared future. And when we finally did, everything changed.

CHAPTER SIXTEEN

Immortal Beloved

Oil lamps smear a lustrous shine across the top hat's silken expanse. The statement in gentility is worn by a man who knows the soot-dust tang of this northern land well. Until he experienced London's permeating miasma, he believed the coal smoke of Sunderland could have been Lucifer's acrid reek. But that was years – and a lifetime – ago. Now returned, he smiles out at the crowd assembled to listen to his message. Local industrial magnates, rough-handed intellectuals and social revolutionaries alike fill the hard benches. Even some miners are here, rubbing at stubborn grit in the corners of their eyes, as if the hewn black gold sweats from their very bodies. Networks of tiny railways like capillaries bring ton upon ton of dense inky lumps from the Great Northern Coalfield to the Sunderland docks, from where it journeys onwards, fuelling hearths, kilns and furnaces. And vast ships, engines glowing like hell-fire in their iron bellies. He knows ships, the hard biscuits and salt horse; holds stuffed with spoils of empire and so recently, bodies as black as his own.

Removing his hat, Samuel Jules Celestine Edwards clears his throat, feels the heavy truth he carries. Its weight is the hope of generations compressed in his past, like the ancient tropical forests crushed into carbon. Speaking lifts his burden, and beams light outwards from him. As before every lecture, his mind's eye blinks back to a bright island embraced by the limpid Caribbean, to his parents and their stories: his inheritance of sorrow and pride. He refocuses and begins to proclaim, on natural selection and strange skulls which preach that no race should tread upon another. Of a future where man's deep shared history is his path to salvation.

A year after the Feldhofer discovery and 7,200km (4,500mi.) across the Atlantic, a black boy was born. Raised in a British Caribbean colony, Samuel Jules Celestine Edwards' enslaved parents had been freed less than three decades earlier. Aged just 12 he became a stowaway, spending the 1870s travelling the world, before landing in the northern English industrial city of Sunderland. By the 1890s he'd gained a theology degree, was studying medicine and had become a respected and hugely popular evangelical and temperance lecturer. As if that were not enough, he was also a biographer, the first black editor in Britain (of *Fraternity*, the journal of the Society for the Recognition of the Brotherhood of Man) and founder of a magazine, *Lux*.

Staunchly socialist and anti-imperialist, Edward's lecture tours drew massive audiences as he spoke eloquently on emancipation and anti-colonialism.* More particularly, he noted the way that evolutionary science was racist, promoting comparisons of black people to baboons. In an 1892 article entitled 'The Negro Race' – written in-between the Neanderthal discoveries at Spy and Krapina – he made the profound observation that, rather than indicating black or Aboriginal peoples were separate sub-human races, the increasing numbers of hominin fossils implied precisely the opposite. *All* peoples of the earth had a common origin, and were therefore united in equal capacity for intelligence, civilisation and humanity.

---

* In 1891 he revisited Sunderland, where Hall Nicholson – coal miner, local secular speaker and my great-great-grandfather – was very probably among the audience.

Phantasms and Fantasies

Edwards was of course correct, but far ahead of the times. It was the eminent prehistorians who refused to see what was in front of them. While many Indigenous cultures' origin tales include enormously ancient, even eternal, lineages of forebears, for Western intellectuals the deep time represented by Neanderthals took longer to comprehend. The first inklings began with John Conyers, a seventeenth-century apothecary with an antiquarian passion. He indulged this by picking through London's guts during massive rebuilding after the Great Fire. It was he who, in 1673, had collected the Gray's Inn handaxe found by a Mr Lilly in gravel pits near the old spring Black Mary's Hole. Conyers recognised it as a humanly made object, but although he understood basic stratigraphy, noting Roman pots were found beneath more recent remains, the real age of the gravels and handaxe was not within his imagination.

Critical evaluations of the natural world begun in Classical times had somewhat petrified through long centuries of Judaeo-Christian tradition that only permitted a few millennia of history. But there was always a fascination with time and ancient remains, especially fossils.

The paradox of living forms frozen in stone even seems to have attracted the Neanderthals, picking up a shell from the rubble of an ancient sea and carrying it, rubbed red with ochre, across the Italian hills. Far later cultures rationalised fossils by embedding them into existing world views: huge bones in caves were dragons or cyclops. Stone tools, when recognised as formed by hands, were the work of elves. Only four years after the Gray's Inn discovery, the natural historian and chemist Robert Plot reasoned that massive bones matching no known creatures must be proof of ancient giants; they were in fact dinosaurs. The notion of a world filled with vanished races of humanoids alongside extinct animals was deeply embedded.

Nearly two centuries passed between Gray's Inn and the Neanderthal Big Bang of 1856. Those years covered immense social, economic and technological transformations, shattering Western society's view of the universe, while electromagnetism, radiation and wireless communication may as well have been magic to the average person. During the eighteenth century time itself was stretching, as the contemplation of rocks held geologists' minds over great yawning chasms that could never fit biblical chronologies. By the 1850s, growing comprehension of the planet's vast antiquity and acceptance of primate fossils were signposts to the idea of ancient humans. Yet for all that, nobody truly expected the Neanderthals. To Thomas Huxley, illustrious British biologist, they cracked open a Pandora's box wherein lay almost unimaginably deep origins for our lineage:

> *Where, then, must we look for primæval Man? Was the oldest Homo sapiens pliocene or miocene, or yet more ancient? … we must extend by long epochs the most liberal estimate that has yet been made of the antiquity of Man.*[*]

Today we know that all of human history separating you and the ochre brushed on the walls of Lascaux 15,000 years ago could be re-run more than twice over before you reached the last Neanderthals. It would spool past 20 times before the first of their kind.

The meaning of Neanderthals was, and is, myriad. We have no record of the Feldhofer foreman's thoughts on realising those bones were not bear; nor an account of Captain Flint's feelings as he presented the Forbes Quarry skull. An exception to the nineteenth-century dry scientific descriptions comes from Charles Darwin's reaction to the same fossil. He found it 'wonderful', and many decades later, exhilaration was experienced by new generations coming face to face with the Neanderthals.

During the 1908 Le Moustier 1 excavation, Klaatsch kept a vivid personal journal. Alongside professional impressions of the fossils, it describes petty team rivalries, champagne-fuelled evenings filled with moonlit speculations about ice age hunters, and accounts of himself brooding late into the dark. The digging and the skeleton electrified him, the journal slipping into present tense and exclaiming 'What teeth!' The ancient boy really got under his skin, as he notes that the night after painstakingly reassembling fragments – 'the most responsible technical task I have ever attempted' – he dreamed about the skull.

---

[*] Thomas Huxley, 1863. *On Some Fossil Remains of Man*. Proceedings of the Royal Institution of Great Britain 3 (1858–1862): 420–22.

Neanderthals infiltrated imaginations beyond the scientific. Within two decades of their debut, primitive fantasy novels began appearing. They satiated public appetites to experience these captivating creatures, if only in the mind's eye. Blending with the nascent science fiction genre, it's notable how often themes of hostility and combat run through stories of Neanderthal encounters. J. H. Rosny's 1911 *Quest for Fire* depicts violent encounters, the original French title being *La Guerre de Feu* (The war of fire).

By 1955, amid the wreckage of two world wars, *H. sapiens* evolved into the ultimate aggressors in William Golding's *The Inheritors*. It is us, not them, who are the insatiable predatory species, unable to perceive the sentience and compassion in Neanderthals. The general trend over following decades expanded on this, but it wasn't until the 1980s and Jean Auel's hugely successful *Earth's Children* series that Neanderthals were allowed to love, and be loved.

Auel's work is prescient in other ways. Most strikingly, her bold speculation about intimate inter-species relations was viewed as somewhat fringe, but 30 years later, genetic science showed she'd been right.* Hybrid babies would have seemed even more outrageous to nineteenth-century prehistorians, though they'd have also been intrigued. Had a modern time traveller stood up after William King's 1863 species-naming lecture and announced that the strange 'pithecoid' people from the Neander valley were, at the molecular scale, present in the room, they'd have faced ridicule or riots.

But even then the Neanderthals were uniquely mind-expanding, even stimulating imaginations to travel into the future. In 1885 a clergyman with intellectual aspirations named Bourchier Wrey Savile published a treatise against evolution and natural selection, entitled *The Neanderthal Skull on Evolution, in an Address Supposed to Be Delivered A.D. 2085, with Three Illustrations*. After a lengthy preface we find the skull itself as narrator in a late twenty-first-century soiree, on stage at Victorian London's principle concert venue: St James's Hall, Piccadilly. Opening in charming terms – 'unaccustomed as I am to public speaking' – the skull assures its (assumed male) audience via

---

* Auel's biggest influence may have been in inspiring future generations of archaeologists.

verbose religious arguments that it can, after all, claim no human kinship. It politely wishes to 'part as "friends"', then vanishes.* While much may change between now and 2085, we're unlikely to see a total volte-face in our understanding of Neanderthals as long-lost cousins.

Wrey Savile would not have approved of Neanderthals' place as ancestors carried within, but he was right that there is desire for a relationship with them. This is manifest in the transfixing power of reconstructions. The earliest known visualisation is an ink sketch by Huxley, doodled during a meeting. Rather ape-like, nonetheless it has a charming vitality. The first serious attempt at a reconstruction in 1873 was surprisingly modern, with a hafted weapon and canine companion. By the early twentieth-century contrasting visualisations emerged, tracking differing views of human origins.

The overtly simian reconstruction of the La Chapelle 'Old Man' was first published in a French paper and reproduced in the *Illustrated London News* in 1909: bent over, absent-mindedly hefting a wooden club, and with decidedly ape-ish pelt and feet. This representation set the primitive tone that's been so tenacious; supposedly produced in collaboration with Boule, its overly ape-like look may reflect his rejection of *H. sapiens* as possible descendants of Neanderthals. Two years later the same newspaper featured a contrasting vision. Commissioned by British prehistorian Arthur Keith, who saw Neanderthals not as dead-end failures but direct ancestors, it's a far less threatening image. A man with a sizeable but tidy beard sits by a blazing fire carefully making a tool; together with the domestic scene, a thinking, modern mind is hinted at by a necklace.

Reconstructions evolved further into the twenty-first century, with increasing levels of anatomical accuracy and gentrification. The artefacts depicted changed, reflecting richer appreciation of their culture from archaeology. But more fundamentally, their poses and especially faces are infused not with misery but intelligence, dignity and contentment.

---

* It's not much of a spoiler to say that Wrey Savile ends the book with a tried and tested literary mechanism: 'I suddenly awoke and found it was all a DREAM!'

Sculptures are the most arresting of all: circling a body, gaining a measure of presence, looking into a face almost summons presence in a way most illustrations cannot. They too have metamorphosed from dejected creatures to people expressing confidence, joy and love; even in one case, a cheeky sparkle. Appropriately in an age obsessed with celebrity, a 2018 reconstruction at the Musée de l'Homme, Paris, acknowledged their A-list status: dressed by clothing designer agnès b, this was Neanderthals literally refashioned in our image.

Viewing models is, however, quite different to encountering actual bodily remains. Just as with famous archaeological entities like the Nefertiti bust or the Pompeii casts, genuine Neanderthal fossils exert an eerie magnetism that stills the heart. There's an urge to crouch down, tune out the flowing museum crowds and gaze full into those fleshless visages. In their glass-case afterlives, sometimes visiting continents where their feet never trod, Neanderthals are confronted by thousands of times more people than they ever saw during life.

Part of the reason our Neanderthal obsession never waned is the media. Offering a potent cocktail of science and social – even moral – concerns, Neanderthals were perfect content for nineteenth-century newspapers rolling off presses to readerships in the millions.[*] If anything, today they're even more popular. Internet searches for Neanderthals have long outstripped the more generic human evolution, and contemporary media happily caters to public desire.

Yet science can get distorted: the flood of new data is hard enough for professionals to keep on top of. It's not helped by persistent framing around two themes: cognition or extinction. Variations on 'Neanderthals not so dumb after all!' or 'They were still dumber than us!' miss-sell the nuanced and interesting reality based in process, context and variation.

But the furore over the first Neanderthal genome's revelation of direct ancestry was a fair reflection of how researchers felt. That

---

[*] There was even an entire late nineteenth-century newspaper cuttings industry that operating like an internet search service, where for a price you could specify keywords and an army of clippers would search out articles and post them to you.

discovery dramatically shifted public feelings* probably more than anything since their discovery. No longer abstract cavemen, now they're personal. DNA research tends to be presented in black-and-white terms, with a simple and powerful message. But much else that's just as exciting barely gets attention, because the data itself is complex and difficult to communicate.

This means that many stubborn clichés about Neanderthals remain, for example that they lacked sophisticated technology and were unable to innovate. The other reason they persist, however, is that rather than discussing them for their own sake, Neanderthals are cast as a foil for ourselves. In this sense they've always represented the ultimate 'Other'; the shadow in the mirror. To behold them is to encounter a multi-spectral reflection of our hopes and fears, not only for their apparent fate, but our own.

Power in the Past

Let's return to Samuel Edwards. Why didn't the learned scholars of the day reach the same conclusions about human evolution as the son of once-enslaved Africans? Quite simply, because they were invested in a world of hierarchies that placed them at the top. Since the eighteenth century, scientists had not only taken the measure of the world, but also its peoples. Linnaeus was literally looking in the mirror when in 1758 he first named *H. sapiens*, regarding himself as the type specimen.†

Right from the start, the Neanderthals were embedded in scientific justifications for white racial supremacy. Spurious notions that skull size reflected intelligence and even moral capacity became mainstream in the nineteenth century, and were used to justify first slavery then colonialism. Ethnography and prehistoric archaeology positioned hunter-gatherers as 'savage', in the sense of animalistic. Even Darwin – who should have known better, understanding that diversity can come from common origins – viewed the Indigenous population he

---

* Online searches for 'Neanderthal extinction' dropped off.
† Later editions used a geographic sub-species classification for humans, defining non-white races in universally negative terms.

encountered in Tierra del Fuego, Chile,[*] through a lens of supposed animal impulses and violent natures. And despite Alfred Russel Wallace's comprehension that hunter-gatherers and the average Victorian gentleman had identical-sized brains, he puzzled over why 'thoughtless' beings should need such computing power.

Contemporary interpretations of the Feldhofer skull make it obvious how such attitudes both influenced understanding of, and were sustained by, the Neanderthals. The first German newspaper reports in 1856 made direct racial comparisons, confidently describing the skull as of the Flathead people from North America. Schaaffhausen likened the anatomy to 'Negros' and Aboriginal Australians, as did Huxley, who casually described the former as 'brutal' and referred to the brain of Sarah Baartman, known as the 'Hottentot Venus' and thoroughly dehumanised as a publicly autopsied museum specimen. William King's perspective was similar, though he preferred the Andamanese as the most 'degraded' human race, 'standing next to brute benightedness'. Only the previous year, skulls of Andaman islanders murdered by colonists were being shipped over to Britain for anatomical collections.

The pernicious effect of such comparisons was to promote ideas that non-white populations represented primitive branches of the genus *Homo*. Around the same time that Samuel Edwards was writing his forward-thinking article on human evolution, Ernst Haeckel – influential biologist, correspondent and protégé of Darwin – classified 'lower races' closer to beasts and claimed colonialism was justified because such lives were less valuable.

As the new century turned, prehistory promoted itself as nobly centred on advancing knowledge, yet Neanderthals continued to be drawn on in bolstering racist beliefs. Claims that black people lacked the refined range of facial expressions seen in Caucasians fuelled debates as to whether or not Neanderthals could smile. While discussing the La Chapelle discovery, Boule had no compunction in stating that Aboriginal Australians were the most primitive of all humans.

---

[*] At the time referred to generically as 'Fuegians', they may have been Yaghan.

Hierarchies of anatomical and cultural advancement in past peoples were created by archaeology, supported by illustrious scholars like Egyptologist Sir Flinders Petrie providing ancient skull measurements. This directly fed poisonous notions of competition and racial purity that were foundational in eugenics. They're reflected in some fiction, such as H. G. Wells' 1921 *The Grisly Folk*, which subtly positions the eradication of an almost parasitic, animalistic hominin race as vital to human survival. Individuals directly associated with Neanderthal fossils were part of the scientific framework: in other publications, Klaatsch placed Aboriginal people on a separate hominin branch stemming from Oceanian and Asian primates, and Wienert, another anthropologist who studied Le Moustier for Hauser worked for the Race and Settlement Office, which controlled the 'purity' of SS staff and co-edited the *Zeitschrift für Rassenkunde* (Journal of Racial Science). The original Feldhofer find was within the Reich, and the Nazis forced the museum to close in 1938 because it didn't sufficiently distinguish the racial origin of the 'Germanic people'.

After the horrors of the Second World War a growing rejection of racially based science began. Neanderthal research began to shift away from obsessions with cranial metrics towards behaviour. But the effects lingered: prominent anthropologist Carlton Coon describes Caucasians as the 'Alpha' race and Aboriginal Australians as 'Omega' in his 1962 book *The Origin of Races*. It's only during the past four decades that theories of a common human origin in Africa became really mainstream, and confirmed genetically. The human-ness of Neanderthals was now debated in collective opposition to *H. sapiens*. This matters because heritage is fundamentally about identity, and archaeology provides material evidence legitimising or disproving contested pasts.

For Neanderthals, it's mostly been creationists – whether Christian or other faiths – who have tried to mould the fossils to their views. Bizarrely, one of the first people to gain access to the Le Moustier 1 remains after the reunification of Germany was an American creationist, Jack Cuozzo. A dentist by profession, his study on the teeth formed a chapter in the publication on the skeleton. Yet Cuozzo's belief in revealed scripture led to his explanation of Neanderthal anatomy within a biblical framework,* and a suspicion of scientists. This culminated in a *Da Vinci Code*-esque book, with accusations of conspiracy against him.

Notwithstanding Neanderthals' reputation as the losers of the Hominin Games, they haven't escaped nationalist manoeuvrings. The French animosity towards Hauser's work at Le Moustier wasn't just about sales of antiquities, but also his foreign identity. Following the Franco–Prussian war a generation earlier, the public were keen to reclaim parts of Alsace and Lorraine, so a prestigious French scientific 'possession' becoming German property rankled. Hauser himself was referred to as German despite being Swiss, and was subject to rumours of espionage. He fled when First World War mobilisation began in 1914, which left the excavation records in disarray. Hauser's enduring reputation as a thief of French heritage is not entirely fair, however, since Peyrony was also making a tidy profit selling things, but to the Americans.

Franco–German prickliness may persist as a comedy trope, but Le Moustier 1's national ties can still raise twenty-first-century hackles. Specialists from nine countries collaborated to produce the definitive 2005 publication on the skeleton; a testament to international research. Yet one reviewer bristled that none of the articles were in French, despite it being Moustier 1's 'native' country. The next year, a UNESCO-supported conference took place celebrating the 150th anniversary of the Feldhofer discovery. The museum unselfconsciously referred to itself as 'Home of the most famous German'.

Many hundreds of kilometres away, others are possessive of 'their' Neanderthals. The new excavations at Shanidar came at the invitation of the Kurdish regional government. Fighting figuratively and literally for their existence as an independent nation, Shanidar signifies deep cultural history for the Kurds. Foreign scholars – not even Palaeolithic specialists – report being asked during TV interviews to confirm that the Shanidar Neanderthals are, in fact, the 'first Kurdish people'.

---

* He claims that the Moustier 1 skull's shape was due to being bombed, which isn't the case as it was not damaged, and also asserts that because fossils are vastly rarer than lithics, this casts doubt on interpretations of the Palaeolithic.

While repatriation requests for human remains and cultural objects taken in the past few hundred years without consent are increasing, it's difficult to see how any particular community could claim ancestral connections and therefore ownership of Neanderthal bones. Nevertheless, the first repatriation request has already happened.

In 2019 the Gibraltar authorities requested the return 'home' of both skulls from the Rock. The choice of wording is emotive, and also probably inaccurate. The original living ranges of the Devil's Tower child and the Forbes' Quarry woman would likely have been far beyond the tiny area of Gibraltar. Though their remains rested there, they may well have spent most of their lives in what's now Spain, or even the Mediterranean seabed.

Human origins as a discipline is rooted in the belief that it pursues questions of global relevance, on behalf of all people. But Western interests are often privileged, shown by the fact that a significant amount of past and current research comparing Neanderthals to hunter-gatherers – including studies mentioned in this book – use skeletal material from Indigenous communities. At least some arrived in Western museums by dubious or outright unethical means, and there are very likely living relatives for some of the bones currently packed in plastic and pincered by callipers. It is important for the discipline to constructively engage with this legacy, so that Neanderthal research does not continue to negatively impact Indigenous communities.

Thinking Outside the Cave

While Indigenous bodies – and most recently, DNA – have been exploited to further Neanderthal research, their knowledge and world views were largely deemed irrelevant to scientific understanding of the past. But the insights of hunter-gatherer communities show how vital it is to centre perspectives that don't originate in largely urban Western scientific traditions. Very different explanations for what we see in the archaeological record are possible from people with skills that many scientists lack. A recent project modelled the collaborative possibilities, inviting expert Ju/'hoan San trackers from Namibia to examine physical traces within European Upper Palaeolithic caves.

Their knowledge identified new tracks, and gave fresh interpretations of what was happening in these places.

No such partnerships yet exist for Neanderthal archaeology, but it's still possible to draw on Indigenous perspectives in rethinking human origins. Even the question '*Where* do we come from?' isn't a universally shared starting point, since many Indigenous cultures instead see original peoples as having arisen and existed outside a linear time. In some ways that increasingly chimes with recent evidence for massive meta-populations in Africa and Eurasia, out of which we and the Neanderthals emerged: shifting kaleidoscopes of communities separating and merging over many hundreds of millennia.

Emphasising Indigenous notions of interconnectedness, unity and relations are other ways to de-emphasise Western dominance in thinking about Neanderthals. Our biology literally limits how we perceive reality: if your eyes could perceive infrared, you'd see yourself glow. But other things can also affect the way humans see the world.

Hunter-gatherers will notice things others don't – tracking being a case in point – while some cultures easily classify shades of green that Westerners struggle to even distinguish. Perceiving Neanderthals' motivations that don't fit typical explanations is one way to try and apply these ideas. They may well have followed rationales matching energetics or cost-benefit equations, but their decisions would have been based in emotions, as is the norm for all humans.

Beyond this, Neanderthals potentially also shared broader perspectives with Indigenous hunter-gatherers, whose cosmologies are often based around relational ideas. This isn't about clumsy cut'n'paste analogies, but questioning the objectivity of assumptions most researchers already use.

Using such ideas, we might reimagine Neanderthal interactions with other animals. Current interpretations are structured around themes of dominance, exploitation and conflict; life as struggle *against* nature, and animals as unthinking, unfeeling commodities. In stark contrast, relational frameworks emphasise the similarities between human and non-human. Hierarchies exist, blood is still spilled, but a relational world is filled with communities based on recognition of common personhood, of which humans are members, not masters.

Human survival is not in conflict with creatures, but entwined in relationships *with* them.

Suddenly, Neanderthal hunting and subsistence looks rather different. For hominins born into an intensely social existence, the world from the beginning is filled with entities and mutual obligations. Assuming the creatures you interact with also have minds is logical and even adaptive, since hunting skills require attention to bodies, habits and motivations.

What Neanderthals thought and felt about animals in such a world extends outwards from their calorific value. Many of their caves and rockshelters have clear phases when Neanderthals were absent, but animals lived there. At Cova Negra this 'other-time' was populated by wolves and rodents gnawing at discarded remains of Neanderthals' meals, while generations of bats lived in fluttery darkness, their winter-frozen bodies falling to the floor. A relational perspective allows us to not just consider what Neanderthals made of the traces of previous hominin occupants, but of others creatures too.

Conceiving of animals as thinking and feeling beings would also mean that alongside a fine appreciation for the material qualities of their bodies, Neanderthals might have valued other aspects. This offers us a context to explore how their focus on hunting particular species went beyond simple availability.

Interest in birds and especially raptors is an obvious possibility, but also the systematic bear hunting at Biache-Saint-Vaast, some 200,000 years ago. Across multiple levels of river deposits, many thousands of bones show that Neanderthals hunted and thoroughly butchered at least 107 Deninger's and brown bears. Though flatness is the overwhelming feature of northern France, Biache-Saint-Vaast is right at the point where the Scarpe River flows north from hills to the Flanders plains. Since bears will den in slopes with reasonably soft ground, hunting during hibernation may have been possible even without caves. But most are adult males, which doesn't really fit the pattern seen in denning caves like Rio Secco. The river itself might have provided an ambush locale, especially if adult males were distracted by fishing.

But however they were obtained, bears aren't that easy to find, and while they represent a lot of meat and fat, hunting them is more dangerous than other more abundant prey like horses or giant deer. These species were present at Biache-Saint-Vaast, but Neanderthals aren't as interested in them, preferring aurochs and bear. Heavy bear fur may well have been a motivation, and cut marks do support this. But bear hunting is *least* common during the colder phases here. In the absence of clear economics a socially motivated explanation for the Biache-Saint-Vaast bears was proposed, but it was very Western: Neanderthals were intentionally selecting dangerous prey to gain prestige.

However, it's equally possible that this was about something relational between hominins and bears. Intriguingly, even if many Indigenous cultures consume bears – some like the Natashquan Innu even naming their land as 'the place where we hunt bear' – conceptions of these creatures as strongly linked to personhood and humans also exist, including in Naskapi, Tlingit, Iroquoian and Algonquian peoples.* During the Palaeolithic, bears shared Neanderthals' habits of moving into the earth to live, leaving their bones and claw marks in the same caves. With this in mind, there's another weird thing about Biache-Saint-Vaast: high numbers of butchered skulls. Bears weren't arriving here as entire carcasses, but if furs and fat were the main interest, why carry extremely heavy heads? Eyes, tongue and brain could easily have been removed elsewhere.

There's endless evidence that Neanderthals selected what they carried based on quality. In animals, that includes relative size and richness of body parts, yet here and in other places heads are more common than expected, especially for big animals. If Neanderthal social relationships were created, renewed and negotiated through sharing of food, perhaps animal body parts were also doing this in other ways.

This raises the phenomenon of fragmentation, which was everywhere in Neanderthals' lives. Soft hammers and retouchers fulfilled functional needs, but looking through a relational window they may have had deeper significance. First, their sheer numbers in some sites – more than 500 at Les Pradelles – is striking, and seems an overabundance compared to other places with similar activities and

---

* Bears are long-lived and can walk on two legs, sit and suckle tumbling babies.

numbers of animals. Though most seem to be made on fresh bone, others were probably carried between places. They can even be thought of as objects representing one point on a cycle of interacting materials: stone tools cut flesh and take apart skeletons, then bone tools sharpen and form stone edges. This recursion may have had deeper resonance for Neanderthals, and potentially connected to how they chose particular species and even skeletal parts. This brings us back to Biache-Saint-Vaast, where despite bears being the second most frequent species overall, hardly any of the hundreds of retouchers came from their bodies.

The watching, breathless chasing, bloody unmaking, carrying and keeping of animals and their bodies was likely central to Neanderthals' understanding and emotional responses to the world. It's interesting then that though stone is much more common as a material, the vast majority of intentional markings and engravings are on bone. Again the species marked don't match the main food sources: think of the raven at Zaskalnaya or the hyaena at Les Pradelles, which was old and potentially brought from elsewhere. Les Pradelles is a place full of material interactions and cycles: hide working, bone knapping, abundant retouchers, and from the same layer as the engraved hyaena bone, the intensively processed Neanderthal bodies.

Death and the responses to it are another area that's well overdue for fresh interpretations. Even without continuing obsessions over the absence of neatly dug graves, discussions of body processing tend to be limited to nutritional cannibalism or violent scenarios that say more about our assumptions than any solid data.

Western world views regard a butchered Neanderthal as being *brought down* to the level of prey: captured, consumed, discarded. But imagine if Neanderthals understood themselves as living among other intimately known entities, whose actions make sense in the same way their fellow group members do. Perhaps when bodies were touched, moved, taken apart, the borders shifted between furred, feathered, smooth-skinned. With a relational perspective, fragmenting and eating of the dead might be less about endings, and more about cycles and connections within sites or across landscapes. Sometimes left as whole individuals, others joined the life-sustaining rhythm of bodies and blood and fat.

When we imagine interactions revolving around more than antagonism, fear and struggle, there's even potential to see the story of Neanderthals and humans another way. Instead of *H. sapiens* as colonisers pushing into fresh lands ripe for exploitation, a different story suggests itself. A world that opens *itself* up, paths unfurling as seasons change, offering new opportunities. Unfamiliar lands and creatures to meet; new-yet-ancient peoples becoming partners in a never-ending dance.

Flip the mirror, and the Neanderthals weren't powerless and waiting for extinction, but intuitive and astute, seeing incomers not as an existential threat but as an opportunity for connection. There was not an end, but many meetings, joinings, transformations; a way to survive, to be reborn.

Vanishings

However we conceptualise it, the vanishing of Neanderthals has loomed golem-like over virtually every aspect of how we've researched, portrayed and dreamed of them. Narratives of their failure – and our success – have dominated. The rest of this book should have made it clear that, in fact, there is no obvious or simple answer for why we are here and not them (in corporeal form at least). There were no giant asteroids or global volcanic Ragnaroks like those that probably took out most of the dinosaurs.

In these final pages, we can reconsider what happened. First: bodies. There's relatively little evidence for *specific* features that gave us dramatic advantages. Differences in walking were marginal, although running less so. Neanderthals' beefiness did not come at the expense of fine grip; just as much as early *H. sapiens*, Neanderthals had the hands of artisans. For virtually every anatomical negative, there are plausible counterbalances.

Was it behavioural? Sometimes the Neanderthals are claimed to have been too much like picky pandas, with restricted big game diets that couldn't cope with change. But at other times they've been accused of not being selective enough. Some of the newest isotopic research shows that Neanderthals and early *H. sapiens* in Europe shared a distinct hominin niche, far more similar to each other than to furred predators. They both ate mammoth, thereby culling another theory for potential inferiority. In certain environments a big game habit was probably the best strategy, but elsewhere and else-when Neanderthals were perfectly capable of hunting small game and collecting plants when it suited them.

But perhaps other challenges did them in. They appear to have disliked full-on glacials, but was the climate during the last few millennia up to 40 ka worse than what they'd weathered before? At Geißenklösterle Cave, Germany, between 45 and 40 ka eagle owls and kestrels enjoyed many small furry meals. As rodents are often sensitive to climate, the raptors' leavings tell us that while true tundra species do become more common at the end of the Middle Palaeolithic, they only really dominate after the last Neanderthals were gone. Perhaps if it wasn't intense cold, maybe the wider *impact* of MIS 3 is what made it different.

While hunter-gatherers can successfully adapt to extremes, instability can be catastrophic. After coping with the heat of the Eemian, Neanderthals rode out a series of ups and downs in climate during the later MIS 5, which may even have been a time of expansions and cultural diversity. But after the subsequent MIS 4 glaciation thawed, the climate got jittery. From 55 ka, MIS 3 degenerated into a jagged, fitful frenzy of stadial–interstadial cycles, sometimes plunging from not too bad to truly bitter within a lifetime. That doesn't mean Neanderthals lived in permanent crisis. Summers still saw sunburned faces after halcyon days on the flower-speckled steppe. But consistent uncertainty would have amped up risks, and this is visible in their core prey species: horse and mammoth were being forced to rapidly alter their diet, and were probably behaving in new ways that impacted traditional hunting tactics.

Yet climate chaos can't be the full story. Recent research tracking what happened to hyaenas during that crucial period came to a surprising conclusion. They were even more affected than Neanderthals by the huge decline in prey during the MIS 4 glaciation, but recovered from the crash in the temporary warmish conditions afterwards as steppe-tundra and even open forest loaded with herbivores developed. Neanderthals mirror this, expanding as things warmed up again, even recolonising Britain, and revealing a burst of technological diversity.

The golden age didn't last, though. Both Neanderthals and hyaenas came under pressure from declines in prey as MIS 3 got colder. The carnivores hung on with their tenacious jaws along with cave bears in south-western Europe until about 31 ka, but in contrast, Neanderthals, who had previously outcompeted them, never made it past 40 ka.

Something else was going on. Based on the demographics, at least some Neanderthal groups included elders, whose wisdom and experience probably acted as sources of disaster mitigation. But if the situation deteriorated beyond common memory, fleeing might have been the only option for survival. If the southern lands were already filled with other Neanderthals, then those from the north, used to relying more on big game, might have had no refuge. And there was an added factor no previous generations had encountered in large numbers: *H. sapiens*.

Hominin populations overall likely grew during early MIS 3. The genetics tells us that Neanderthals were definitely encountering *H. sapiens* around this time, with interbreeding happening in multiple phases. Even if relations were largely friendly, competition for resources would still have been at its most intense in our collective history, just as the climate instability really kicked off around 45 ka.

Finishing this book in the late spring of 2020, it's impossible not to wonder if a terrible contagion might have been added into the mix, jumping from us to them. Obviously invisible on skeletons or in genomes, nonetheless what seemed like a fringe concern over the past decades no longer appears so unlikely.

Though some Neanderthal lineages were less genetically isolated than others, overall the wider population had been slowly withering for hundreds of thousands of years. For all their cleverness, flexibility and resilience, the archaeology does suggest they had weaker and smaller social networks made up of small groups that rarely came together in large gatherings. Long-distance lithic movements get more extreme and more common as the Upper Palaeolithic develops, and crucially, things other than stone begin to be carried far. Shared *symbolic networks* reflecting connections with far-flung communities are what define the post-Neanderthal world. Being welcome at the fires of friends many valleys away might make the difference between infants getting by on milky dregs, or tiny hollow bodies being laid down in cold crevices.

Small tragedies over thousands of years could ramp up as localised genetic pools became cut off and sluggish. In contrast to long histories of genetically small worlds and within-group reproduction visible (though not universal) in Neanderthals, so far no early *H. sapiens* genomes point to similar processes. But there's a paradox: if Neanderthals tended to be cut off socially, it's remarkable there was so much contact and interbreeding not just with us but also with Denisovans.

Climate meltdown, plus a much more crowded continent, could have provided the stage for our persistence and the passing of the Neanderthals. Think back to those European macaques who went extinct around the same time. They'd probably also been treading a fine line in energetic terms: trying to cram food, travel and social relations into short northern days is hard, maybe impossible, in harsh winters. With bodies costlier to run than ours, teetering on the edge in extreme conditions would be very dangerous for Neanderthals.

A perfect storm of different stresses may have together been overwhelming. Crucially, populations and species can vanish through factors that have nothing to do with cleverness, but that simply come down to time and babies. The fate of the last Neanderthals who did not join with us may have been more of a whispered goodbye than a war cry; a quiet vanishing accompanied by the crooning of mothers in the night.

We may never discover the precise details of what happened; how could we, when every Neanderthal's story from the Atlantic to the Altai would be unique? But we do already know something: the belief that Neanderthals were a failed early release hominin, on a road to nowhere, has tainted perceptions of the archaeological record for well over a century. The initial mtDNA results showing no interbreeding was always only a partial account, but we gladly accepted this as evidence. Try a thought experiment: what if we'd only found the first Neanderthal fossil in 2010, with the genetic tools to immediately see where we fitted together? They would have been known from the start as a branch of the family we were only temporarily separated from, and at the same time our thousand–times–great–grandparents. What would they have been called? And what if Denisova had been the first fossil site, and we'd known from the outset that first-generation Neanderthal hybrids were possible?

Things look different now that we know the blood feeding neurons crackling like fireworks in 6 billion living brains – yours, as you read this page – carry the legacy of the Neanderthals. That the vast majority of living people are their descendants is, by any measure, some sort of evolutionary success. Speaking in terms of extinction no longer feels right, but at the same time, we didn't assimilate them on equal terms. Our bodies aren't identical to early *H. sapiens* people, but nobody alive *really* looks like the Neanderthals. The fact that hybrids existed, lived, loved and raised their own children is the most persuasive argument for our closeness at every level. Not only did we find each other attractive, but some level of cultural communication must have been involved.

Everything is clearer in hindsight of course, but if the past decade has taught us anything, it's that nobody expects the next Neanderthal revelation.

Until incredibly recently, the world was sparkling with hominins: Neanderthals, Denisovans, the diminutive Indonesian *H. floresiensis* 'hobbits' and other tentatively named Asian populations like *H. luzonensis*, while in Africa *H. naledi* must be the vanguard for other as-yet-unidentified populations. Even for Neanderthals, researchers are only just comprehending that there's a mass of 'unknown unknowns' out there too. The great challenge going forward will be in combining the ever-growing but very different kinds of evidence: how genetics connects to physical diversity, and making sense of them both in relation to the cultures they produced.

Fundamentally, the long obsession over the Neanderthals' fate reflects our deep dread of annihilation. Extinction is frightening; even the syllables slam up against each other. Is it a coincidence that as our species belatedly wakes up to what may be its greatest threat, apocalyptic fiction becomes all the rage? In the face of obliteration, we desire comforting parables where we are always the Ones Who Lived. What's more, we want to feel special: most of the stories we've told about Neanderthals have been narcissistic reassurances that we 'won' because we're outstanding, destined to survive.

Yet the Neanderthals were never some sort of highway service station en route to Real People. They were state-of-the-art humans, just of a different sort. Their fate was a tapestry woven from the lives of individual hybrid babies, entire assimilated groups, and in remoter corners of Eurasia, lonely dwindling lineages – endlings – who left nothing behind but DNA sifting slowly down into the dirt of a cave floor.

Futures

The twenty-first-century advances in genetics that allow us to recover Neanderthal DNA from dust can seem like magic. But they face us with dilemmas that edge into the realms of science fiction. Enormous curiosity about what made the Neanderthals different and how interbreeding affected us is driving research using multiple approaches. Extrapolating from genomes and databases of medical histories is one, but the only way to know for sure is to observe DNA in *living* organisms.

With this in mind, biological experimentation splicing Neanderthal genes into mice has already begun, while 'Neanderthalised' frogs have been studied to determine if they experience pain differently. But we must ask whether causing suffering in sentient creatures for the purpose of human origins research is appropriate.

If that isn't problematic enough, there are now multiple projects building Neander-oids: small clumps of gene-edited human brain cells. Taking nine months to grow, they're not true brains, lacking consciousness or known processing capacity. Yet they spontaneously develop varied structure, are capable of internal electrical connections, and in 2019 – contrary to initial claims they would be isolated from stimuli – reports emerged that they'd been connected to robots controllable through signalling. These quadruped machines can actually walk, and one team plans to create a feedback system of input signals to track neural development.

All these projects raise profound concerns. No open discussion within the scientific community on ethical issues has taken place, and as the research itself is not yet published, it's happening in the dark. In the absence of any professionally agreed code of practice, we're collectively inching forwards on an undirected trajectory towards self-aware Neander-brains.

Doubts over all this aren't simply about morals, but about effective science. Environment and social context have enormous impacts on

DNA function, and since we can replicate neither the Pleistocene world nor Neanderthal society in a test tube, exactly what gene-editing studies can achieve is arguable.

Without action, even more shocking futures may come to pass. It's entirely possible that an unregulated lab will decide to put Neanderthal genes in primates, 'just to see'. Once that red line is crossed, the risk of someone trying to create a human–Neanderthal hybrid baby is real. Some genetics experts have already half-joked about cloning using surrogate human mothers, and the 2019 unsanctioned genetically edited human babies show it's no longer out of the question. Such experiments would risk serious health issues, and it's entirely uncertain if 'nearly humans' would receive legal protections or rights.

If our ancient ancestors become objects with no consent in their own study, we are accepting humans – albeit of another sort – as scientific playthings. And for what purpose? The earth hosts many creatures that possess sentience, intelligence, self-awareness, even culture. Not only do we fail to show much interest in really communicating with elephants, crows, cetaceans or primates (other than chimpanzees, though only on our terms), but the scale of abuse we commit against them is a damning indictment of what we could permit to happen to Neanderthals, despite knowing what – and who – they are.

Beginnings, endings and uncertainty are at the heart of all this. The great nineteenth-century upheavals in comprehensions of the universe included not only a gigantic inflation of time, but also of space. Slow realisations over centuries that the earth was just one sphere in a solar system suddenly accelerated to understanding that our sun was itself not unique but one among untold numbers. What was then the world's largest telescope focused on fuzzy 'spiral nebulae' at the edges of vision, and found yet more galaxies, downy with stars. The cosmos in all four dimensions had grown almost beyond the limits of the mind. Within four decades of the Feldhofer discovery, ideas about life on the moon, encircling other stars or in the far future were flourishing. By 1878 there were flying saucer stories; in 1893 H. G. Wells was imagining what humans would look like in a million years' time. Four years later *War of the Worlds* was published, and in 1909 – the same year as Peyrony was digging out the leg bones of the

La Ferrassie 1 Neanderthal – the first radio projects listening for alien signals began.

Tuning into the voices of the Neanderthals somehow would be a dream come true. But in some ways they already represent our first encounter with ET: intelligences not from off the earth, but out of our time. The original Pleistocene meetings between them and us must have been something to behold, but 40 millennia later, the gulf between the Victorians and the entities whose bones they excavated evoked similar awe as any notion of aliens.

Moreover, the profound questions this forced in the nineteenth century about where humanity came from and where it might be going are exactly those still asked today, whether by prehistorians, SETI* researchers or science fiction authors. What is sentience, intelligence, creativity, self-consciousness?

Such themes also intersect the newest Other entity that arose during the twentieth century: artificial intelligence. The ways we permit Neanderthals to be human – assessing their competence as hunters, knappers and artists, or imposing restrictions on which practices around death hold meaning – resonate with the tests for self-awareness in AI systems. True machine consciousness may soon be within reach, and we now believe that a fifth of all stars have a potentially habitable earth-mass planet. Although the journey to such a place would be on the multi-millennial scale, one day we might reach them. If so, we will not travel alone: along with a voice in the machine, there will be the Neanderthals inside us.

The magnificence and desolation of deep time in earth's history terrified and mesmerised early geologists. To this unspeakably old planet populated by endless terrible lizards, monstrous-jawed fish and ammonite swarms, Neanderthals brought a kind of solace; humanity's origins were part of this greatest story.

And the revelation of their existence came not at the hand of divinity, but amid the noise and muck of the Industrial Revolution.

---

* Search for Extra-Terrestrial Intelligence.

Our first-discovered Neanderthals were pulled out of the ground by quarrying, mining, infrastructure and urban sprawl; even war. The black powder explosives that tore apart the Feldhofer Neanderthals were developed for munitions, while both Gibraltar skulls were uncovered by men only on the Rock for military reasons. In today's digital biotech age, thousands of bone fragments can be painstakingly tested for hominin biomarkers. The child of a Neanderthal and Denisovan is recovered not from the ground, but tiny collagen filaments and numbers on a screen.

It seems fitting that, as the first hominin species we (re)discovered, Neanderthals are the one we know most intimately and are now closer to than ever. After more than 160 years, we have finally begun viewing them on their own terms. Successful, flexible, even creative: all can justifiably be applied. More than anything, Neanderthals were survivors and explorers, pioneering new ways to be human, expanding themselves through space and even in time. They experimented with new ways to fragment, accumulate and even metamorphose material substances. Long-burning aesthetic embers and bright eruptions of symbolic engagements are there in collecting special objects, marking things and places, exploring what it meant to be dead.

Let's finish our shared journey through these pages by letting your guard down. Push against the impossible, and perform a quantum shift back in time to the Pleistocene. Close your eyes and pick a world: a grassy plain under cool winter sun; a warm forest track, soft loam underfoot; or a now-sunken rocky coast, gulls' cries salting the air. Now listen, step forward, she's here:

> *When you are close enough, press the skin of your palm against hers. Feel her heat. The same blood runs under the surface of your skin. Take a breath for courage, raise your chin, and look into her eyes. Be careful, because your knees will weaken. Tears will come to your eyes and you will be filled with an overwhelming urge to sob. This is because you are human.*[*]

Neanderthal. Human. Kindred.

---

[*] Prologue, *The Last Neanderthal*, by Claire Cameron; quote by permission.

Epilogue

This is the decade of the Neanderthals. Generations have stared at the great bony monuments to their own existence, trying to re-flesh them in the mind's eye. We want to see those broad feet and legs that stretched and climbed rough hills or crouched behind leaves; those arms and hands that hefted boulders filled with future tools, or still-warm horse thighs marbled with delicious fat. After so long grasping brief seconds of connection, the visceral, electrifying thrill to finally know they're still with us – in the rushing beat of billions of hearts, in babies coming squalling into the light – has not yet worn off. But it's always been their skulls that haunted us. Huge faces familiar and also strange, behind which subtle brains once rested, processing a vanished world beheld by voided eye sockets.

And yet. All things pass.

Like a famished wolf in the hollow of a tree … They are like the river and the fall … nothing stands against them.[*]

This is William Golding's vision of humanity spreading across the world, seen through the eyes of his gentle Neanderthal protagonist Lok. Its chilling vision was published 99 years after the Feldhofer discovery, and 40,000 years – almost 10 times less than the span of Neanderthals' existence – since the world shrank to just one sort of human: us.

Carried within our bodies, today the Neanderthals face another crisis. Earth, where we exist in a frighteningly thin atmosphere like honey smeared on an apple, has long strained against the increasing load we place upon it. Our shared fascination for material properties has metastasised into a tumour of creation and consumption, as our clever fingers fashion ever more things from stone, iron, plastic.

As I finish this book from home lockdown in 2020, existential questions abound. The COVID-19 pandemic has overtaken the world in barely a month, vastly sped up by millions of flights connecting every corner of the globe. Slower, temporarily forgotten but even more serious is the climate crisis.

Since the current interglacial began around 12,000 years ago, we've largely basked in the kind climates of a world where ice caps are dormant.

Without the Industrial Revolution there probably would have been a few thousand more golden years before the mercury began juddering downwards. Instead, the massive release of CO_2 – outstripping anything from the entire Pleistocene and beyond – has delayed the next ice age indefinitely.

What's happening is unprecedented. Over the next millennium – roughly 30 generations – we are heading into a world hotter and more dangerous than any previous hominin survived. The Eemian 120,000 years ago was on average just a degree or two warmer than today, yet along with hippos in the Thames, sea levels were 5 to 7m (15 to 22ft) higher. Coasts where picturesque cottages and teeming cities now

---

[*] From *The Inheritors*, William Golding.

stand were swamped. And that's with far lower CO_2 levels than we've already reached.

In the absence of immediate, drastic action, the most up-to-date climate models put us on track for a terrifying future. Polar ice caps are at genuine risk of disappearing, and if so, oceans would rise by 20m (65ft) or more. In the past year the Great Barrier Reef has withered, the Arctic, Amazon and Australia have all been ablaze, and heat records have been breaking like waves, one after the other.

Over the ancient Eurasian steppe superhighway where Neanderthals once trod, Pleistocene corpses melt out from vast yedoma[*] ice peats – mammoth feet, wolf heads, entire infant cave lions – like some ghastly outriders of doom. The Great Thaw might even be how we meet Neanderthals for a third encounter: somewhere, still ice-clenched by 50,000-year-old muds and permafrost, a body surely lies.

We might console ourselves with the knowledge that Neanderthals survived similar extremes of climate change. As glacials expired, the land itself must have seemed to disintegrate, as old permafrost bubbled up into lake-speckled bogs running horizon to horizon. Hillocks appeared and disappeared like gigantic seasonal fungi, forests staggered and drowned, vast craters opened up. Entire mountainsides liquefied like ice cream as soil, plants and everything slid off, despoiling local ecosystems and once-lucid rivers – the infrastructure of life – ran heavy with sediment as the land was sloughed off. All this, and they held on.

But a Eurasia with maybe a few hundred thousand souls is very different to today's teeming millions. Neanderthals could move to try and escape hard times. We have no guidebook for the destination our sprawling, industrialised, unimaginably complicated civilisation faces. What's been shockingly proven by COVID-19 is that, even with technological buffering, we're on a course for uncertainty and ever-greater instability.

This future of blistering sun, suffocating cities, flood, tempest and maybe more pandemics is like a bison thundering towards us. If we do not move fast, our children's children will be impaled. And bleeding from them out onto the ground will be the last Neanderthals.

---

[*] Yedoma comes from the Siberian Nenets for 'no reindeer', being places where they had to walk on foot.

Acknowledgements

Writing Kindred was just as daunting and difficult as I'd feared when I started work on it some eight years ago; attempting to write a definitive account on the subject of your great passion is an immense privilege, but also a millstone. Doubts over accuracy were magnified, and even considering such a project felt presumptuous at times. And yet it also felt vital and important to try, because of the Neanderthals themselves. They're the first thank you I owe, never failing to be fascinating, confounding, surprising and impressive. They've always revitalised me, even when the writing process itself dragged or stalled.

Talking of writing, a special shout-out goes to the reviewer who rejected my first academic article in 2009, stating that my examination of the social contexts and cognitive implications of Neanderthal birch tar technology 'would fit better in a Jean Auel novel'. They made me determined to find a way to keep writing 'New Agey discussion of the feelings of archaic Palaeolithic humans' while remaining committed to the archaeology.

On that note, I want to proudly acknowledge the debt I owe to Jean Auel. The great trouble she took to try and represent tiny details of Palaeolithic life fired up my nascent childhood interest in prehistory, and in many ways her depiction of Neanderthals was prescient. Other novelists who have brought the Palaeolithic to life for me include Elizabeth Marshall and Claire Cameron; to the latter I'm grateful for permission to use a quote from her book *The Last Neanderthal*. In addition, I am grateful to many other writers of varying sorts whose demonstrations of skill have inspired me. A tiny selection include Gavin Maxwell, Richard Fortey, Kerstin Ekman, Primo Levy and Nan Shepherd. Special appreciation goes to the other Bloomsbury Sigma authors who have all been encouraging and supportive, especially Jules Howard, Kate Devlin, Ross Barnett and Brenna Hassett.

There are so many archaeological colleagues without whom I would not have reached this point. I'm grateful to Robert Symmons, Richard Jones and Naomi Sykes for getting me into the whole archaeology thing in the first place; to John McNabb for my Masters training and his support since, and to the many institutions who made my PhD on the late Neanderthal archaeology of Britain possible. I thank Beccy Scott, Matt Pope and others for the opportunity to work on the Neanderthal artefacts from La Cotte de St Brelade following my PhD.

Just a few names of scholars and thinkers whose work helped me consider Neanderthals more deeply include Clive Gamble, Tim Ingold, John Speth, Louis Liebenberg, Zoe Todd, Vanessa Watts, Kim Tallbear, Donna Haraway and the Yolŋu Aboriginal community including Bawaka Country.

Many of my closest postgraduate peers helped me develop my thinking about prehistory and archaeology, and were supportive of my writing too: my thanks go to Ana Jorge, Christina Tsoraki, Erick Robinson, Nick Taylor, Geoff Smith, Karen Ruebens and Becky Farbstein.

I started to write Kindred just after my postdoctoral fellowship at the PACEA laboratory, Université de Bordeaux. Many colleagues there were welcoming, encouraging and inspiring, but most of all I am deeply grateful to Brad Gravina for the intellectual wrangling over Neanderthals, the regular fires being lit under me and his warm friendship that has continued long after we occupied adjacent offices.

I am also massively thankful towards my wider professional network, particularly when I decided to emerge from an academic chrysalis into a more free-flying writing and creative career. All of the following I've learned from and enjoyed enlightening archaeological discussions and debates about Neanderthals with, often via social media: John Hawks, Alice Gorman, Julien Riel-Salvatore, Chris Stringer, Will Rendu, Colin Wren, Annemieke Milks, Marie Soressi, Jacquelyn Gill, Tom Higham, Kate Britton, Catherine Frieman, Jacq Matthews, Paige Madison, Jenni French, Andrew Sorensen, Hanneke Meijer, James Cole, Radu Iovita, Clive Finlayson, Ben Marwick,

Manuel Will, James Dilley, Shanti Pappu, Michelle Langley, Antonio Rodríguez-Hidalgo, Patrick Randolph-Quinney, Caroline VanSickle Joseba Rios-Garaizar … and many, many others who I am forgetting.

While writing the book, I often had very little income and so I'm enormously thankful to the Society of Authors for an Author's Foundation Grant in 2016, and Contingency Fund grants in 2018 and 2020 that made a huge difference. Equally, I am very grateful to editors from *The Guardian*, *Aeon* and other places for publishing my writing.

Enormous gratitude goes to Jim Martin at Bloomsbury Sigma for offering me the chance to write such a mammoth book, and keeping faith even when my critical thinking didn't extend to realistic estimates of how long it would take. My editor Anna MacDiarmid at Sigma also has my grateful thanks for consistently being understanding, and for her ability to deliver a chivvying coated in calm optimism. Working with Myriam Birch made the copyediting a genuine pleasure despite time pressure. At Bloomsbury, first Kealey Rigden and then Amy Graves and Alice Graham handled the publicity and marketing magnificently. Working with Alison Atkin as the artist for the chapter-opening images was wonderful; she was always open to trying even my most odd ideas and captured the emotions of particular scenes exactly as I hoped. Grateful thanks to author Jen Marlow for kitchen-table writing companionship. Feedback from the kind people who read sections or chapters of the manuscript was invaluable: thank you to Brad Gravina, Angela Saini, John Hawks, Geoff Smith, Brenna Hassett, Tori Herridge and Suzanne Pilaar Birch.

The latter three women all also deserve infinite gratitude for Being There as my personal, nearly 24-hour anarchist feminist cheerleading collective. Since 2013, however monumental or minor the crisis, they've offered unconditional support, wisdom and help. And for every tiny step forward or triumph, their big-ups gave me the confidence I needed to keep going; not just with *Kindred*, but in choosing to take a path out of academic research.

Most of all, it's my family who have been like a mountain to cling to, never mind a rock. I wish I could give the book to my grandfather Sam who shared my passion for history, and my grandmother Dorothy who loved literature and poetry (her grandfather Hall Nicholson, who was a miner and speaker for the Sunderland Secular Society in the 1890s, has a mention in the last chapter). My other grandfather Neville's quiet devotion to classical music sparked an appreciation that has also sustained me while writing, and I continue to be inspired by my grandmother Jean, whose fierce and unwavering love means no bad reviews will ever matter.

My brother Jack and my parents Rosalynd and Peter have always been there for me and always will be, unconditionally supporting me through undulations, tumbles and ascents. Their love, faith and pride mean more than I can express.

Lastly, I want to say thank you to my husband Paul. No words are adequate for what I owe you.

Kindred is for my children, and the 3,000-odd generations of mothers before me, tangibly connecting us to the Neanderthals. They're still here in every fibre of my being, and in that of my two little girls (who sometimes enjoy hearing about the 'Anderthals mummy is writing about, but would quite like for her to come and play now).

Mid Wales, June 2020

Index